Meeting Diversity in Ergonomics

Meeting Diversity in Ergonomics

Pikaar, Koningsveld, and Settels

ELSEVIER

Amsterdam • Boston • Heidelberg • London • New York • Oxford • Paris
San Diego • San Francisco • Singapore • Sydney • Tokyo

Elsevier
Linacre House, Jordan Hill, Oxford OX2 8DP, UK
Radarweg 29, PO Box 211, 1000 AE Amsterdam, The Netherlands
30 Corporate Drive, Suite 400, Burlington, MA 01803, USA

British Library Cataloguing in Publication Data
A catalogue record for this book is available from the British Library

Library of Congress Cataloging-in-Publication Data
A catalog record for this book is available from the Library of Congress

ISBN: 978-0-0804-5373-6

For information on all Elsevier publications visit our web site at http://books.elsevier.com

Transferred to Digital Printing 2009

Contents

Preface

It is a pleasure to introduce this collection of excellent papers that have been developed by selected authors who represent a cross-section of the ergonomics domain. These authors were selected from the International Ergonomics Association (IEA) Congress in Maastricht, 2006 and requested to extend their work to provide a broader perspective of their research and to provide linkages between research areas of the domain. This book assists in the extension of the Congress theme to reflect the diversity of ergonomics research and application.

The papers selected for this publication reflect aspects of this diversity with a common theme of studying and considering the needs of people in a range of applications.

As the IEA celebrated 50 years since it is initial conception in The Netherlands in 2006 it is worth acknowledging the breadth of ergonomics as reflected in this publication. Research papers have been provided on the specific physical capabilities such as with a gloved power grip and hand tool design, musculoskeletal disorders, and hospital architecture through to "pleasurable" products.

Cognitive ergonomics has been a growth area with papers on process control displays, CCTV, software design and virtual reality. Advances in ergonomics methodologies focusing on user centred approaches and embracing economics and marketing are part of the ongoing links between ergonomics and the holistic approach to hazard analysis and resolution.

The application of ergonomics in product, systems and process design is a focus of the IEA EQUID program which was discussed during this Congress. The application of ergonomics by the designer utilizing a user centred approach is an area for development and promotion by the IEA.

As the 16th President of the IEA I would like to specifically acknowledge and thank the Organizing Committee from the Nederlandse Vereniging voor Ergonomie (NvvE) being Ernst Koningsveld, Ruud Pikaar and Paul Settels. Without their commitment and leadership this IEA Congress would not have been such a wonderful success.

Yours Sincerely,
David C Caple
IEA President

Part I
Marketing Ergonomics

Meeting Diversity in Ergonomics (2007)
Pikaar, Koningsveld, and Settels (Eds)

1

Meeting Diversity in Ergonomics

E.A.P. Koningsveld

TNO – Ergonomics for Innovation, P.O. Box 718, NL-2130 AS, Hoofddorp, The Netherlands

P.J.M. Settels

Health & Safety Department, ING – HR Netherlands

R.N. Pikaar

ErgoS Engineering & Ergonomics, P.O. Box 267, NL-7500 AG Enschede, The Netherlands

Abstract. The field of ergonomics ranges from Occupational Safety and Health (OSH) issues to human performance, efficiency and productivity. Actually, ergonomics deals with activities of humans or groups of people, and so all aspects relate to human factors. The objective of this book is to provide an overview of recent trends in the field of ergonomics. Based on experiences at the 16th congress of the International Ergonomics Association (IEA), a selection of excellent authors wrote contributions that focus on new trends, and join together to form a vision of the future of ergonomics. As is proper, the editors start with a reflection on the past of the discipline. They place this reflection in the perspective of developments in society, in work and in technology. Developments in the content of our work are pointed out, as well as important steps towards more professionalism in ergonomics and human factors.

Keywords: diversity, ergonomics, human factors, human performance

INTRODUCTION

The development of ergonomics followed two different approaches: one has its origin in health and safety (OSH), the other in human performance. For decades these two lines co-existed. In recent years some discrepancies between these two principles grew. Societal changes towards more efficient work, cost containment and an increasing international competition are factors on the one hand. These caused, for instance, a growing interest in the cost–benefit aspects of ergonomics [1]. On the other hand, the professionalism of experts in ergonomics and human factors was enhanced. Courses in ergonomics and human factors, which started within a diversity of faculties, such as psychology, human kinetics and industrial engineering, had to improve their qualities in order to survive in the ever-changing academic structures. The creation of certification systems for professional ergonomists, and especially the core competencies that were defined in that respect, has been important.

Besides, the world of human activities is constantly changing. New trends in information and communication technology have changed the work of many dramatically.There is an ongoing process of new trends in the organization of work. Worldwide we see a migration of people, resulting in a growing diversity within the workforce. There is a trend towards keeping ageing workers at work, and towards making work accessible for people with handicaps of any kind.

For ergonomists and human factors experts, these trends are challenges for their contribution to 'sustainable performances of people'. With this statement we want to express that ergonomics and human factors endeavour to create work, workplaces and organizations that are safe and preserve people's health, and that allow them to perform well during all their working life.

THE DEVELOPMENT OF ERGONOMICS

Ergonomics to promote safety and to protect health

Health protection and providing a safe working environment have been important matters for many years. Working can be unsafe and overstraining, resulting in large numbers of early dropouts from the working processes. This used to be reality in most of the work worldwide, and is still the case in large parts of the world.

Scientific knowledge about capacities of humans and actual workloads has been extended greatly over the past 50 years. Solutions for hazardous workloads and unsafe conditions have been developed, tested and introduced. Nevertheless, today you will still not have to search for long to find adverse working conditions, even in the most industrially developed parts of the world. Shareholders and management of companies seem to be blind to the long-term benefits of safe and healthy work environments.

The consequences of poor working conditions are serious. Workers suffer from health impairment and experience a diminished quality of life. Their power to earn money decreases, and their independence is affected. For the employer, the resulting costs are high. The employer is held progressively responsible for the effects; in most cases as a legal body, but several cases show that the employer can be held responsible as an individual. At the societal level, the costs are a real burden. Several European studies show that the societal costs are in the order of two to three per cent of the Gross National Product (GNP) [2, 3]. Health effects, resulting in lost working days, occupational disability and medical treatment, are responsible for the majority of these costs. The quoted Dutch study provides a good overview of cost categories and their values, but does not include all effects. It turned out to be difficult to validate the financial effects of poor working conditions on companies' performances, so these costs are not included. Some experts estimate that these may be another two to three per cent of the GNP. Others, including the International Labour Organization [4] point at the importance of better working conditions for companies performances. There is a growing understanding that working conditions are important for the competitive strength of a country or an economic region like North America or the European Union [5].

Many individual companies and organizations are cost-driven; but according to Oxenburg [6] and Marlow [7], cost–benefit considerations related to occupational health and safety are scarcely out of the egg.

The rapidly increasing costs of health care in the western world emphasize the need for prevention of health-impairing work and the creation of safe working conditions. It is not easy to express lost healthy lifetime and individual harm in financial terms, and so these are not included in studies about the costs of poor working conditions. However, these factors are becoming important matters in the scope of politicians, employers and workers, and their representatives.

Ergonomics to promote human performance

The development of ergonomics related to human performance has its roots in studies on physical performance. Employers were highly interested to increase human performance, and thus in the maximum capacities of humans. As the military was a large employer and battle actions were highly demanding, the armed forces formed a major target group for ergonomics studies. Later, the scope changed from maximum performance to 'optimal', indicating the workload that workers can endure for a lifetime of full-time working weeks. Today such studies are still important for many trades in which physical efforts are required. In addition, ergonomists investigate how to design workplaces that allow maximum performance without negative effects on physical workloads and discomfort [8].

From the 1940s, ergonomists have been involved in cognitive, mental and organizational performances. As several ergonomics publications indicate, during World War II, more planes crashed on their way to the battlefields or on the return flight than were shot down by the enemy. A vision of 'human-centred technology' was born, and has become important. Today, the quick pace of development of new technology makes this even more important.

Yet a large gap still exists between the designers of information and communication technology and the way humans expect how to deal with technology, or how they want technology to work. Yet, as early as the 1970s, studies about controlroom operator's behaviour and design recommendations were published by Edwards and Lees [9]. The European Coal and Steel Community put much effort into this field. Later, mistakes in controlroom-operated industries resulting in serious accidents generated studies with an impact on ergonomics. Examples of such accidents with proven human factors misfits are: the Three Miles Island nuclear incident (1979), the Bhopal catastrophe (1984, release of a toxic compound) and the Flixborough disaster (1974, an exploding chemical plant). Reliability assessment became an important, though underestimated, task. Authors like Kirwan published several books on the matter [10], also focusing on the mental workload and the internal process representation of the human operator.

Other studies on human performances based on mental processes focus for instance on drivers' tasks in cars, trains, ships, and even in bicycle riding. Recent studies warn for instance of the hazards of dual tasking, such as when using a mobile phone while driving. Organizational design and management has become a central topic in ergonomics. Subtopics, such as socio-technical design and participatory ergonomics, get much attention at conferences. This field of interest is not completely free of hype. Every decade new gurus or visions arise, many of which deal with human performance. Shapiro [11] has written a rather critical review of work organization design and management views, entitled *Fad Surfing in the Boardroom: managing in the age of instant answers*. Much organizational hype passed by, such as total quality management, systems re-engineering, six sigma, or learning organizations. Of course, each of these has its values. But Shapiro states that at the end, the organization consultants are the only ones who benefit.This is probably because many of these approaches don't address the actual production tasks at the shop floor. And that is basically what ergonomists and human factors specialists do. This raises two questions:

- Isn't management interested in their most important and valuable asset: the human resources?
- Why does ergonomics (or human factors) <u>not</u> reach the boardrooms?

Regarding the first question, there is a trend that shareholders stress short-term financial performance, which may very well conflict with durable human performance. Ergonomists need to be proactive. Attacking the short-sightedness of management seems like a plausible approach, but will not work. It is better to find ways to prove the short-term benefits of ergonomics. Ergonomists have adequate knowledge of the relation between people's health and their performance. They can very well argue the importance of these factors for companies' health and performance, both in the short and long term. According to Gallaway [12] it is time to do so; don't wait until tomorrow.

Dul and Neumann [1] analysed the drivers behind business strategies. Obviously, ergonomics, in and of itself, is not a strategy or business goal. However, they show that ergonomics is a potentially important feature of strategy formulation and implementation, since attention to ergonomics can contribute to many different aspects of business performance (profit, cost minimization, productivity, quality, flexibility).

Referring to the second question, 'why does ergonomics not reach the boardroom?', we can see that ergonomists tend to pay only little attention to trends in management styles. A current trend to 'manage as a coach' focuses particularly on questions of human performance: how to get workers to perform well, and how to motivate them to continuously develop and innovate, in both themselves and their work. Professional businesses may learn from the field of professional sports. The role of the manager changes from a hierarchic boss to a facilitating coach. However, does this interesting approach reach the ergonomics society?

Four generations of ergonomics

In his keynote address at the IEA 2006 congress, Boff [13] distinguishes four generations of ergonomics and human factors, and so gives another view of the development of the discipline(s). The first generation focused on adapting the 'physical fit' of equipment, workplaces and tasks to match human capabilities and limitations. The second generation was spawned by the growth of complexity of work environments and systems, and focuses on systems rather than on workplaces and tasks. Complexity is an unintended by-product of combining people, technology and work. The third generation is characterised by enabling humans to perform better, instead of designing better work environments. The fourth generation, according to Boff, aims to maximize human effectiveness. Information, biological and nanotechnologies are enabling the ability to redesign our basic human factors: how we think, how we feel, how we look, how we age and how we communicate. This generation of ergonomics looks forward to the future. We we will return to this point later in this chapter.

THE DIVERSITY IN ERGONOMICS

The previous section clearly shows the diversity in ergonomics. Diversity has been the basis for the 16th world congress on ergonomics of the IEA in 2006. The congress theme was *Meeting Diversity in Ergonomics* [14]. The combination of diversity and ergonomics expresses the broad scope of ergonomics and the global variety within the field.

- Ergonomics in itself is diverse. It is an engineering activity, but also a product development cycle; for instance, to design for a population as diverse as possible, taking into consideration psychological, socio-technical and organizational matters.
- Ergonomics is driven by the ambition to optimize the combination of prevention and performance, but can also be driven by legislation and standardisation, which are more corrective (or curative) by nature.

Within the group of ergonomists and human factors specialists, a large diversity can be found, as they are:

- scientific researchers: mostly in-depth research in one or several topics;
- professionals in applying scientific results in ergonomic design projects;
- practitioners, who may be full-time and as such are educated ergonomists, or specialists from other disciplines, who apply ergonomics knowledge and principles.

There is a large diversity in the ergonomics fields of interest, in the areas of application, and, of course, in humans themselves:

- humans with their anthropometrics, capacities and limitations;
- scope, such as the physical workload, mental workload, environmental factors, organizational design and management;
- areas of application, such as manufacturing, assembly, process industry, office work, rehabilitation, hospitals, schools, transport, agriculture, sports;
- methodology: research methodology, engineering practices, ergonomic design projects;
- worldwide cultural differences: for example, industrialised versus industrially developing countries.

The challenge for ergonomists' research and consultancy is to meet the diverse requirements in the target groups.

A NEW MILLENNIUM: NEW TRENDS IN ERGONOMICS?

Changing attention

The level of attention given to the two ergonomics approaches, i.e. OSH and human performance, has fluctuated over the decades. Today we may question whether well-based scientific knowledge is required for many of the practical questions asked of ergonomists. For decades, we have known that loads with a mass of 23 kg or more are unacceptable for manual lifting and carrying under good conditions. In many cases a quick expert's guess will be enough to decide whether intervention is required or not. Of course, a well-performed task analysis or job assessment (or whatever methodology is favoured to investigate) is required to evaluate situations that are close to or just beyond limits of physical workloads. More importantly, the analysis of such work situations may reveal the best possible way to solve the problem.

A second example: in 1994 in the Netherlands a group of ergonomists, bio-mechanical experts and occupational physicians drafted a first instrument to evaluate work situations for the risk of repetitive movements [15]. They based their short checklist on collective experts' insight, as a scientific basis was still lacking. The instrument focused only on use in occupational health care practice, and turned out to be an utilizable tool for the period until new research results became available.

Today we know that, next to physical ergonomics, organizational design and management matters are important, as are the aspects of mental workloads. There is no argument but that both OSH and human performance are extremely important for a healthy and competitive society. In August 2006, Finland organized a European Productivity Conference [5], as part of its time as chair of the European Union. This was in no way an ergonomics conference. Nevertheless, in many presentations, experts and politicians stated that attention to health, safety and meeting humans' capacities in work are the predominant factors in enhancing productivity. Productivity and innovation are essential to meet the Lisbon treaty's goal of March 2000: that Europe become the strongest economy in the world.

At the 16th congress of the IEA, a large diversity of ergonomics approaches was in evidence. Indeed, the protection of health and safety and the enhancement of human performance and productivity both got full attention. Research papers and reports of practical applications and interventions alternated through the whole program. Compared to previous IEA congresses, the balance between *scientific* research papers and *applied ergonomics* papers seemed somewhat shifted in favour of the latter. Much of the research now focuses on 'how to apply scientific knowledge in specific fields of application'. For example, a notable change occurred in the field of work-related musculoskeletal disorders (WMSD). Though over 80 papers were presented, the main emphasis shifted from theory development to application. See this book's chapter 'Physical Ergonomics and Musculoskeletal Disorders: What's hot? What's cool?' for further details [16]. Papers have been presented on WMSDs in harbour crane operators, poultry slaughtering, fishing industry, several manufacturing industrial applications, health care, and in addition workshops on practical analysis tools, such as OCRA (Occupational Repetitive Actions Index), attracted a large audience.

A growing area of interest is health care and patient safety. The approximately 70 papers, including keynotes on patient safety related to systems design and ergonomics by Buckle [17] and medical monitoring displays by Sanderson [18], constituted a symposium on their own. Focusing on application, all kinds of ergonomics topics were covered: architectural design of hospitals Villeneuve [19], patient handling, reliability and patient safety, and product design. For example Jalote-Parmar et al. [20] describe the design of surgical workflow driven information systems.

Another remarkable change is the applied research on topics related to IT. It is very encouraging to notice that, for the first time, some ergonomists have been able to catch up with the rapid developments in this area. Answers could be given to ergonomics design questions regarding video-wall/large screen systems by Groot [21], CCTV control centres by Wood and colleagues [22], and software interaction design by Mulder [23]. In these three cases, the authors give practical design guidelines, thus helping engineers in the development of interactive systems, i.e. control centres.

Future of ergonomics

Meetings supported by the IEA can play an important role in the future of the discipline of ergonomics. To that end, this book reports major results of symposia attracting large audiences. Moreover, the editors created a publication space for invited papers of significant contributions to the IEA 2006 congress.

Where the main scope of conferences nowadays is to report ongoing and completed research, it is even more important that conferences are used to identify which basic knowledge is lacking, and which trends can be seen, or are approaching. If ergonomists are aware of newly raised questions, and of technological, societal and cultural developments, they may be more successful in contributing to new work situations, new products and to developments in working life.

When one wants to explore the future, one should start to learn from history. After the ages of mechanization, industrialization and digitaliation, ergonomists – and particularly certified ergonomists – must anticipate the next great generation of development, even though it may be impossible to identify the first signs for the next 10 or 15 years. When we think back to the development of aeroplanes as tools in war, the basic knowledge of human capacities for such tasks was only developed after serious problems arose. In the 1970s, ergonomists only became involved in human–computer interaction after users suffered from eyestrain. Even then, most ergonomists focused only on visual impairments; almost 20 years later, the quick development of arm, neck and shoulder complaints by computer workers took them by surprise. So it was not until the mid-1990s that ergonomists started research into work-related upper body complaints.

Another example of a trend can be found in production and assembly work. Over the past century, this trend started with the attention to energetic workloads and exhaustion from work.

Via low back pain, the attention moved towards upper body complaints. Which body part will be the next predominant item for ergonomists? How to deal with new trends in productivity, organizational matters, human performances in complex (technology-based) systems, an ageing population in relation to ICT, and so on?

Conferences should focus on the identification of trends like these, and not only be a platform or a market place for a thousand different presentations, never mind how interesting each individual paper may be. Boff's fourth generation [13] may serve as an example. Boff states that this generation 'is emergent and focused on applications of pharmacology, biotechnology and genetic medicine, seeking to stall, reverse and modify effects of disease or ageing. The human factors and ergonomics implications derive from the inevitable consideration of these techniques to enhance capabilities and overcome the limitations of otherwise normal individuals'. Completely new questions arise, for which knowledge should be developed. Human factors and ergonomics experts can anticipate ethical questions as well. We encourage the human factors and ergonomics society to start discussions on this matter.

This brought the editors to the idea behind this book: 'to provide some major trends in ergonomics and in human activity as the basis for new trends in ergonomics'.

ERGONOMICS: EDUCATION AND CERTIFICATION SYSTEMS

The content of ergonomics may be important, and skills and competencies are essential to act successfully as a professional ergonomist. The IEA promoted the development of educational programs, e.g. by defining minimum criteria. Over the years IEA has been publishing directories of educational programmes in ergonomics, and a version is available on the Internet.

Since the late 1980s, certification systems for ergonomists were developed and introduced in different parts of the world. A worldwide generic system would have been good for the profile of ergonomics, but efforts to develop such would certainly have failed, due to differences in cultures, backgrounds and local regulations. The fact that many European countries agreed on one system (Centre for Registration of European Ergonomists (CREE), can be considered as a large step. Today, certification systems for ergonomists exist in Australia, Canada, Europe, Japan, New Zealand and the USA. In other parts of the world certification is being developed.

In 2001 the IEA succeeded in issuing a document on the core competencies of ergonomists with worldwide acceptance, as well as criteria for IEA endorsement of certifying bodies [24]. And in July 2006 the IEA's Council decided positively about a renewal of the IEA Code of Ethics (see Appendix).

The IEA is now developing criteria and requirements for ergonomic design of products, work systems and services. The proposed standard will be called: Ergonomics QUality In Design (EQUID) [25]. Such a standard is most important for the profile and status of ergonomics, and provides a backing for professional ergonomists, and their clients.

The commercial aspects of the work of ergonomists get attention as well. In 1989 the Dutch Ergonomics Society organized an international conference on Marketing Ergonomics [26]. Even today, attendees report that participating in that conference has been important for their further development as a professional. In May 2006, the New Zealand Ergonomics Society chose the same topic for their annual conference [27]. Next to commercial competencies, the ability to pinpoint the benefits of ergonomics in core business indicators, and to convince stakeholders with economic or other convincing arguments, has become increasingly important recent relevant publications are written by Oxenburg [6], Marlow [7]; Koningsveld [2, 28] and Dul [1].

ABOUT THIS BOOK

The aim of this book is to outline major trends in ergonomics and in human activity. This can provide a basis for new trends or actualization of ongoing developments in ergonomics. Contributions on the scientific state of the art as well as papers based on practical experiences were selected. The papers of the 16th congress of the IEA are the basis for the book's content. The editors selected experts who organized outstanding sessions, or sessions on topics in which notable developments got attention. Some chapters deal with new trends in ergonomics. We asked the authors to write chapters based on their sessions, extended with experiences in discussions, with recent developments in the field of ergonomics, and with an eye on the future. The focus is on application, not on in-depth scientific research projects.

As mentioned earlier, Dul and Neumann discuss in the next chapter *The strategic business value of ergonomics*. This is followed by a chapter on *Ergonomics in Engineering Projects* by Pikaar, based on several cases of applied ergonomics. The participation of endusers plays a major role in the ergonomic approach to engineering. Imada elaborates this topic further in his overview *The value of participation in Ergonomics.* The last, somewhat provocative, paper of Part One, *Marketing Ergonomics*, is written by Gallaway, and presents a plan of action for promoting ergonomics.

Part Two is largely based on some of the summary presentations of the closing session of IEA 2006. Overviews are presented on physical ergonomics, organizational design and management and ergonomic work analysis.

The third part of the book focuses on Comfortable Products and is introduced by a paper written by Bonapace and Dejean. Papers concern a design methodology for icons on process control displays (by Liang) and several tools for product development, i.e. a people's database (Porter et al.) and the applicability of Virtual Reality (Sharples et al.).

Part Four highlights the link between research (results) and everyday practice, for example on CCTV systems, software design and information systems in surgery, all mentioned earlier in Part Three. Other outstanding papers of IEA 2006 in this part concern driver support systems (Thompson), and hospital architecture (Villeneuve et al.).

TO CONCLUDE

Beyond the content as such, we would like to motivate ergonomists and their societies to foresee the trends in ergonomics for the next 10 to 15 years in good time. In that respect it is advisable that ergonomists in research and application collaborate in the anticipation on human factors problems and challenges that may arise.

The development of ergonomics as a profession, and of ergonomists as professionals, deserves more attention. What is the value of our contribution to society, to our clients, to the workers? How can we successfully 'sell' ourselves and our discipline? Which profile do we choose: the health prevention expert or the human performance expert, a combination, or another profile?

Several chapters show successful interventions. In our field, a successful intervention is not only an intervention that is implemented; the intervention must benefit core business values or process values. It needs to contribute to a positive balance between efforts and effects, and between costs and benefits to be made by the client organization and its workers. This book aims to inform ergonomists and scientific researchers about ergonomics theories and recent research findings. Researchers will find the latest trends. The professional ergonomists, involved in the application of knowledge in practice, will find data, knowledge and experiences that they can apply in their own practice. Ergonomics Societies and educational centres may find indicators to update their policies. Finally, management and other interested clients of ergonomists can get an actual insight into the broadness of the expertise of ergonomics, and the contribution of ergonomics to society.

REFERENCES

[1] Dul, J. and Neumann, W.P. The strategic business value of ergonomics. This volume.

[2] Koningsveld, E.A.P., Zwinkels, W., Mossink, J.C.M., et al. (2003). The Dutch national costs of poor working conditions (in Dutch: Maatschappelijke kosten van arbeidsomstandigheden van werknemers in 2001). Werkdocument 203, Ministry of Social Affairs and Employment. Den Haag, The Netherlands, 2003.

[3] Koningsveld, E.A.P. (2005). Participation for understanding: An interactive method. *Journal of Safety Research – ECON Proceedings*, 36, 231–6.

[4] Forastieri, V. (July 1999). *Improvement of Working Conditions and Environment in the Informal Sector Through Safety and Health Measures*. Geneva: International Labour Office.

[5] Anon. European Productivity Conference EPC 2006. Proceedings can be found at: http://www.eanpc.org/EPC_agenda.php.

[6] Oxenburgh, M., Marlow, P., and Oxenburgh, A. (2004). *Increasing Productivity and Profit Through Health and Safety*. London: Taylor & Francis.

[7] Marlow, P. (2006). Making ergonomics a usual business tool; Marketing ergonomic interventions. *Proceedings of the Annual Conference of the New Zealand Ergonomics Society*, pp. 39–42. New Zealand: Christchurch.

[8] Rhijn, J.W. van, Looze, M.P. de, Tuinzaad, G.H., et al. (2005). Changing from batch to flow assembly in the production of emergency lighting device. *International Journal for Production Research*, 43, 3687–701.

[9] Edwards, E. and Lees, F.P. (1974). *The Human Operator in Process Control*. London: Taylor & Francis.

[10] Kirwan, B. (1994). *A Guide to Practical Human Reliability Assessment*. London: Taylor & Francis.

[11] Shapiro, E.C. (1996). *Fad Surfing in the Boardroom: Managing in the Age of Instant Answers*. Knoxville, USA: Capstone.

[12] Gallaway, G.C. An operational and marketing infrastructure for the human factors and ergonomics professions – a call to action. This volume.

[13] Boff, K.R. (2006). Revolutions and shifting paradigms in human factors and ergonomics. *Applied Ergonomics*, 37 (4), 391–400.

[14] Pikaar, R.N., Koningsveld E.A.P., and Settels, P.J.M. (2006). *Meeting Diversity in Ergonomics*. Proceedings IEA 2006 Congress. ISSN 0003-6870, Elsevier Ltd.

[15] Huppes, G., et al. (1994). *Repetitive movements: Definitions, extent of the problem, risk evaluation instrument* (in Dutch: Repeterende bewegingen, definities, omvang, instrument voor risico-inventarisatie). The Netherlands: Sdu Publishers.

[16] Beek, A.J. van der and Ijmker, S. (2007). Physical ergonomics and musculoskeletal disorders: What's hot? What's cool? This volume.

[17] Buckle, P., Clarkson, P.J., Coleman, R., Ward, J. (2006). Patient safety, systems design and ergonomics. *Applied Ergonomics*, 37 (4), 491–500.

[18] Sanderson, P. (2006). The multimodal world of medical monitoring displays. *Applied Ergonomics*, 37 (4), 501–12.

[19] Villeneuve, J., Remijn, S.L.M., Hignett, S., et al. (2007). Ergonomic interventions in hospital architecture. This volume.

[20] Jalote-Parmar, A., P.M.T. Pattynama, R.H.M. Goossens, et al. (2007). Workflow integration matrix as a part of user-centric design methodology for surgical workflow driven information systems. This volume.

[21] Groot, N. de and R.N. Pikaar (2006). Videowall information design: Useless and useful applications. In *Meeting Diversity in Ergonomics* (Pikaar, R.N., Koningsveld E.A.P., Settels, P.J.M. Eds.). Proceedings IEA 2006 congress. ISSN 0003-6870, Elsevier Ltd.
[22] Wood, J. (2007). CCTV-Ergonomics: Case studies in rail and road transportation. This volume.
[23] Mulder, H.W.B. (2007). Making software work. This volume.
[24] See: http://www.iea.cc/events/education.cfm.
[25] See: http://www.iea.cc/events/equid.cfm.
[26] Pikaar, R.N. and White, T.N. (1990). Marketing ergonomics: Profile of an interdisciplinary field. *Ergonomics*, 33, 245–50.
[27] Purcell, N. (2006). *Anon. Marketing for Ergonomist.* Proceedings of the 13th Conference of the New Zealand Ergonomics Society. New Zealand: Christchurch.
[28] Koningsveld, E.A.P. (2006). *Selling ergonomics: Tricks or competencies?* Proceedings of the 13th Conference of the New Zealand Ergonomics Society. New Zealand: Christchurch.

Appendix – IEA CODE OF CONDUCT FOR ERGONOMISTS

Code of Conduct for Ergonomists
Approved by
the Council of the International Ergonomics Association
on July 9, 2006

INTRODUCTION

High standards are important for the effectiveness and credibility of the ergonomics profession. Codes of conduct help to set out the standards required of ergonomists in terms of professional practice and research ethics. Federated societies of the International Ergonomics Association (IEA) and professional certification organizations should demand their members to abide by a code of conduct. Any reported violations must be addressed through disciplinary procedures.

The IEA seeks to promote the ergonomics discipline and the consistent application of standards in ergonomics practice within and between countries. The IEA Code of Conduct provides a model document for federated societies and professional certification organizations that are developing or revising their own codes.

ETHICAL PRINCIPLES

The IEA Code of Conduct for Ergonomists is based on fundamental ethical principles relating to:

- beneficence – do good
- veracity – truthfulness, accuracy and integrity
- autonomy – respect for persons
- justice – fairness

PROFESSIONAL CONDUCT

In the conduct of their profession, ergonomists shall:

(1) maintain at all times personal integrity, objectivity and respect for evidence.
(2) not lay false claim to educational qualifications, professional affiliations, characteristics or capacities for themselves or their organizations.
(3) refrain from making misleading, exaggerated or unjustified claims for the effectiveness of their methods, and they shall not advertise services in a way likely to encourage unrealistic expectations about the effectiveness and results of those services.
(4) conduct themselves in their professional activities in ways which do not damage the interests of the recipients of their services or participants in their research and which do not undermine public confidence in their ability to perform their professional duties.
(5) limit their practice to those areas of ergonomics for which they are qualified by virtue of training and/or experience, and endeavour to maintain and develop their professional competence. Any work taken outside the competence must be conducted only with proper professional supervision or they shall give every reasonable assistance towards obtaining the required services from those qualified to provide them.

(6) always value the welfare of all persons affected by their work, protecting the privacy of individuals and organizations and follow ethical principles when conducting or reporting on research involving human participants.
(7) not use race, ethnicity, disability, gender, sexual preference, age, religion, or national origin as a consideration in hiring, promoting or training in any situation where such consideration is irrelevant.
(8) avoid all situations that contain elements of conflict of interest, and provide full disclosure of those conflicts to all potentially affected parties.
(9) take all reasonable steps to preserve the confidentiality of information acquired through their professional practice and to protect the privacy of individuals or organizations about whom the information was collected or held. Furthermore, they shall not divulge the identity of individuals or organizations without permission from those concerned.
(10) neither solicit nor accept financial or material benefit from those receiving their services beyond what was contractually agreed. Furthermore, they shall not accept such rewards from more than one source for the same work without the consent of all parties concerned.
(11) when becoming aware of professional misconduct by a colleague, that is not resolved by discussion with the colleague concerned, they shall take steps to bring that misconduct to the attention of appropriate authorities in the professional organizations to which they belong.
(12) take all reasonable steps to ensure that those working under their supervision act with full compliance to this code of conduct.

RESEARCH

Ergonomists who conduct research should follow ethical guidelines including:

Conduct of research

All ergonomists shall comply with the Geneva Convention and Helsinki Accord in treating both human and animal participants, in addition to obeying national and local laws, and generally accepted procedures within the scientific community. In particular, ergonomists shall:

(1) where there is any potential for harm, seek and act on guidance from a competent ethics advisor or committee.
(2) identify all potential sources or causes of harm in the research they are conducting. These hazards must then be effectively managed, including compliance with any requirements of the ethics advisor, to ensure that the risk of harm to participants is minimized.
(3) ensure that participants are fully informed of the outcome of the risk assessment and of any requirements identified by the independent ethics advisor before seeking informed consent.
(4) obtain prior written informed consent from human participants. Information must be provided in writing and orally to human participants in plain and clear language indicating the terms of participation, particularly about any hazards involved. Occasionally there may be exceptions in which the human participant is not able to consent. In such cases prior informed consent should be obtained from a person with (preferably legal) responsibility for the participant.
(5) empower human participants to terminate their involvement in the research at any time without prejudice.

(6) terminate any research process or experiment immediately if the participant's exposure to hazards exceeds commonly accepted thresholds. Further, if necessary, medical treatment must be provided.
(7) keep the identity of human participants confidential unless permission is obtained from the participants.
(8) not coerce anyone to participate in research nor use undue monetary reward to induce participants to take risks they would not otherwise take.
(9) ensure these ethical guidelines are followed by collaborators, assistants, students, and employees.

REPORTING OF RESEARCH

In pursuit of their profession, ergonomists who are engaged in research and scholarly activities have an obligation to report their work to the scientific community. In particular, ergonomists shall:

(1) ensure the integrity and accuracy of the data recorded before reporting results and conclusions to the scientific community.
(2) identify original sources (i.e. not plagiarize) and give credit to those who have contributed on a professional level to the work.
(3) pay special attention to the communication of research findings so as to facilitate their practical application.
(4) maintain the highest degree of objectivity when they are reviewing or editing works of other colleagues. In particular, they must ensure that their objectivity is not impaired by their own views even if the data and results reported conflict with their own previously published work.

Meeting Diversity in Ergonomics (2007)
Pikaar, Koningsveld, and Settels (Eds)

2

The Strategic Business Value of Ergonomics*

Jan Dul

Department of Management of Technology and Innovation, RSM Erasmus University, Rotterdam, The Netherlands

W. Patrick Neumann

Department of Mechanical and Industrial Engineering, Ryerson University, Toronto, Canada

Abstract. The value of ergonomics extends beyond health and safety. This chapter presents: (1) our views on the potential of ergonomics to contribute to the strategic goals of an organization, and (2) the results of a workshop held at an international ergonomics conference on this suggested potential. We do not see ergonomics itself as a strategy or business goal. Rather, we see it as a potentially important feature of strategy formulation and implementation, since attention to ergonomics can contribute to many different aspects of business performance. Ergonomics can contribute to the ultimate business goal of profit, or to intermediate business goals like cost minimization, productivity, quality, delivery reliability, responsiveness to customer demands or their flexibility.

In order to test our views we discussed them during a workshop with experienced researchers and practitioners from around the world. During the workshop, participants raised issues including the role of different business functions, the organization's context, the interaction of ergonomics with different business strategies, factors influencing the process of uptake of ergonomics, and the problem of lack of knowledge amongst engineers, managers and ergonomists. The workshop participants were generally supportive of the thesis that ergonomics can contribute to an organization's strategic goals beyond an exclusively health and safety focus. Achieving this may require the ergonomists to take on new roles and to see ergonomics as a means to support organizational development rather than an end in itself.

Keywords: strategy, business, economics, paradigm shift, future of ergonomics

INTRODUCTION

The International Ergonomics Association (IEA) describes ergonomics (or human factors) as 'the scientific discipline concerned with the understanding of interactions among humans and other elements of a system, and the profession that applies theory, principles, data and methods to design in order to optimize human well-being and overall system performance' [1]. This implies that ergonomics contributes to the optimization of both human well-being (a social goal) and total system performance (an economic goal); that ergonomics is broader than just occupational health and safety; and that it includes issues like workplace design, job design, work organization design, etc.

However, most ergonomics research and advice primarily deals with the area concerned with human well-being; in particular, the prevention of musculo-skeletal disorders and other

* Parts of this chapter have been previously published in Dul and Neumann [38] and Neumann and Dul [24].

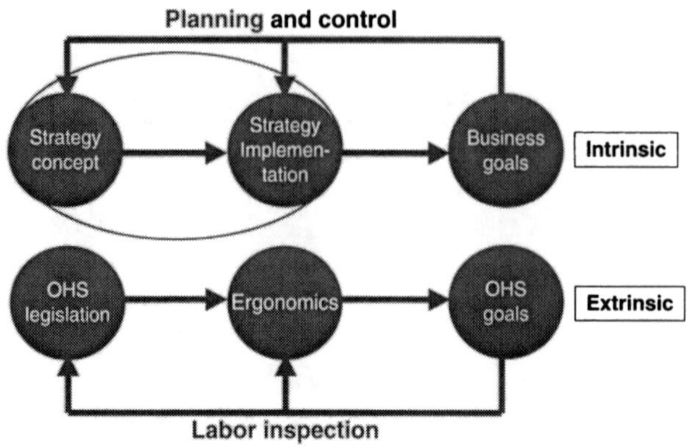

FIGURE 1. Upper part: The relationship between strategy (strategy concept and strategy implementation) and business goals. Lower part: The present isolated position of health ergonomics.

occupational health and safety goals. Furthermore, in several countries ergonomics is closely linked to occupational health and safety legislation. Under these circumstances companies may perceive ergonomics as an extrinsic element (lower part of Fig. 1), and not as part of the strategy, business goals and planning and control cycles (upper part of Fig. 1). The current trend in western governmental policies, namely, to reduce command-control legislation and to increase support for voluntary initiatives, is a threat to ergonomics as a health and safety perspective, because we do not believe that organizations will then spontaneously start ergonomic initiatives.

In our opinion and the opinions of others, the position of ergonomists in organizations is not very strong [2, 3], and ergonomics is not well integrated into companies' system design processes [4–7]. Perrow [3] argued that in the USA there are not many ergonomists working in companies, that they have no control over budgets and people, and that they are seen as protectors of workers; for example, not blaming human errors on the workers but on the designers and managers of the systems. There is no reason to believe that the situation in other countries is very different. Hendrick [8] added that ergonomists, wrongly, presume that others are convinced of the importance of ergonomics. Helander [9] listed seven common reasons that ergonomics is not implemented. He noted, among other things, that people think that ergonomics is to design chairs, ergonomics is common sense, and that organizations first design the technical system and then consider ergonomics. We suggest a new direction for ergonomics, using its full potential in organizations, without being exclusively dependent on health and safety considerations (see Fig. 2). We consider 'strategy' and 'business goals' as

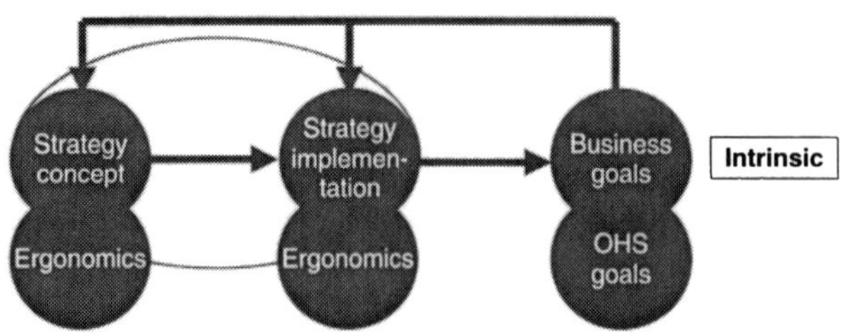

FIGURE 2. Business ergonomics to strategy and business goals.

useful connection points to internalize ergonomics in organizations, because strategy has top management priority and is normally intended to be broadly communicated and implemented in the organization. This raises the question: 'how to link ergonomics to strategy?'

ERGONOMICS AND STRATEGY

We distinguish three 'strategic arenas', which we use as a starting point for linking ergonomics to strategy. Each strategic arena represents a different set of stakeholders that might benefit from ergonomics:

- Corporate strategy
- Business function strategies
- Cross-functional strategies

In the corporate strategy arena, ergonomics must show that it can add value to the corporate business strategy for realizing competitive advantage. In this arena, the top management of the organization is involved, as well as external stakeholders including shareholders. In the business function arena, depending on the business function (e.g. product design, production engineering, marketing, human resource management, middle managers and the employees representing the business function will be primary stakeholders. Here ergonomics must show that it can support the chosen strategies, and the corresponding performance indicators of the functional field.

Cross-functional strategies involve two or more business functions, and hence several corresponding middle managers and employees from these business functions will be primary stakeholders. Below we present some possibilities to link ergonomics to specific strategies in each arena.

CORPORATE STRATEGIES AND ERGONOMICS

Cost and differentiation strategies

Porter [10] suggests two basic corporate strategies that can be distinguished: a differentiation strategy and a cost strategy. In the differentiation strategy, the company produces and delivers products or services with unique features to attract consumers. User-centred products, created by ergonomic product design, can be such a feature [11]. In the cost-strategy, a company competes on the basis of the cost of the product or service. By ergonomic design of the production system, including ergonomic job and workplace design, or human work elimination by mechanization or automation of inefficient, unhealthy or hazardous tasks, the costs per unit can be reduced and labour productivity increased [12]. Reducing costs and increasing productivity is an ongoing activity in most organizations.

Resource based view

According to the resource-based view (RBV) of the firm [13] a company can outperform other companies by the way the company combines its technical, human, and other resources. When people are considered to be a key resource, it is important to maximize their capabilities and knowledge and to prevent its outflow by using ergonomics. The RBV attempts to reach sustained competitive advantage by choosing and developing resources that are valuable, rare, costly to imitate, and exploitable by the organization. By improving ergonomic job and workplace design, ergonomics can contribute to the maximization of the use of valuable, rare, and costly human resources, and hence to the maximization of sustained competitive advantage and to economic performance above normal.

BUSINESS FUNCTION STRATEGIES AND ERGONOMICS

There are many different business functions in a company to which ergonomics can be linked. Here are just a few.

Product design

Product design and engineering can benefit from the application of ergonomics both to the design of the product for the end user and to the design of a product that is easy to produce [14]. All too often, products are not designed to accommodate the physical or mental characteristics of the target customer. Better design, with attention to the user, can result in more desirable products [15].

Design for Assembly (DfA), or Design for Manufacturability (DfM) [16], is an approach by which the ergonomics of assembly and manufacturing is considered in the product-design stage. By considering production ergonomics in this phase it is possible to avoid the high costs associated with corrective ergonomics processes, with little extra investment in the design phase. It is in the early stages of design, while the concept is still malleable, that changes can be implemented with the least cost [17, 18]. Embedding ergonomics considerations early in design can therefore be much more cost effective than retrofitting.

Helander and Nagamachi [16] show for several cases that DfA provides a double benefit: it reduces production cost as well as occupational health and safety (OHS) problems, just because assembly has been simplified.

Production engineering

Industrial work has become increasingly repetitive and monotonous. The resulting problems of unmotivated and injured workers have long been known. Attention to ergonomics can support alternative designs that may result in systems, such as long-cycle parallel assembly-flow systems, with superior performance [19].

Automation is another strategy by which performance may be increased and exposure to repetitive monotonous work decreased, although it is also important to address the tasks remaining for operators, not just the tasks that are automated away [20]. Integrating ergonomics into production engineering, so that solutions are optimal for both productivity and operator well-being can be difficult due to a 'clash of perspectives' between engineers and ergonomists [21]. Other barriers to integration that have been reported by engineers include lack of time, knowledge, methods, tools and the absence of demand from customers and management [22].

Corporate communication/marketing

In marketing communication the company's competitive product (or production) characteristics can be communicated to the customer. Positive product characteristics of ergonomically designed products like functionality, usability, health and comfort, can be communicated to the customer. With respect to production ergonomics, 'fair trade' products are an example in which the communication may target the consumer aware of their availability. A barrier here remains the extent to which consumers are prepared to differentiate products based on the working conditions of their manufacture, and the extent to which credible information on the working environment is available.

Ergonomics can form a part of a company's 'corporate social responsibility' and 'corporate sustainability' platforms [23] in a society that is placing increasing demands on companies to be more than money-making organizations. Such advertising of ergonomics as part of 'harmless product' or 'harmless production' campaigns can offer the potential consumer a better product, made in better working conditions, for a better world.

Human resource management

Good working conditions present one strategy for attracting and retaining high-quality employees. The need to attract people to manual assembly jobs in Sweden was one of the driving forces of production system innovation away from traditional Tayloristic-line production toward new more productive and attractive solutions [24]. HRM departments have long been held responsible for employee welfare, even though they tend to have little responsibility for work system design. The gap between human resources and operations management (OM) has been noted and presents a challenge for the design of work systems that are motivating and productive [25].

While many HR strategies exist, we mention only 'High Performance Work Systems' (HPWS) as one of those that incorporates elements of involvement and employee empowerment consistent with existing 'participatory' ergonomics approaches, as well as job design. HPWS have shown themselves capable of increasing organizational performance [26], but appear to operate on the HR side of the HRM-Operations Management gap [25]. Ergonomics, with its focus on both human well-being and system performance, could help bridge this gap.

CROSS FUNCTIONAL STRATEGIES AND ERGONOMICS

Most of the common and well-known management models, fads and hypes fit into this cross functional category. Lean Production, Business Process Re-engineering, Downsizing, Total Quality Management, and the Service Profit Chain are examples of broader strategic concepts affecting different functions in the organization, usually accompanied by a specific set of tools to implement the strategy. For these strategies to be successful, several business functions must work together to realize an effective implementation. This implies a need to co-ordinate among a complex network of stakeholders to ensure that all functions are 'pulling in the same direction' towards strategic objectives. This poses both challenges and opportunities for capitalizing on the application of ergonomics in the organization.

Downsizing, lean production, business process re-engineering

The ergonomics and health and safety communities have tended to view strategies such as downsizing, lean production, and business process re-engineering negatively – as a health threat. Vahtera et al. [27] have found risk of musculo-skeletal disorders to increase by 5.7 times during 'corporate downsizing'. The individuals' perception of the downsizing process itself also appears to affect health [28]. Landbergis et al. [29], in their review of available literature, noted increased negative health outcomes are often associated with the adoption of lean manufacturing approaches. While it is tempting to look at these results and say: 'Strategy X is bad ergonomics', this is perhaps not the right conclusion. In our view, strategy includes both a concept (a strategy plan) and its implementation. The extent, and the way in which a strategy is realized in practice, may vary with the gap between strategy and practice being apparently a more important indicator of (poor) performance than the strategy itself [30]. It is difficult therefore to determine the ergonomic consequences of production strategies directly without considering the specific implementation for each case. There may be a gap between the strategic concept and its implementation that is leading to poor ergonomics, and compromising the effective realization of the strategy.

Total quality management and the service profit chain

'Total quality management' is a general term for improving business processes by incremental improvements, involving 'all' business functions. For the implementation and management of this strategic concept, specific tools can be used. Many European organizations use the EFQM model (European Foundation for Quality Management). In this model, nine criteria for quality are considered including two for people ('people enablers' and 'people results'). Ergonomics can be applied as part of a people-enabling approach, and therefore can contribute to people results and total quality.

Quality has become an important competitive domain that has been seen to have links to ergonomics [31]. For example, Axelsson [32] found that jobs with poor ergonomics were ten times more likely to have quality deficits than jobs with good ergonomics, and Yeow and Sen [33] found a reduction of $574 000 in rejection costs with less than $1100 in modifications and training, which led to a 5.2% reduction in customer side deficits. These examples illustrate how application of ergonomics can contribute to strategic quality goals in manufacturing.

Heskett et al. [34] proposed the Service Profit Chain (SPC) model that relates employee satisfaction to customer satisfaction, and further to financial performance of a service organization. Empirical studies suggest that the relationships between employee satisfaction, customer satisfaction and business performance exist if the employee-customer contact is more important [35]. Ergonomics can contribute to employee satisfaction, and therefore to the strategy concept of the SPC [36]. A multiple case study showed that managers in service-based warehouses decided for ergonomics improvements because of the expected effect on customer satisfaction, and not because of health and safety [37].

THE WORKSHOP

In order to 'test' the ideas outlined in the previous section, we arranged a workshop at an international conference to discuss them with experienced researchers and practitioners from around the world. This allowed us to hear of examples reflecting the presented framework, and to refine this frame to incorporate the experience and understanding of the participants (many of these examples are now incorporated in the conceptual frame presented in this Chapter).

During the workshop '*The Strategy – Ergonomics relationship: Exploring and combining available knowledge to improve performance*' held at the Human Aspects of Advanced Manufacturing: Agility and Hybrid Automation (HAAMAHA) Conference, San Diego, USA, 2005 we discussed the above views. The workshop was attended by 12 conference participants (see acknowledgements), and included both researchers and practitioners with ergonomics, engineering and business administration educational backgrounds. The workshop consisted of a brief introduction of our views [38], followed by a two-hour 'open forum' round-table discussion in which participants were encouraged to discuss aspects of the topic they felt were most relevant. Both the authors, as well as two 'volunteers' amongst the participants, took notes. The summary of the workshop (presented below) was sent to the participants to confirm that their comments were not misrepresented.

WORKSHOP RESULTS

The discussions during the workshop often focused on examples from the manufacturing industry, although also examples from process industries and office-type organizations were discussed.

Participants mentioned four important business functions relevant to the thesis: finance, marketing, human resources (HR), and operations. Of these, the operations function was seen as the most critical to reach, since it is here that influence over the work system, and therefore ergonomics, is highest. While some participants questioned the utility of work environment as a marketing strategy, examples of 'best practices' awards for good workplaces exist and, in Denmark for example, work environment ratings are made public and can thus influence a company's image. A challenge to such efforts is a reliable 'certification' process that could verify, or even score, a company's production processes. The Red Stripe beer company was named as an example of having used work environment as part of their advertising campaigns in conjunction with a broader corporate social responsibility (CSR) agenda. There was a call for more evidence that application of ergonomics principles can provide financial, and not just health, benefits to the company. Finally, HR was not seen as commonly connected to ergonomics, although a number of examples exist of companies taking a broader interest in their employee's well-being. Ergonomics could support the notion that employees are the 'intangible assets' of the company.

Company context was seen as important in terms of its uptake of ergonomics. Wealthy companies with long-term perspectives might be more likely to include ergonomics as part of its social strategic thinking. Companies operating in a survival mode, with a short-term focus, may not see the social goal of ergonomics as a critical aspect the company must satisfy its business goals even though it may want to satisfy its social goals. Companies with high investment intensity (very expensive systems) may have more problems adopting ergonomics due to the 'rigidity' caused by the high cost of change. The 'time horizon' of manager decision-making will change with the company context – the nature of managers' decision-making will influence their perception of ergonomics. Company size was also seen as important: while large companies can afford specialists to help comply with occupational health and safety (OHS) regulations (and perhaps reap other benefits from ergonomics over time), smaller companies may have no idea how to begin incorporating ergonomics into their development processes. Once in the organization, the ergonomist can also give attention to the economic goal of ergonomics.

A number of corporate strategies and their ergonomics connection were identified over the course of the workshop. Firstly, ergonomics was seen as easier to 'sell' to companies that had a strategic philosophy that included employee well-being. Similarly, CSR agendas, previously mentioned, can provide support to the ergonomics agenda. 'Lean manufacturing' was discussed as two sided: while operator participation is critical, the work itself often remains Tayloristic. Participants noted a trend in continuous improvement processes (a 'lean' component) to focus on both quality and ergonomics in the improvement process. This parallels a trend towards the integration of management systems such as ISO 9000 (quality), ISO 14001 (environment), and OHSAS 18001 (work environment) [39]. This integration, and particularly the links between ergonomics and quality, may support a more natural and effective application of ergonomics in future development processes. 'Downsizing strategies' were noted as problematic, since the ergonomist may be 'downsized' early as the company tries to 'focus on the essentials'. Finally, benchmarking approaches were seen as a possible strategy that could help motivate companies to improve their own utilization of ergonomics.

The adoption process by which a company might take up ergonomics emerged as a thread in the workshop discussions. How, participants asked, should ergonomics 'grow' or evolve in the company? Can we learn from the successes of the quality movement? Some participants were sceptical that ergonomics could come from a strictly bottom-up approach – management support is needed. Obtaining such support may need some kind of 'ergonomics champion' inside the organization, especially at senior levels. Unfortunately, managers at these levels are mostly accountants, lawyers, or engineers, with little understanding of ergonomics. The group also made several observations with regard to ergonomics uptake. A newly hired manager had imported ergonomics thinking into her new company's processes. Companies had been seen to adopt ergonomics only to the extent required by OHS law – nothing more. Participants generally agreed that ergonomics should be included in the long-term planning of the organization and that integration at all levels of the organization was needed – although there is a need for tools that would support consideration of ergonomics in decision-making. An intervention strategy was discussed, which started by examining a company's mission statement and long-term business plan to identify how ergonomics could contribute to the company reaching its goals. In this way ergonomics could move beyond its legislated OHS 'pigeonhole' to achieve higher levels of application in the organization.

Knowledge and training were mentioned as potentially important factors for ergonomics adoption. The extent to which engineers receive training in ergonomics appears to vary across countries, with many engineers receiving almost no education in ergonomics. While the ergonomics community has developed and presented in textbooks and handbooks information on how to apply ergonomics in workplace design, the field of ergonomics is not widely known or understood in the engineering community that does design work. Management-training programmes generally provide no education in ergonomics – which almost certainly influences the manager's decision-making with regard to the application of ergonomics in their organization. Similarly, ergonomists seem to be receiving little training in engineering and management, which can also hamper the integration of ergonomics into engineering design processes and management decision-making. A cautionary note on ergonomics education for engineers was that if the organization does not expect the application of ergonomics, then the expertise of the ergonomically trained staff would not be used. Here again, context plays a role in determining the extent to which knowledge might be applied inside the organization. Employee's knowledge also represents a part of the company's 'intangible assets' or 'business intelligence' that can play a vital role in adapting to continually changing market conditions. Although case studies are lacking, ergonomics was seen to have potential in retaining valuable personnel who can contribute knowledgeably to an organization's development.

From discussions in this workshop the following research priorities were identified:

- Good studies demonstrating the 'win-win' (productivity and health) effects of good ergonomics practice and that provide insight into the process challenges faced in the adoption of ergonomics by the organization.

- Development of an 'Ergonomic Index' allowing an organization to judge its performance, and potentially gain certification similar to ISO9000.
- Studies of how customers respond to (process) ergonomics as a marketing tool.
- Studies of managerial and engineering attitudes and decision-making processes with regards to ergonomics.
- Broad surveys examining how companies manage or apply ergonomics in their organizations.

Finally, we add that a number of participants agreed that there is a need to engage the management community in such discussions as to how ergonomics can be applied to contribute to corporate strategic goals.

DISCUSSION

While the workshop participants appeared to generally agree with the thesis presented, it is possible that deeper reservations were not raised in this context. There remains a sense that ergonomics is a health issue, supports companies' social objectives and can be driven by the creation of local/national legislation. While we do not discount the need for legislation (in particular, when no economic benefits can be expected), we see great potential for ergonomics to contribute to organizations' core (economic) goals, while also furthering social (health) goals [38]. Seeing ergonomics as contributing to organizational goals may facilitate successful implementation. This requires a paradigm shift and a change in thinking about what 'ergonomics' is (more than just a health issue) and represents (a potential contributor to strategic goals) for the company. For some health and safety professionals this may be a foreign way of thinking, and that such a move could weaken the health and safety (H&S) structures in the organization. We argue the opposite – attending to both the strategic performance goals and the health goals simultaneously, as in the definition of ergonomics, may actually improve the effectiveness of H&S efforts and increase the credibility of ergonomists with management decision-makers. By paying closer attention to the performance aspects of ergonomics, and to the strategic needs of their client organizations, ergonomists may be able to open new doors and avenues of application, and thereby improve the effectiveness of ergonomics in the design of work systems that are both humanly sustainable and economically productive.

CONCLUSIONS

Our analysis suggests that ergonomics can contribute to many different company strategies, and can support the objectives of different business functions like production, marketing and HRM, and cross-functional strategies like TQM. In these terms ergonomics is seen as a tool or a means, rather than an end in itself. Discussions within the ergonomics community showed that linking ergonomics explicitly to business strategies and goals is feasible, has been done at times, yet remains a great challenge for the ergonomics discipline. For many ergonomists it means a paradigm shift, which requires a repositioning from an exclusive H&S focus to one that also includes strategic business objectives. However, by contributing to the shared goals of business performance, ergonomists may also be better able to reach their traditional objectives of well-being and health and safety.

ACKNOWLEDGEMENTS

This work has been supported by Erasmus Research Institute of Management of the Erasmus University Rotterdam, and the SMARTA theme of the Swedish National Institute for Working Life. We are thankful for the workshop participants who included: Mark Boocock, Tony Vitalis, Stefano Marzani, Arne Bilberg, Anabella Simoes, Pedro Ferreira, Damian Graham, Canan Ceylan, Ole Broberg, Roberto Montanari, Enda Fallon, and Reem Aekadeem.

REFERENCES

[1] IEA. (2000). *The Discipline of Ergonomics*. International Ergonomics Association. p. 1.

[2] Frick, K. (1994). *Från sidovagn till integrerat arbetsmiljöarbete: arbetsmiljöstyrning som ett ledningsproblem i svensk industri*. ISBN 91-86576-25-9. Stockholm: Arbetslivscentrum.

[3] Perrow, C. (1983). The organizational context of human factors engineering. *Administrative Science Quarterly*, 28 (4), 521–41.

[4] Jensen, P.L. (2002). Human factors and ergonomics in the planning of production. *International Journal of Industrial Ergonomics*, 29 (3), 121–31.

[5] Imbeau, D., Bellemare, M., Courville, J., et al. (2001). Ergonomics in a design environment. In *International Encyclopedia of Ergonomics and Human Factors* (W. Karwowski, ed.). London: Taylor and Francis, pp. 1233–5.

[6] Neumann, W.P. (2004). *Production Ergonomics: Identifying and Managing Risk in the Design of High Performance Work System*. Lund Techn.

[7] Skepper, N., Straker, L., and Pollock, C. (2000). A case study of the use of ergonomics information in a heavy engineering design process. *International Journal of Industrial Ergonomics*, 26 (3), 425–35.

[8] Hendrick, H. (1996). Good ergonomics is good economics. *Human Factors and Ergonomics Society*. Santa Monica, CA, USA.

[9] Helander, M. (1999). Seven common reasons to not implement ergonomics. *International Journal of Industrial Ergonomics*, 25 (1), 97–101.

[10] Porter, M.E. (1985). *Corporate Advantage: Creating and Sustaining Superior Performance*. Free Press.

[11] Dul, J. (2003). The strategic value of ergonomics for companies, in Human Factors. In *Organisational Design and Management VII* (H. Luczak and K.J. Zink, eds). Aachen, Germany: IEA Press, pp. 765–9.

[12] Put, M. *The Contributors of Ergonomics to Cost Leadership*. Master Thesis. Rotterdam: RSM Erasmus University.

[13] Barney, J.B. (1991). Firm resources and sustained competitive advantage. *Journal of Management*, 17 (1), 99–120.

[14] Broberg, O. (1997). Integrating ergonomics into the product development process. *International Journal of Industrial Ergonomics*, 19 (4), 317–27.

[15] Vicente, K.J. (2004). *The Human Factor: Revolutionizing the Way People Live with Technology*. Taylor & Francis.

[16] Helander, M. and Nagamachi, M. (1992). *Design for Manufacturability: A Systems Approach to Concurrent Engineering and Ergonomics*. Taylor & Francis.

[17] Alexander, D.C. (1998). Strategies for cost justifying ergonomic manufacture and assembly. *Manufacturing Engineer*, 77 (5), 221–4, Improvements, *IIE Solutions*, 30 (3), 3035.

[18] Miles, B.L. and Swift, K. (1998). Design for manufacture and assembly. *Manufacturing Engineer*, 77 (5), 221–4.

[19] Ellegård, K.D., Jonsson, T., Engström, M.I., et al. (1992). Reflective production in the final assembly of motor vehicles – en emerging Swedish challenge. *International Journal of Operations and Production Management*, 12 (7).

[20] Neumann, W.P, Kihlberg, S., Medbo, P., Mathiassen S.E., et al. (2002). A case study evaluating the ergonomic and productivity impacts of partial automation strategies in the electronics industry. *International Journal of Production Research*, 40 (16), 4059–75.

[21] Kirwan, B. (2000). Soft systems, hard lessons. *Applied Ergonomics*, 31, 663–78.

[22] Broberg, O. Integrating ergonomics into engineering: empirical evidence and implications for ergonomists. *Human Factors and Ergonomics in Manufacturing* (in press).

[23] Hardjono, T. and De Klein, P. (2004). Introduction on the European Corporate Sustainability Framework (ECSF). *Journal of Business Ethics*, 55 (2), 99–113.

[24] Neumann, W.P. and Dul, J. (2005). Workshop report: Ergonomics contributions to company strategies. In: *Ergonomics as a Tool in Future Development and Value Creation*. Proceedings of the 37th Annual Conference Nordic Ergonomics Society (NES), 10–12 October 2005. Norway: Oslo.

[25] Boudreau, J., Hopp, W., McLain, J.O., and Thomas, L.J. (2003). On the interface between operations management and human resources management. *Manufacturing and Service Operations Management*, 5 (3), 179–202.

[26] Appelbaum, E., Bailey, T., Kalleberg, A.L., and Berg, P. (2000). *Manufacturing Advantage: Why High-Performance Work Systems Pay Off*. Cornell University Press.

[27] Vahtera, J., Kivimäki, M., and Pentti, J. (1997). Effect of organisational downsizing on health of employees. *The Lancet*, 18 October, 1124–8.

[28] Pepper, L., Messinger, M., Weinberg, J., and Campbell, R. (2003). Downsizing and health at the United States Department of Energy. *American Journal of Industrial Medicine*, 44, 481–91.

[29] Landsbergis, P.A., Cahill, J., and Schnall, P. (1999). The impact of lean production and related new systems of work organization on worker health. *Journal of Occupational Health Psychology*, 4(2), 108–130.

[30] Rho, B.H., Park, K., and Yu, Y.M. (2001). An international comparison of the effect of manufacturing strategy-implementation gap on business performance. *International Journal of Production Economics*, 70, 89–97.

[31] Drury, C.G. (2000). Global quality: linking ergonomics and production. *International Journal of Production Research*, 38 (17), 4007–8.

[32] Axelsson, J.R.C. (2000). Quality and ergonomics: towards successful integration. In *Linköping studies in science and technology*. Dissertations, 616, University of Linköping.

[33] Yeow, P.H.P. and Sen, R.N. (2003). Quality, productivity, occupational health and safety and cost effectiveness of ergonomics improvements in the test workstations of an electronic factory. *International Journal of Industrial Ergonomics*, 32 (2), 147–63.

[34] Heskett, J.L, Jones, T.O, Loveman, G.W., et al. (1994). Putting the service – Profit chain to work. *Harvard Business Review*, 72 (2), 164–74.

[35] Dean, A.M. (2004). Links between organisational and customer variables in service delivery – Evidence, contradictions and challenges. *International Journal of Service Industry Management*, 15 (3–4), 332–50.

[36] Hogenes, E., Dul, J., and Haan, G. (2006). Human centered designed work environments at Interpolis. In: Pikaar, R.N., Koningsveld, E.A.P., Settels, P.J.M. (eds) *Meeting Diversity in Ergonomics*. Proceedings IEA 2006 Congress. ISSN 0003-6870, Elsevier Ltd.

[37] Janssen, K., Van de Vecht, H., and Wong, Y.W. (2004). *Employee and Customer Satisfaction are the Most Important Motives for Solving Ergonomics Problems*. Bachelor Thesis. Rotterdam: RSM Erasmus University.

[38] Dul, J. and Neumann, W.P. (2005). *Ergonomics Contributions to Company Strategies*. Proceedings of the 10th International Conference on Human Aspects of Advanced Manufacturing: Agility and Hybrid Automation, 18–21 July 2005. San Diego: USA, HAAMAHA, 2005.

[39] Matias, J.C.D. and Coelho, D.A. (2002). The integration of the standards systems of quality management, environmental management and occupational health and safety management. *International Journal of Production Research*, 40 (15), 3857–66.

Meeting Diversity in Ergonomics (2007)
Pikaar, Koningsveld, and Settels (Eds)

3

New Challenges: Ergonomics in Engineering Projects

R.N. Pikaar Eur. Erg

ErgoS Engineering & Ergonomics, P.O. Box 267, NL 7500 AG Enschede

Abstract. This chapter presents experiences of ergonomists in applying a system ergonomic approach in real life engineering projects. The aim is to show what is useful in system ergonomics, and what isn't. First the impact, or perhaps a lack of impact, of ergonomics in an engineering environment is considered. The methodology of system ergonomics is discussed briefly and illustrated by a case study of an emergency control centre. Several other cases are presented, which have at least two factors in common:

(1) comparable system ergonomic approaches, and
(2) the involvement of ergonomists until or including the implementation phase.

Each case leads to some lessons learned. Practitioners evaluating work and designing, or implementing, solutions may develop best practices. These best practices should be included in the theoretical framework of system ergonomics. Questions may be raised on whether the ergonomics community is studying the most relevant issues from an engineering point of view.

Keywords: marketing system ergonomics, socio-technique, design project, applied ergonomics, engineering project, control centre, case studies

MARKETING ERGONOMICS

Ergonomics

Ergonomics (or human factors) is described as fitting tasks, workplaces and interfaces to the capacities, needs and limitations of human beings. The aim of ergonomics is to optimize safety, health, comfort and efficiency for the human in the work system. The tools which are used and the production systems which are controlled are numerous and varied. Due to the variety of tools and differences between users in terms of, for example, body size, muscular strength and cognitive abilities, favourable human task matches will not arise as a matter of course. Therefore, designing human-machine systems is a complex task [1], characterized by the need for an interdisciplinary approach.

A succinct definition of ergonomics is *user-centred design* or *user-centred engineering* [2], expressing a focus on the human being, and at the same time emphasizing prevention by design. (Note: such a definition would also cover other words related to ergonomics and frequently used as well, such as human factors, design for all, or participatory ergonomics. In this article 'ergonomics' will be used as an overall term.)

Unsafe, unhealthy, uncomfortable or inefficient work situations can be avoided by taking into account the limitations of human beings during design. Ergonomics contributes to the prevention of inconveniences and, to a considerable degree, improves system performance in terms of an increased productivity (Vink et al. [3]). Vink et al. give several references to support the claim that good ergonomics is good economics. In addition, Vink discusses four cases in an effort to convince ergonomists to apply a positive approach to the outcomes of their work. At the same time Vink is worried about the negative connotation of ergonomics. Dul and Neumann [4] discuss the same topic: 'Most ergonomics research and advice primarily deals with the well-being goal of ergonomics, in particular the prevention of musculoskeletal disorders, and other occupational health and safety goals. Furthermore, in several countries ergonomics is closely linked to occupational health and safety legislation. Under these circumstances companies experience ergonomics as extrinsic... We do not believe that organizations will then spontaneously start ergonomic initiatives.' Dul suggests linking ergonomics to business strategies and goals, such as:

- reducing costs and increase productivity
- maximizing the use of valuable, rare and costly human resources
- designing products for (easy) assembly, design for manufacturing
- integrating ergonomics into production engineering
- corporate communication; i.e. ergonomically designed products and/or the company's corporate social responsibility.

As a final remark, Dul and Neumann suppose: 'for many ergonomists it means a paradigm shift, which requires a repositioning from health ergonomics to business ergonomics. However, by contributing to the shared goals of business performance, ergonomists will also better be able to reach their traditional objectives of well-being and safety.' To be more successful in the market, ergonomists should focus on the business performance, or more precisely on investment projects: the design, redesign or extension of production systems. It is by definition that ergonomists are not the only profession involved here.

Working for many years as a professional ergonomist in Europe, the author has made comparable observations, inevitably leading to the conclusion that occupational health and safety issues are not the real ergonomic issues. The value of ergonomics is beyond health and safety. It is illustrative to reread some of the opening statements, made in 1989, at the International Conference 'Marketing Ergonomics' [5]: 'Is ergonomics a quality tool to use in building a profitable and well-run organization? Ergonomists think so, but often they are not very convincing... they have a way of saying "the answer to your question depends on the situation at hand... being not very to the point".' In addition it was stated: 'If one aspect above all others is lacking in the expertise of many good ergonomists, it is the marketing of their profession'. Verhoeven [6], a senior project manager, expressed the importance of ergonomics by stating: '...ergonomics are an integral part of system management... Professional skills, good cooperation and sufficient motivation are absolute requirements to achieve the common goal to finish a project within the goals set for time, money and overall quality which includes good ergonomic solutions in hardware and organization.' In 2006, the general feeling hasn't changed much, according to the general opinion at the triennial IEA Congress in Maastricht. However, some progress, particularly in process control and regarding standardization, is indicated (and elaborated under 'Case study: 112–emergency control centre' and 'Ergonomics in projects' below).

Ergonomist/Human Factors specialist

It may not be very clear what an ergonomist is or does by profession. The International Ergonomics Association (IEA) describes ergonomics as: 'the scientific discipline concerned with the understanding of interactions among humans and other elements of a system, and the profession that applies theory, principles, data and methods to design in order to optimize well-being and overall system performance'. Apparently, there is a distinction between (a) scientific research in human sciences and (b) professionals applying the results of this basic research in practice. The scientists are experts in one research area (for example, medical sciences, biomechanics, behavioural sciences, cognitive sciences, industrial design). The professional ergonomist needs to know something about all of these areas. He also needs a background in one or several engineering sciences. Because this is almost impossible, the professional ergonomist relies heavily on methodology, in particular for analysis, design and engineering, in other words the process of systems design.

Aim of this paper

De Looze and Pikaar [7] assume a gap between scientific research on human factors and on design theories on one hand, and the needs of professionals in engineering and design on the other hand. Is the ergonomics community studying the most relevant issues from a practical point of view? Closing the gap between practitioners and researchers remains a challenge. Two steps should be taken:

(1) organize access to the best practices developed in the field and
(2) organize research programmes with potential societal and market value.

There is one important thing scientists are missing: case material. Professional ergonomists have a tremendous amount of case material. After 25 years of working in the field, the author would be able to present over 100 cases. Some are included here.

Practitioners, evaluating work and designing or implementing solutions, may develop good or even best practices. Publishing a report on a successful, or perhaps an unsuccessful, project is seldom part of the work contract. It is not a standard line of business if one is not affiliated to scientific research. In addition, getting a project report published may easily fail, because this type of work is not commonly accepted in the international journals. Hence writing this article has been an opportunity taken to publish about projects.

DESIGN AND ENGINEERING

Project

Projects concern building a new production system, extending an existing system or automating (or mechanizing) existing systems. It doesn't make any difference whether the system is a large production system, an office building or merely a small facility, such as a bridge master's guard house. A project is characterized by a physical result, something that actually works. A project always requires the input of several disciplines:

- management (project owner, project managers)
- end users or their representatives
- engineering disciplines (mechanical, instrument, building, software)
- (building) contractors
- others, such as an architect, ergonomist and a personnel manager.

Usually a project goes through several project phases, starting with a feasibility study, via several design steps, to (detailed) engineering and implementation (see 'Methodology –Systems Ergonomics'). The ergonomics in projects concerns all human-centred activities during a project: analysis, design and detailed engineering. The ultimate goal is to help optimizing the (new) work system, by introducing a user centred approach. This should be the linking pin between a company and ergonomics (refer to Dul and Neumann [4]).

Socio-technique

The idea of work system optimization or joint design of the social system and the technical system is an old one. Socio-technical theory has at its core the notion that design and performance of new systems can be improved, and indeed can only work satisfactorily, if the social and the technical system are brought together and treated as interdependent aspects of a single work system. Clegg [8] argues that the application and diffusion of socio-technical principles and practices have been disappointing. A joint design and optimization of both the social and the technical system seems to be rare. It is speculated there are two reasons for this:

(1) In practice, engineering (or technology) is leading.
(2) Neither engineers nor management accept a design methodology (or approach) guided by or developed by others, such as social scientists or, for that matter, ergonomists. Hence, why not follow the leadership of engineers?

Does this imply that an ergonomic input in projects is doomed to fail? Looking at literature on system ergonomics and socio-technical theory, it appears that most of the research has been done in manufacturing of consumer goods (the well known work-structuring projects at Philips), or car manufacturing (at Volvo [9]). Apparently, less work is published on projects in process industries, though much work has been done here, particularly on the integration of human factors in design projects. Unlike manufacturing, process control and, for example, software design are not very visible or physical. You cannot touch the design result. There is no manual handling involved. The focus is on cognitive skills, which are far more difficult to understand.

The interest of process industries and power generation (nuclear) industries in ergonomics may be driven by safety and reliability issues. The influence of the human factor in chemical processes or power generation has been recognized widely. For example, Rijnsdorp [10] integrated ergonomics into his standard engineering work *Integrated Process Control and Automation*. Rijnsdorp also stated in many publications, that if one wants to realize ergonomics in system design, the ergonomic activities should be integrated in existing company procedures or standard engineering approaches. Formalizing an ergonomic design procedure will not be successful; you should adapt to company procedures.

Standards

In industry, standards play an important role, particularly in contracting work. It is relatively easy for the project owner to prescribe standards if they are available. Most ergonomic standards concern general requirements for designing products, workplaces or processes. General requirements are not easy to use for non-specialists, and therefore usually ignored with the silent consent of both project owner and contractors. Examples of frequently prescribed standards are ISO 6385 (2003) *Ergonomic principles in the design of work systems*, EN614-1 (1995) *Safety of Machinery – Ergonomic design principles* or ISO11064 – Part 1 (2000), *Ergonomic design of control centres – Principles of the design of control centres*. On a general level the requirements in these standards are valuable for ergonomists. Once the standard is accepted, it implies that an ergonomic task analysis shall be the starting point for the project, because this is a mandatory requirement. Which profession is capable of performing task analysis adequately?

Dul and de Vries [11] mention a lack of specific technical standards as compared to general standards. They assume that simple specifications of technical requirements can more easily be understood and used by designers and engineers. In the field of process control rooms such a technical guideline has been operative since 1998 [12]. We know, because every now and then ergonomic questions reach our desks, i.e. we are invited to participate in a project. Several observations regarding this guideline can be made:

- The Ergonomic Guideline [12] provides guidance for layout, workstation and interaction design, also including some more 'difficult' topics, such as workload assessment and job design. This guideline is accompanied by a general document, the so-called Engineering Guideline [1], giving general ergonomic principles and discussing the ergonomic inputs required at each step of a project.
- The user is confronted with several warnings.
  - Engineers, will be able to implement ergonomic guidelines, in particular for standard situations. However, a professional ergonomist may be able to do the same job in less time.
  - Some design tasks should be carried out by experts, namely the situation analysis, task allocation, job design and managing user participation.
  - Don't follow legislation, because it sets minimum requirements for health and safety, which differs from optimal design. Optimizing work conditions is a complex trade-off between costs, hazards, productivity and quality of working conditions.
- Specific guidelines may only last about 10 years. In the case of process control rooms, it is evident that, for example, display technology rapidly changes. Before applying 'simple' guidelines to a changing technology, engineers should have some understanding of universal ergonomic principles. Studying these principles would be a good idea anyhow!
- The guidelines have been written within one year. In parallel, the author participated in the development of ISO 11064, the aim and content of which is comparable to the guideline [12]. Efforts on ergonomics standards are primarily made by ergonomists, and hardly funded by industry [11], which is contrary to the general approach to standardization. Somehow, working on standards is inefficient. For this particular topic it took almost 20 years. As a consequence standards are always 'late'.

METHODOLOGY – SYSTEMS ERGONOMICS

A successful project, including a system ergonomics approach, is possible provided ergonomists adapt to company strategy and standard engineering procedures. A general project structure for human centred systems design was published amongst others by Rijnsdorp [10], Pikaar et al [13] and Kragt [14]. While studying at the Ergonomics Group of Twente University, the author remembers clearly the efforts made by Prof. Rijnsdorp to match (several) commonly used industrial engineering procedures with a system ergonomics approach. After some smaller projects in industry, the first big challenge has been applying system ergonomics in the Exxon FLEXICOKER Consolidated Central Control Room project (1981–1986). This project is well documented and evaluated ([10], [13], [15]). The project included also job design, local work organization design and an extensive user participation program. The schematic of Fig. 1 has been a scientific product of this industrial project. Fig. 1 emphasizes the integration of technical and organizational design, as suggested by Singleton's *Ergonomics*

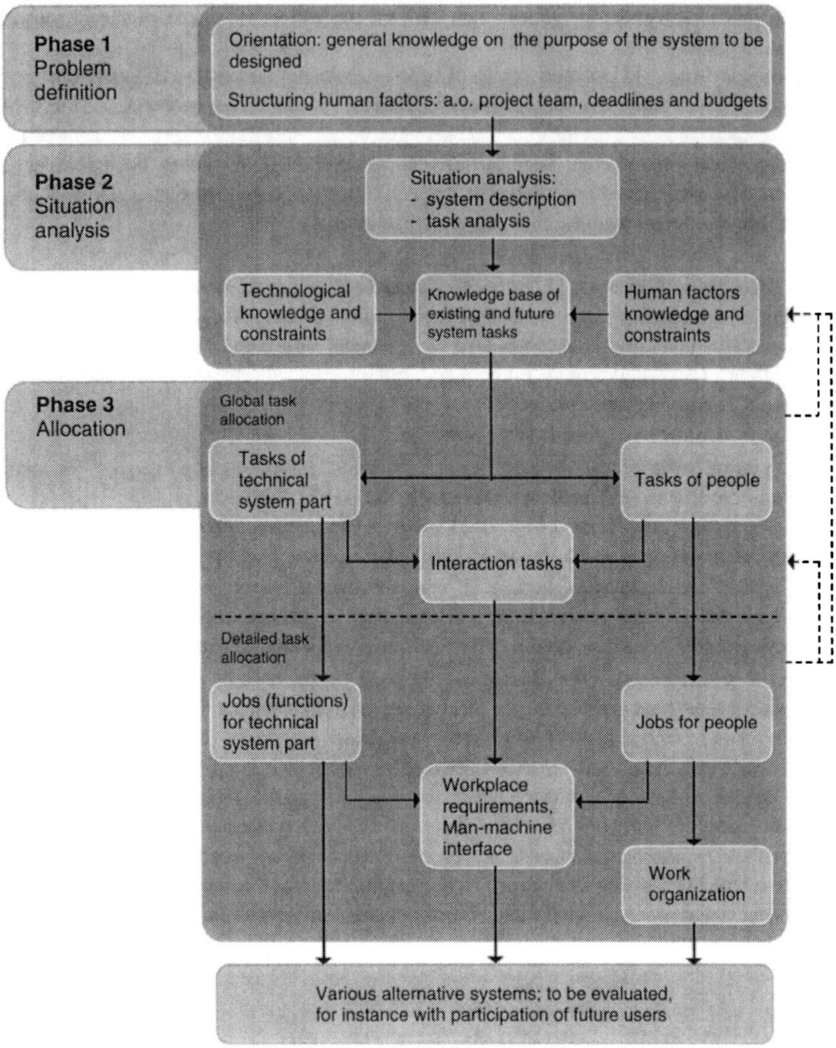

FIGURE 1. General project structure for system design from an ergonomic point of view.

in System Design [16]. The ergonomic phases, problem definition, situation analysis and task allocation could successfully be related to major industrial engineering phases, i.e. design basis, design specification and detailed engineering and construction [17].

Ever since the FLEXICOKER project, the approach has been debated many times with colleagues, project managers, engineering groups and so on. The approach has been the basis for many quotations for ergonomic projects (see 'Ergonomics in Projects' below). However, there are also some weak spots:

- The task allocation steps are difficult to substantiate because there is little guidance (theory, or a proven scientific basis) on this topic. Task allocation is difficult to explain to engineers and project managers. The word 'task allocation' typically belongs to the human factors vocabulary. Also, the structure of the allocation phase as given in Fig. 1 is rather complicated. Finally, job design is usually not associated with ergonomics.

- In practice, one will experience difficulty explaining why so much time has to be spent on a task analysis of an existing situation, while working on a 'new' project.
- Apparently, Fig. 1 stipulates that the largest part of the ergonomics work is done before the actual workplace design is even started (it is the last step in the design approach). This is not in line with the general expectations.

Though our ergonomic approach to design didn't change over the years, the presentation did. One could say that the marketing of the ergonomic contributions to projects has changed. Fig. 2 [1] gives a current overview of the presentation of ergonomic engineering steps related to a general industrial project procedure.

Some changes made to the original schematic of Fig. 1 are:

- The general project phases and the ergonomic engineering steps are presented in two parallel flows. Related to each project phase, a typical ergonomic input or activity is specified. Even if the ergonomic input starts some project phases later than indicated in the schematic, every ergonomic engineering step still has to be taken, preferably catching up with the main project as fast as possible.
- Phase 3 – ergonomic step 3 is a condensed version of the Allocation Phase; typical ergonomics terminology to be avoided in real projects.
- The last box has been considerably expanded; there is a lot of work to do in ergonomics engineering during detailed engineering, construction and commissioning.

Remijn [18] compares typical phases in architectural design (build environment, Fig. 3) and an ergonomics approach (see Fig. 4). Refer to the chapter by Mulder [19] for a comparison to an ergonomic design approach for software development.

Short explanation of design steps

This explanation concerns the theory as presented in Fig. 2. Practice is illustrated under 'Case study: 112–emergency control centre' below.

(1) Feasibility: The feasibility step typically encompasses a review of human factor assumptions.
(2) Problem definition: This step starts with a general description of the project and the purpose of the system to be designed. The outline of the design steps have to be negotiated with project management, including design constraints.
(3) Situation analysis: The aim of the situation analysis is to gain insight in existing and future tasks. A situation analysis includes activities such as:
    (a) collecting formal documents, specifying the existing system
    (b) analysing the existing situation by observations and interviews about work tasks, problems the users experience and wishes they might have for the new situation
    (c) gathering relevant knowledge on the new system.

Ergonomists have many tools available for an analysis, as can be found for example in Wilson & Corlett [20], and the book by Meister [21]. A careful selection has to be made. Within a project, there is only a need for detailed knowledge on tasks and topics relevant

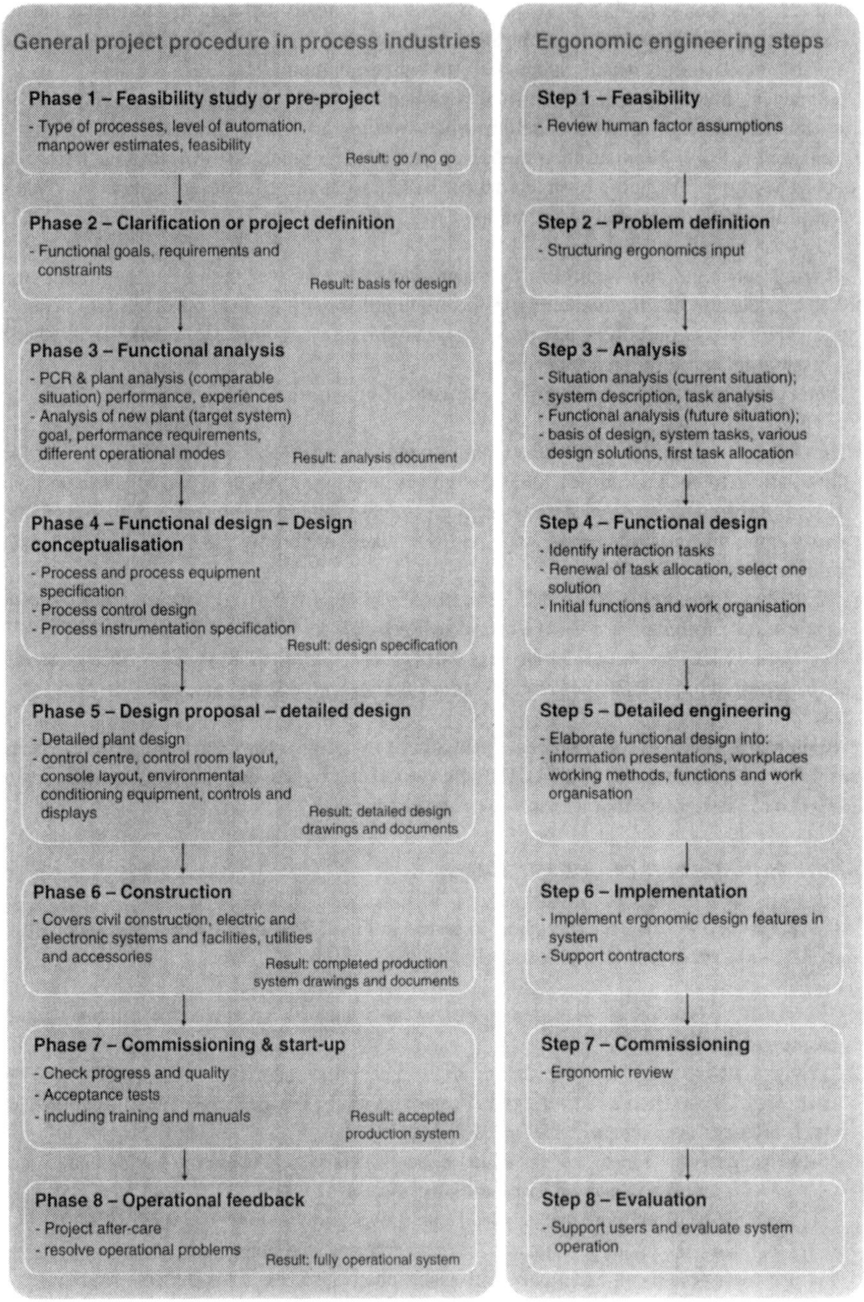

FIGURE 2. General project structure.

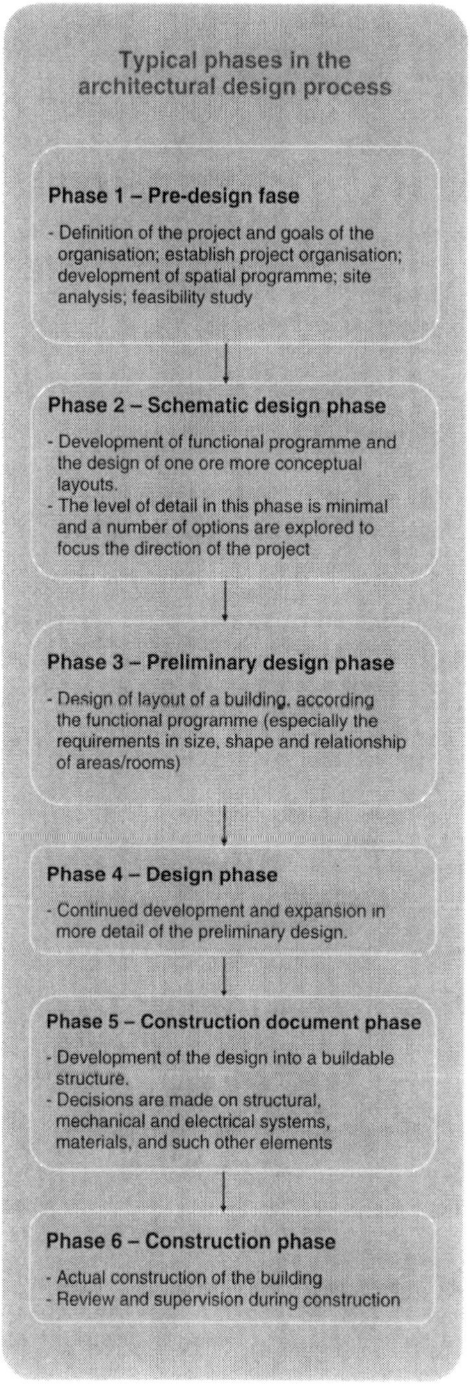

FIGURE 3. Typical phases in architectural design process.

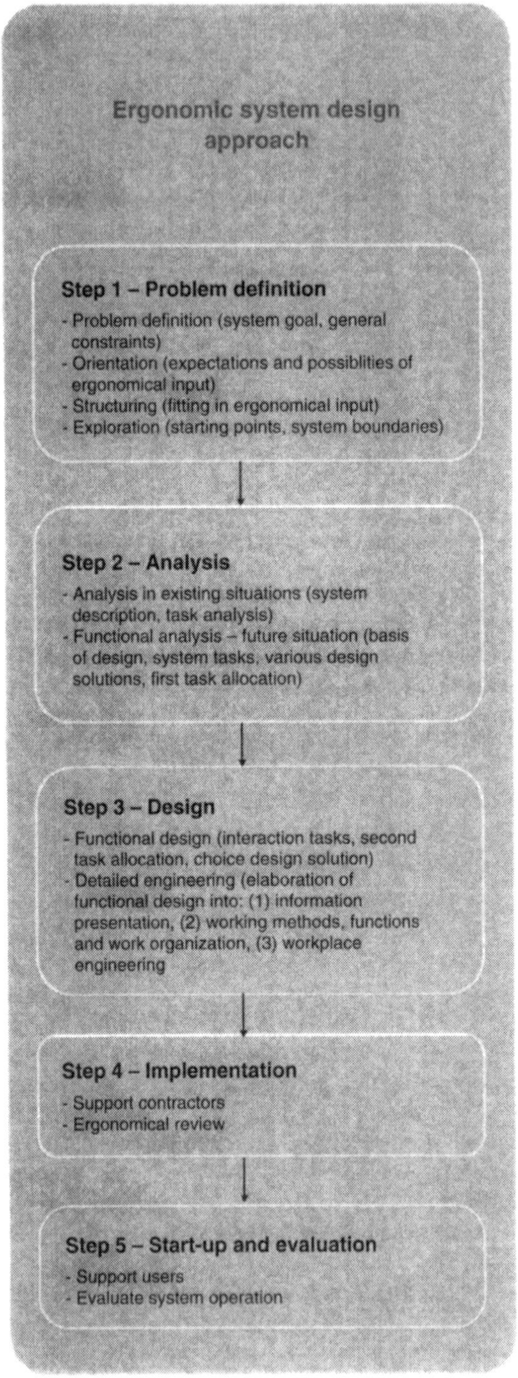

FIGURE 4. Typical ergonomic design steps, related to building projects.

to the project. For example, there is no need assess manual lifting situations in a logistics department, if a project concerns full mechanization of packaging.

(4) Functional design specification: In theory, functional design specification is about the allocation of system tasks. An allocation procedure includes a discussion on the level of automation, job requirements and the design of a work organization. Following this, a programme of functional design requirements has to be drafted, which concern amongst others things:
    (a) the allocation of tasks to workplaces
    (b) the layout of a system
    (c) the shape and size of workstations (including equipment)
    (d) environmental requirements (noise reduction, lighting levels).
(5) Detailed design (also indicated as detailed engineering): On the basis of a set of functional design requirements, various design solutions can be developed. Choices have to be made, which implies weighing all aspects involved, including ergonomics. Basic decisions regarding the detailed engineering may be based on 3D drawings, mock-up evaluations or prototyping.
(6) Implementation: During the construction phase the production system is realized. Typically, this will start with the production of workshop drawings and building site drawings. For example, from an ergonomic point of view, assistance in making workshop drawings for dedicated furniture may be required.
(7) Commissioning: Once the system is finished, formal handover (commissioning) of a working (and tested) system to the project owner will be organized. Typically an ergonomist could be involved in the review of all workplace-oriented parts of a system.
(8) Evaluation: Ideally, though not common procedure, an evaluation of the running system, for example resulting in operational feedback on design, engineering and management of the project, should be organized.

User participation and project ergonomics

User participation is the systematic consultation of those involved in the system to be designed. Experienced users dispose of a wide and detailed knowledge of system tasks and have practical experience that is not always documented or known to designers, engineers or staff. The main problem is to find ways to involve them efficiently in the design process.

User participation is strongly related to ergonomics, i.e. human-centred design. However, participation of future users is not listed anymore in the suggested ergonomic engineering steps, because it is not the same as ergonomics. User participation could best be approached as a third procedural line: the user participation steps (as suggested in Fig. 5). Participatory ergonomics (Vink [3]) is about participation and not necessarily about (good) ergonomics. In practice, the ergonomist will be a candidate for managing participation because of his user-centred background. However, any other person with management qualities could do equally well in this job (also refer to the case studies under 'Ergonomics in Projects' below).

Important steps in user participation are related to the situation analysis and the evaluation of various design solutions. During the analysis users meet the ergonomist and get acquainted with human factors issues. The evaluation of design alternatives gives them an opportunity to contribute to the design of their new workplaces and to influence an important decision making process.

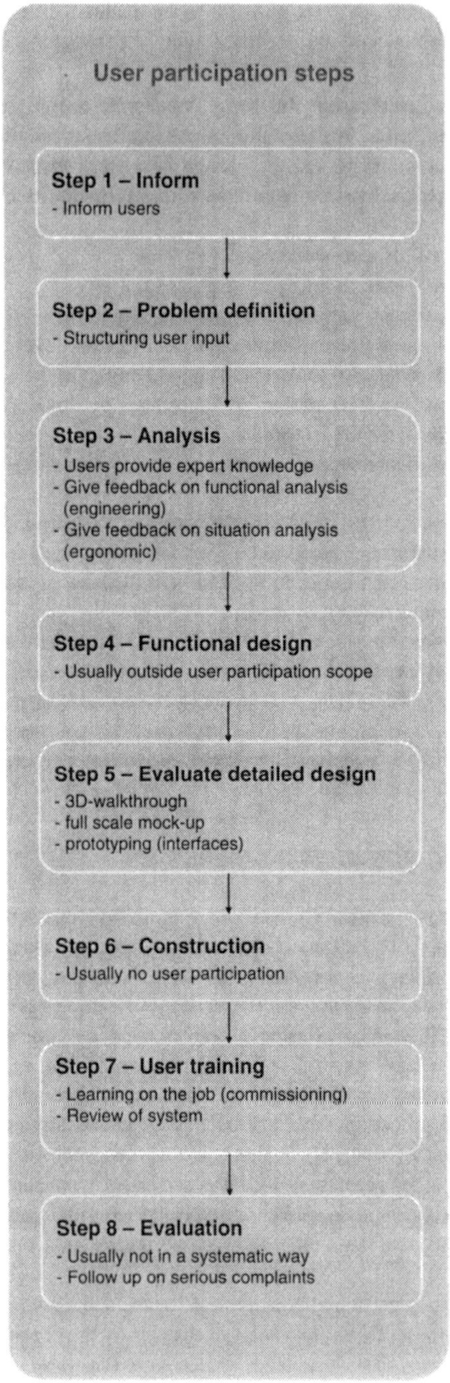

FIGURE 5. Steps in user participation; compare these steps to the phases given in Figure 2.

CASE STUDY – 112–EMERGENCY CONTROL CENTRE

In this section a recent complex design project is presented. An extended version of this case can be found in Wood [22]. This case is of particular interest because a new feature was added to the traditional ergonomic steps. Namely, a senior ergonomist was given the role of managing all building, construction and interior design activities, including ergonomics.

Project scope

The project concerns the redesign of an existing four-storey office building in order to accommodate an integrated regional 112–Emergency Control Centre for police, ambulance and fire brigade ('112–Centre'). One $500m^2$ floor was dedicated to the 112–Centre itself. Three other floors accommodate technical equipment, rack rooms, staff offices and three command centres for handling calamities or planned large-scale operations, such as a high risk football match. The full scope of the project included an investment of €11 million, half of which was related to technical equipment, 20% to project management and consultants, and 20% to (interior) civil, mechanical and electrical work, and interior design including dedicated furniture.

Before the project start, a situation analysis in the existing police emergency control centre was carried out within the scope of another project. One could say that by coincidence, the ergonomists were able to support the search for, and the selection of, a suitable building for the new regional 112–Centre, by providing some clear requirements. This process had been rather convincing and resulted into an assignment to manage the building activities as well as all ergonomics activities, including user participation. The project meant several challenges for the ergonomists:

(1) to show that it is advantageous to have an ergonomist as a design and building manager (interior, workplace, environment)
(2) to show that an ergonomics approach to control room design for industrial settings (see 'Methodology – Systems Ergonomics') applies equally well in a non-industrial organization
(3) to integrate three work organizations (i.e. police, ambulance and fire brigade), with different organizational structures as well as cultures
(4) to work within an environment not used to handling engineering projects within limited budgets and time frames.

Project organization

The project was headed by a general project manager, reporting to three directorates (regional police, ambulance, fire brigade), which in the end are controlled by political bodies. Four sub-projects were defined:

(1) **Accommodation**
Including interior design and all building related activities (internal walls, floor, ceiling, cabling, emergency power supply system, HVAC, special furniture). A sub-project team was managed by an external, half-time, senior ergonomist. The team consisted of an interior architect, an ergonomist, external engineering capacity for mechanical and electrical aspects, and two facility managers of the existing organizational bodies.
(2) **Information and communication technology (ICT)**
The largest area of interest was ICT, including information and communication technology, software, hardware and telecom facilities. Several full-time specialists have been working in this area.

(3) **Organizational development**
After completion of the project, the manager for this sub-project would become director of the 112–Centre. Due to political reasons, it took until nearly the end of the project to fill in this job. Therefore, both the senior ergonomist and the general project manager worked occasionally on this topic (though not systematically), because the best opportunities to realize acceptance of a new working environment occur during early phases of ergonomic activities (analyses) and participatory design [17].

(4) **Other**
Public relations, financial controller, legal aspect.

Project procedure

The project steps were carried out as follows (summary).

- Step 1 – Feasibility: Ergonomists and ergonomics requirements, were involved in the selection of the new building.
- Step 2 – Problem definition: Because an ergonomist was in charge of all building related aspects, there was a high degree of freedom for organizing the human factors input and user participation.
- Step 3 – Situation analysis:
  Techniques used:
  - Observations (inventory and registration of events) and semi-structured interviews at the 112–police emergency centre and in the combined ambulance/fire brigade centre. Each centre was visited during at least five shifts and at different periods of the day.
  - Semi-structured interviews with key persons in day shift.
  - Drafting a report of analysis results for feed back by the user group.
- Step 4 – functional design specification: At the new location, new furniture (consoles), displays and communication equipment would be installed. However, the work tasks and the functionality of the equipment would not change. The functional design requirements concerned amongst others:
  - the allocation of task areas (workplaces) over the building floors
  - the layout of the control centre
  - shape and size of control workstations (including equipment)
  - environmental requirements (noise reduction, lighting levels).

  The requirements have been developed by the ergonomist in close cooperation with the user group.
- Step 5 – Detailed engineering: Based on a set of functional design requirements, various design solutions have been developed. Basic decisions were made during mock-up sessions for the layout as well as prototyping of the workstations.
- User participation: User participation was implemented by a series of meetings with a group of 10 representatives of the control centre users, two line managers, an ergonomist and the interior architect. Several techniques have been used:
  - layout design: brainstorming, scissors/paper design
  - layout design: a full scale mock-up of 25 workplaces; for this purpose the emergency control centre floor, completely stripped, was available
  - detailed design: mock-up of layout and shape of two types of desks
  - implementation: prototype testing of desks.

All activities were structured by short questionnaires, reports by the ergonomist and feedback during user group meetings. The ergonomist reported to the building manager (i.e. the senior ergonomist).

FIGURE 6. Full-scale mock-up evaluation.

FIGURE 7. Full-scale mock-up evaluation.

Results and lessons learned

Many things can be said on the ergonomics of the design. In this case, the following items have been important:

- Avoiding glare from lighting and outside windows at all costs.
- An essential requirement for workstation design has been 'easy communication'. Therefore one should be able to see over the top of display screens towards colleagues within the same grouping of three or six workplaces.
- Boxes to accommodate, cabling, extenders, power supplies etc. in the console, versus sufficient knee room below the desk.

An ergonomist managing building issues contributed much to realizing a human factors approach. Compared to other projects, ergonomics has been much more influential. Lessons learned:

- At the end of the project, during the implementation steps, the ergonomist lost full grip on the project because, as an external manager, he was not allowed to make any financial commitments. As a result and under time pressure, unsuitable types of lighting fixtures, touch screen control panels and some TFT-LCD screens were installed.
- In industrial projects, technology usually is leading. Thus controls, displays, cabling and so on are specified early. Following this, workstations can be engineered in detail by an ergonomist. Here the detailed console design was scheduled ahead of ICT design (or, one could also say, ICT was late on this topic). Part of the equipment was specified, after the workstations had already been delivered or specifications changed without consulting the accommodation manager. As a result, cabling initially didn't fit into the consoles; some changes had to be made by the interior craftsmen.
- A project manager is not the right person to coach, as an ergonomist, the user group (which had been the initial approach). In this case, the other ergonomist took over this responsibility.
- End users and line management were not used to a type of influence based on consensus between all parties involved. In addition, working with three different organizations and cultures made the participation more complicated.
- Project circumstances led to a limited interaction between accommodation and ICT. It is safe to assume that the lack of interaction had to do with:
  - limited budget for external consultancy (a half-time accommodation manager; a larger ICT staff (up to three full timers))
  - external influences on the ICT design (new nationwide communication systems, contractors, politics)
  - insufficient knowledge of ergonomics aspects relevant to ICT, as well as the interfacing of building and ICT aspects; probably the scope of each area was not sufficiently defined.

Nevertheless, it can be concluded that the approach to systems ergonomics, as given in Fig. 2, has functioned well. The human factors approach has been very influential. A general lesson learned is that it requires a full-time involvement instead of a part-time contribution, thus enabling much more communication with the other members of the project team.

ERGONOMICS IN PROJECTS

The author and his colleagues at Twente University and ErgoS Engineering have been involved in over 100 commercial projects during the past 25 years. In this section a summary of 10 cases is given. Refer for a description of each case to Appendix 1. This overview is not yet based on a systematic document research. A more detailed and systematic research is suggested and scheduled for 2007, preferably supported by structured interviews with the project ergonomists, project managers, project owners and end user representatives.

A word on methodology

In general, there will always be a methodological problem regarding case studies: $N = 1$. Usually a project isn't carried out twice, with and without an ergonomics approach. A project, as defined earlier, is always unique. Therefore, pure science or a statistical analysis isn't possible. Even when applying a specific design approach in several consecutive situations, differences will occur, simply because the ergonomist is using previous experiences during his next project. Remember, most work has to be done within actual project deadlines, with incomplete specifications of the system, restrictions in the set of acceptable solutions, a pressure to produce applicable results as soon as possible, and the like [23]. One rarely has the scientific luxury of using e.g. properly counterbalanced experimental designs. Nevertheless, case studies should be taken seriously in ergonomics literature, as the core of the business is in applying it.

Project characteristics

The '112–Centre' case (see above) serves as an example for the identification of relevant project characteristics. What is it we would like to show? The aim of this article is to share experiences of applied ergonomics in projects with other practitioners and scientists, in order to identify the gap between scientific research on human factors and design theories, and engineering. Thus the following items are relevant:

- Type and extent of the ergonomic project, to indicate the type of work done by ergonomists.
- As a starting point, system ergonomics, is taken. What is the applicability of a system ergonomics approach?
- What are the lessons learned?
- If possible, a cost–benefit indication: has it been worthwhile to include ergonomics in a project?

Project selection criteria are:

- The project included the implementation of a new (or redesigned) system.
- Application of a system ergonomics approach, as discussed under 'Methodology – Systems Ergonomics' above.
- The project has been realized/finished, possibly some evaluation has taken place.
- Project documents readily available at the author's desk.

The following characteristics will be helpful in answering the questions above:

- **General project scope**
  - type of industry or organization
  - scope of the (overall) project

  - investment in project (indication)
  - realisation period.
- **Project organization**
  - general project organization
  - position of ergonomist within the project
  - hours by ergonomists and/or made under full responsibility of ergonomist
  - percentage of the project or investment influenced directly by ergonomics
  - type and extent of user participation.
- **Ergonomic topics to be handled**
  - main emphasis, i.e. job design, workplace design, interaction design
  - amount of different jobs or workplaces involved
  - procedural matters, i.e. interaction with other disciplines (architect, engineering), responsibilities, management of user input, and so on.
- **Procedure – ergonomics**
  - project phases that included ergonomics
  - ergonomic engineering steps (eight steps, refer to 'Methodology – Systems Ergonomics' above).
- **Lessons learned**
  - motive to hire ergonomist
  - typical results
  - lessons learned.

Case studies

Below, 10 cases are briefly summarized. Further descriptions of the cases are given in Appendix 1. The projects are:

(1) 112–Centre (presented in 'Case study: 112–emergency control centre' above); 2002–2004.
(2) Consolidated Control Centre for the FLEXICOKER project, 1981–1986.
(3) Integrated control room and crane cabin for waste incinerator, 1992–1994.
(4) Waste incinerator plant (green site project), 1994–1996.
(5) Central control room for existing waste incineration plant, 2002–2004.
(6) Redesign of existing control room for waste incineration plant, 2006.
(7) Central control room for extended waste incineration plant, 2005–2006.
(8) Supervisory control room for off-shore gas production, 2004.
(9) Lock master's control centre, 1995.
(10) Bridge for a large dredging vessel, 2000–2001.

The following characteristics are interesting for as they assist in comparing the cases:

- For all projects the same basic approach of system ergonomics has been applied by the same group ergonomists.
- Cases studies 3 to 7 are related to the same type of industry.
- In five cases (1, 3, 5, 6 and 8) the ergonomist cooperated with the same interior architect. Though the architect had his own 'vision' on the design project, he was also very eager to follow the ergonomic requirements. For cases 1, 5 and 6, the architect was contracted by the ergonomists.

Ergonomic input

Projects range from 120 hours for the smaller projects to 3000 hours in the larger projects. The maximum percentage of the investment, related to ergonomic engineering is approximately 2%. Also, the ergonomically designed dedicated furniture, is usually less expensive than standard furniture delivered by the accepted deliverers. In two cases (6 and 8), about 10% of the project costs are attributed to the ergonomic engineering, because major engineering was done by the ergonomist and the projects were 'limited' to the ergonomic design.

Motive to involve ergonomics

Earlier experiences of the project manager are almost always the driving factor behind hiring ergonomics engineering. These project managers are highly convinced of the benefits it will bring to the project as well as the end results. Because this is a personal motive, the ergonomist should be prepared for a situation in which others involved in the project are not so enthusiastic.

Ergonomic steps and role

In general, ergonomists should be involved during all project phases, though in different roles. There is a tendency to hire ergonomists until detailed design. This role can best be indicated as consultancy, giving good advice. In some cases the ergonomist also does the preliminary stages and even final engineering. Summarizing, the ergonomist is often out of the project once the contractor is building the work environment, workplaces and man–machine interfaces. In several cases we stumbled across problems that could have been avoided easily, e.g. by a review of workshop drawings by a professional. A typical example of a problem is given in Fig. 8. A library of this type of problems may help convincing project owners to involve ergonomists in all project phases.

The most influential role for the ergonomist can be found when they are hired by the (top) project management or even when they are a member of the management team (cases 1, 4, 6 and 8).

Job load and work organization design

In cases 2 and 4, job load assessment and work organization design were major items for the human factors consultant. Carrying out a situation analysis is essential to being able to understand operator workload parameters. The next step, the allocation of tasks and the design of jobs is a difficult one and has to be done very carefully. Amongst others, the impact of job design is very high, because it inevitably dictates the number of personnel required. There is no useful material to be found in ergonomics or ergonomics related literature regarding task allocation and job design.

In automation projects, such as the lock master's control centre (case 9), it may be difficult to win the trust of the end users. The ergonomist (or the employer) has to invest in the situation analysis (spend much more time on site than required to get sufficient insight in the tasks) and also in e.g. a mock-up session, in order to get the support by the end users.

FIGURE 8. Heating radiator located exactly at the best standing position for a lock master.

User participation

In the FLEXICOKER case 2, user input was organized in a working group format. The working group consists of one or two representatives of each shift, a representative of the shift supervisors and an experienced (older) supervisor as the link between operations, project and ergonomist. The working group meets for:

(1) kick off, getting to know each other
(2) discussing the results of the situation analysis
(3) discussing/working on the functional design
(4) building and evaluating a full-scale mock-up of the design
(5) finalizing design.

This model for user participation was applied in the cases 1, 2, 5 and 6 and worked extremely well. In fact designs were made with full consensus between all parties. For projects 4, 8 and 10 this approach was impossible due to various valid reasons that can be summarized as 'no operators available'. At the remaining companies, it proved to be difficult to convince management of the positive results a time-consuming involvement of the end users would have. Nevertheless, successful mock-up studies (i.e. consensus reached) could also be arranged for the cases 3, 9 and 10.

During various mock-up evaluation studies it became clear that neither company management nor project management should participate, or at least not on the same day as the end users. This included the ergonomist, in case he had a management role (case 1).

Specific – case-related – findings

- Projects are usually organised according to engineering disciplines, i.e. civil engineering, automation, mechanical, and so on. Engineering contractors spend much time in defining the borderlines and interface between these disciplines. The ergonomic engineer should not forget to do so too. If not, problems may occur, as shown in case 1. In addition, to be successful, it is essential to know a lot about the production processes and plant design, be acquainted with technical terminology and so on, and, last but not least, be visible and available to all other disciplines within the project, as well as to the user organization.
- Not all decisions during a project are made objectively. Engineering contractors easily use the threat of not meeting deadlines or budgets if they dislike a specific design requirement (see for example case 4). Don't back off in such situations.
- Several cases show that a handover of the workplaces to the end users (and the production organization) should be taken seriously. The users should get acquainted with the intended use/task performance at the workplace (for example: case 4 and 5), which may not always be the task situation they are used to in previous systems.
- The off-shore gas production control room was an example of a small facility and project. To acquire the job of doing an ergonomic design, the ergonomist should be prepared to make a turnkey offer for the full project, or at least be aware of a suitable partner. As a rule of thumb, the relationship between ergonomic engineering and architect/civil engineering during a project is indicated in Fig. 9.

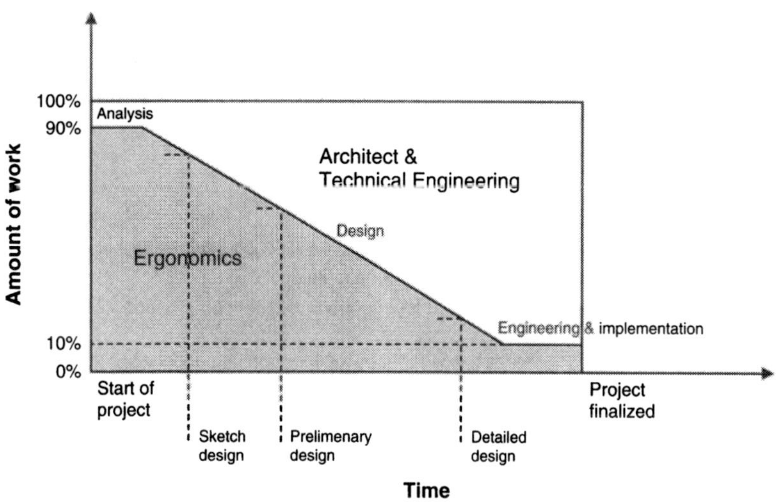

FIGURE 9. Relationship input ergonomic engineering – input architect.

CONCLUDING REMARKS

In the previous sections experiences of applying a system ergonomic approach in engineering projects have been presented. Though this has not (yet) been a systematic research into the process of the participation of ergonomists in projects, some conclusions stand out clearly.

(1) Ergonomists should not be afraid of a negative connotation, as indicated by Vink [3]. Once a project manager has had the pleasure of working with an ergonomic engineer, he will do so in every new project. After several years, management usually doesn't remember the results, but 'the design process': ergonomics had tools to tackle human factors issues effectively.
(2) The cases show that (professional) ergonomics is not about an additional effort or higher project costs; probably the contrary.
(3) Stick to the system ergonomics approach, because it works well and is understood by the engineering community. Always perform a complete situation analysis, if an existing or comparable system is available. You will need the outcomes during functional design or in spin-off activities. If the project owner is not convinced of its use, you may consider doing the analysis (partially) for your own risk (it will pay out later).
(4) It is important to stay in the project during implementation and commissioning. The ergonomics community should compile an overview of cases and examples indicating the benefits of this involvement, in order to be able to convince project management to spend some money on this activity. More or less the same may be suggested regarding user participation, i.e. the use of full-scale mock-ups.
(5) Job load assessment, task allocation (job design) and work organization design are considered human factors/ergonomics. In this area useful and validated design tools are missing and could be developed by the ergonomics community.

The triennial IEA world congresses give excellent opportunities to evaluate ergonomics in practice. The number of reports on actual projects is limited [24], but increasing. The most interesting series of projects are compiled in this book. The author sincerely hopes that his colleagues will find the time and opportunity to publish critical reports of their projects.

The author thanks all his colleagues at ErgoS Engineering and Ergonomics for their contributions to this article and to the cases presented here.

REFERENCES

[1] Pikaar, R.N., et al. (1998a). Ergonomics in process control rooms. *Part 1: Engineering Guideline*. The Hague: WIB International Instrument Users Association.
[2] Ergos, E. (1995). *Human Centred Design (Mens-gericht ontwerpen). Enschede: Ergos Engineering & Ergonomics.*
[3] Vink, P., et al. (2006). Positive outcomes of participatory ergonomics in terms of greater comfort and higher productivity. *Applied Ergonomics*, 37, 537–46.
[4] Dul, J. and Neumann, W.P. (2007). The strategic business value of ergonomics. This volume.
[5] Pikaar, R.N. and White, T.N. (1990). Marketing ergonomics: Profile of an interdisciplinary field. *Ergonomics*, 33, 245–50.
[6] Lenior, T.M.J. and Verhoeven, J.H.M. (1990). Implementation of human factors in the management of large scale industrial investment projects, a management point of view and ergonomics practice. *Ergonomics*, 33 (5), 643–54.

[7] de Looze, M. and Pikaar R.N. (2006). Meeting diversity in ergonomics. *Applied Ergonomics*, 37, 389–90.
[8] Clegg, C.W. (2000). Sociotechnical principles for system design. *Applied Ergonomics*, 31, 463–77.
[9] Berggren, C. (1993). *The Volvo Experience*. London: The Macmillan Press Ltd.
[10] Rijnsdorp, J.E. (1991). *Integrated Process Control and Automation*. Amsterdam: Elsevier.
[11] Dul, J. and Vries, H.J. de. (2006). Differences between ergonomics and other ISO/CEN standards. Lessons for standards development. In: Pikaar, R.N., Koningsveld, E.A.P., Settels, P.J.M. (eds) *Meeting Diversity in Ergonomics*. Proceedings IEA 2006 Congress. ISSN 0003-6870, Elsevier Ltd.
[12] Pikaar, R.N., et al. (1998). Ergonomics in process control rooms. *Part 2: Design Guideline*. The Hague: WIB International Instrument Users' 'Association.
[13] Pikaar, R.N., Lenior, T.M.J., and Rijnsdorp, J.E. (1990). Implementation of Ergonomics in design practice: Outline of an approach and some discussion points. *Ergonomics*, 33 (5), 583–600.
[14] Kragt, H. (1992) (ed.). *Enhancing Industrial Performance: Experiences of Integrating the Human Factor*. Taylor & Francis.
[15] Pikaar, R.N., et al. (1997). Ergonomics in process control rooms. *Part 3: The Analyses*. The Hague: WIB International Instrument Users' 'Association.
[16] Singleton, W.T. (1967). Ergonomics in systems design. *Ergonomics*, 10, 541.
[17] Pikaar, R.N., Thomassen, P.A.J., Degeling, P., et al. (1990). Ergonomics in control room design. *Ergonomics*, 33 (5), 589–600.
[18] Remijn, S.L.M. (2006). Integrating ergonomics into the architectural design processes: Tools for user participation in hospital design. In: Pikaar, R.N., Koningsveld, E.A.P., Settels, P.J.M. (eds), *Meeting Diversity in Ergonomics*. Proceedings IEA 2006 Congress. ISSN 0003-6870, Elsevier Ltd.
[19] Mulder, E. (2007). An integral approach to make software work. This volume.
[20] Wilson, J.R. and Corlett, E.N. (1990). *Evaluation of Human Work*. London: Taylor & Francis.
[21] Meister, D. (1985). *Behavioral Analysis and Measurement Methods*. New York: John Wiley and Sons.
[22] Pikaar, R.N. (2007). Integrated design of an 112–Emergency Control Centre. In *Case Studies in Control Room Design* (J. Wood, ed.) Taylor & Francis.
[23] Pikaar, R.N. (1992). Control room design and systems ergonomics. In *Enhancing Industrial Performance: Experiences of Integrating the Human Factor* (H. Kragt, ed.) Taylor & Francis.
[24] Koningsveld, E.A.P., Settels, P.J.M., and Pikaar, R.N. (2007). Meeting diversity in ergonomics. This volume.
[25] de Groot, N. and Pikaar, R.N. (2006). Videowall Information Design: useless and useful applications. In: Pikaar, R.N., Koningsveld, E.A.P., Settels, P.J.M. (eds), *Meeting Diversity in Ergonomics*. Proceedings IEA 2006 Congress. ISSN 0003-6870, Elsevier Ltd.

APPENDIX 1 – CASE STUDIES

Estimates of the project investment and money spend on ergonomic activities are given in Euros (for the year of the project, i.e. not indexed) and are with +/– 25% accuracy.

CASE STUDY 1 – 112–EMERGENCY CONTROL CENTRE

See for references [22] and 'Case study: 112–emergency control centre'.

Project scope and organization

- organization: regional administration
- project scope: interior design; 1500 m^2 offices, 500 m^2 112–Centre
- investment: €11 000 000; 2% ergonomic engineering
- period: 2002–2004
- management: employee of regional administration;
- project team: ergonomist is member of project team; manages a sub-project team for accommodation which includes architect; a second ergonomist is member of the sub-project team.
- engineering: mainly by internal staff of regional administrations
- hours: 2000 hours, (senior) ergonomist
- users: working group of end user representatives; at first very open minded; gradually becoming more 'political', due to uncertainties in future work contracts.

Ergonomic topics

- main topics: building layout, workplace layout, detailed design.
- workplaces: 10 for call centre; 15 for "112–Centre"; several other, social areas and standard offices
- role: management, ergonomic design, organizing user input.

Procedure – ergonomic steps

Start: at feasibility phase of the project

(1/2) feasibility: yes (building selection criteria)
(3) analysis: 2 existing control centres; 2 × 40 hours on site (all shifts)
(4) funct. design: building layout, all workplaces
(5) tools: mock-up evaluation, prototype testing, 3D drawings
(6) implementation: progress review of manufacturers, building contractor
(7) commissioning: no
(8) evaluation: aftercare (operator chair selection, seating instructions, and risk evaluation (mandatory)).

Lessons learned

- Motive to hire ergonomist: earlier experience of one of the project owners.

- Ergonomist delivered engineering/architectural and management services that was needed anyhow. Dedicated furniture proved to be considerably less expensive than standard furniture used for these centres.
- Lessons learned: refer to 'Case study: 112–emergency control centre'.

CASE STUDY 2 – FLEXICOKER – OIL REFINERY CONTROL ROOM

See for references amongst others [10], [11] and [15].

Project scope and organization

- organization: process industry/oil company
- scope: control room building, control room and workplaces, work organization design (workload assessment)
- investment: €6 million, exclusive instrumentation/computers; €100 000 ergonomic consultancy
- Project was part of a $2000 million extension of a refinery with the s- called FLEXICOKER.
- period: 1981–1986
- management: managed by control centre design manager (member of general project management team)
- ergonomics: external consultant of Twente Universities Ergonomics Group to control centre design manager
- engineering: turn-key engineering contractor; except: ergonomics consultant and interior architect
- hours: 3000 hours, ergonomist, senior ergonomist, professor
- users: working group of end user representatives, a shift supervisor and the production manager; In addition, an experienced (former) shift supervisor was the daily contact between refinery and ergonomists.

Ergonomic topics

- main topics: job design/work load assessment; building, control room and workplace layout including detailed design
- workplaces: five operator consoles; workstations for process automation
- roles: ergonomic design, organizing user input; in addition advice on a scientific level, in particular used by manager to convince the company of ergonomic topics/design.

Procedure – ergonomic steps

Start: at feasibility phase of the project

(1/2) feasibility: yes (choice of one or two control rooms)
(3) analysis: existing control room; 5 × 40 hours on site (all shifts)
(4) funct. design: building layout, all workplaces
(5) tools: layout mock-up evaluation, workplace mock-up evaluation
(6) implementation: review of detailed design by architect and engineering contractor.

(7) commissioning: no
(8) evaluation: full situation analysis, two years after commissioning.

Lessons learned

- The motive to hire ergonomists: earlier experience of production manager; interest of project owner in scientific research on design approach.
- Return on investment in ergonomics within one year (operational benefits) by avoiding unit shutdown during incidents.
- The project has been a major test of the ergonomic system design approach in a (very) complex project.
- The situation analysis and its results were essential (for all parties involved, including the end users) to understand operator workload parameters. As a consequence, discussions on the number of operators for an extended refinery could be done more objectively and therefore were more easily acceptable for all.
- Job design/task allocation and workload assessment tools had to be developed almost from scratch (no useful ergonomics literature at that time available).
- Large-scale mock-up evaluations proved to be very valuable, not only for the end users, but also for the interior architect (who was in fact very cooperative and understood the major ergonomic requirements).
- Critical feedback given by project manager proved to be valuable and challenging (the scientist doesn't meet often this type of feed back).
- There is a (company political) need to convince higher management levels of the importance of ergonomics. In this case, the fact that the team of ergonomists was headed by Prof. Rijnsdorp proved to be important.
- Note: a redesign/upgrading project started in 2006, for which the same ergonomists were hired.

CASE STUDY 3 – EXTENSION OF WASTE INCINERATOR

Project scope and organization

- organization: waste treatment company
- scope: job design, combined control room and crane cabin layout, workplaces, graphic displays
- investment: €150 000 000 for the extension of the plant;
- <0.1% on ergonomic engineering.
- period: 1992–1994
- management: professional management contractor
- project team: ergonomists are consultants of management contractor, which in turn supervises the engineering contractor; thus ergonomic requirements became mandatory for design
- engineering: engineering contractor; interior architect
- hours: 1200 hours, (senior) ergonomist
- users: primarily represented by production management.

Ergonomic topics

- main topics: integration of control room and crane cabin, control room layout, workplace design, graphical user interface

- workplaces: two operator consoles, two crane control workplaces
- role: functional design, frequent check of engineering contractor drawings, and of the interior design made by an architect.

Procedure – ergonomic steps

Start: During problem definition phase

(1/2) feasibility: yes, regarding location of control room and starting point for the work organization
(3) analysis: existing control room and crane cabin (>40 hours on site, all shifts)
(4) funct. design: control room layout, workplaces
(5) tools: mock-up evaluation
(6) implementation: (limited) review of architectural drawings
(7) commissioning: no
(8) evaluation: only a quick look during a follow-up project.

Lessons learned

- The project manager told project owner that 'he would not do the job, if ergonomic consultants were not contracted' (based on earlier experiences [6]).
- Working as a consultant for the project management level gives more impact than working for an engineering contractor. Initially the problem was that neither the project owner nor the engineering contractor liked the ergonomics input, as arranged for by the general project manager.
- The situation analysis showed, as a by-product, that production management didn't function very well, which hindered the project significantly. In fact, the ergonomic consultant was 'used' by the project manager to solve this problem early during the project.
- Visiting the operators and doing the mock up were real 'ice-breaking' activities. Operators liked the possibility to give input, as well as liking the final results very much.

CASE STUDY 4 – WASTE INCINERATION (NEW SITE)

Project scope and organization

- organization: company (owned by regional body of cities)
- scope: new waste incineration plant (no existing organization)
- investment: €250 million; 0.1% on ergonomic engineering
- period: 1994–1996
- management: management contractor; senior ergonomist is member of management team
- engineering: engineering contractor
- hours: approximately a half time job, during 2.5 year of development and construction
- users: not yet contracted!

Ergonomic topics

- main topics: all human factors, i.e. design of the work organization, all workplaces, field work (accessibility for maintenance), graphic displays, mimic panel.

- workplaces: two at control room, two crane cabins, inspection cabin, slag treatment control room, other.

Procedure – ergonomic steps

Start: two months after official project kick off (i.e. feasibility)

(1/2) feasibility: yes
(3) analysis: no, the ergonomist had to rely on earlier experiences (in particular related to case 3) and visits to several other incineration plants
(4) funct. design: work organization, all workplaces
(5) tools: task allocation (based on task estimates), sketches
(6) implementation: frequent meetings with architect/building contractor and engineering contractor; all reported to the management team, i.e. on human factors to the ergonomist.
(7) commissioning; yes, member of commissioning team.
(8) evaluation: risk analysis and evaluation (mandatory) at the end of commissioning; recently (2006) the crane cabin design was evaluated.

Lessons learned

- The motive to hire ergonomists: earlier experience of general project manager; he was a member of the case 3 management team.
- Being a member of the project management team gives ergonomics much impact. Contractors follow your requirements. As a consequence, detailed design of, for example an operator console, is done by the contractor. The result, now based on giving feedback on engineering documents and drawings, may not be as good as possible if done by ergonomists.
- Designing a work, organization was within the scope of the ergonomist. A bottom-up approach was taken: the starting point for task allocation was a list of all tasks, including estimated time needed and frequency of occurrence. Major problem: how to find out about time estimates?
- Not all design decisions are made objectively. Engineering contractors easily use the threat of not meeting deadlines or budgets, if they dislike a specific design requirement.
- To be successful, it is essential to know a lot about the production processes and plant design, be acquainted with technical terminology, and so on. Also, it requires one to be frequently present on the building site (they have to see you).
- The contract ended after commissioning. There was no opportunity to explain to the end users the intended use of the workplaces. For e.g., in this case it was not understood that the crane cabin chair was designed for parallel control of two waste cranes.

CASE STUDY 5 – WASTE INCINERATOR – CONTROL ROOM BUILDING

Project scope and organization

- organization: large semi-state controlled company
- scope: new control room, control room building (replacing the existing worn-out control room)

- investment: €4 million; €80 000 ergonomic engineering
- period: 2002–2003
- management: internal senior (building) project leader
- project team: internal; ergonomist reports to building project leader
- engineering: external for civil engineering/architect
- hours: 800 hours, (senior) ergonomist
- users: working group, one member per operator shift plus one shift supervisor; managed by an older supervisor and highly supported by the project manager.

Ergonomic topics

- main topics: control room building sketch design; control room layout, dedicated furniture layout
- workplaces: two consoles, permit handling, calamity desk, social areas, some offices.
- role: consultant, project engineer.

Procedure – ergonomic steps

Start: feasibility phase

(1/2) feasibility: yes
- (3) analysis: existing control room; 60 hours on site (all shifts)
- (4) funct. design: building layout, all workplaces, partially instrumentation
- (5) tools: mock-up evaluation
- (6) implementation: limited to dedicated furniture (review)
- (7) commissioning: no
- (8) evaluation: no, informal by visits made on site (2005, 2006).

Lessons learned

- motive: project engineer had good experiences with the result of an ergonomic design; he acted a few years as the production manager of the case 4 plant.
- The project ran smoothly until the implementation phase; then it became clear that the civil engineer/architect wouldn't deliver good work. To solve the problems an interior architect was contracted by the ergonomic engineer.
- During the project emphasis was on user participation (for which the operators were not really used), strongly supported by the project manager. During each ergonomic step the operator working group was consulted, i.e. helped designing the new situation.
- A mock-up was completely self-organized by operators (very stimulating).
- During implementation some parties involved, including internal engineering departments, were not interested in ergonomics. Efforts were made to include the graphic design for the operator interfaces and a video wall system. These efforts failed, due to a lack of support.
- Within two years after commissioning, major changes were made to the newly built control room, without any input from users or ergonomists. A carefully balanced layout of the control room was disrupted. The possibility of an extension was foreseen, and suggestions had been made (however not used). This was probably due to the fact that the project manager had left the company in the meantime.

CASE STUDY 6 – WASTE INCINERATOR – CONTROL ROOM REDESIGN

Project scope and organization

- organization: waste incineration company
- scope: re-design and redecoration of existing integrated control room and waste crane cabin (after several extensions of the plant)
- investment: €1.1 million direct costs; including 15% building/construction; 10% dedicated furniture; 10% external consultancy and 20% for new CCTV system and videowall
- period: 2006 (realized within 10 months)
- management: internal; senior project leader (maintenance department)
- project team: three engineers: instrument (internal), electrical (internal), layout and furniture (ergonomists); i.e. the ergonomist is a member of the project team and contractor for interior architect
- hours: 700 hours ergonomist, 100 hours interior architect.
- users: working group, one member per operator shift plus one shift supervisor; excellently managed by project manager.

Ergonomic topics

- main topics: control room layout, dedicated furniture
- workplaces: four at operator consoles, waste crane control, permit handling desk, kitchen/social area, shift supervisor office and visitors area
- role: project engineering, ergonomic engineering.

Procedure – ergonomic steps

Start

(1) feasibility: feasibility study and sketch design were made several years earlier (by another ergonomist)
(2) problem def.: project start at problem definition; project mainly driven by ergonomic steps (not by traditional project phasing).
(3) analysis: limited visits to all shifts (presumably done two years earlier, no systematic analysis report available); 32 hours
(4) funct. design: layout, furniture, instrumentation, interior architecture (look and feel)
(5) tools: 3D, mock-up evaluation
(6) implementation: ergonomist organizes requisition procedure (furniture, civil); review of workshop drawings; fully involved in detailed engineering and construction by contractors.
(7) commissioning: yes, amongst others site acceptance tests
(8) evaluation: planned for mid 2007.

Lessons learned

- In this case, the maintenance department needed additional engineering services, which could effectively be delivered by the ergonomic engineering company (instead of technical-oriented engineering contractors). The ergonomists were found on the internet.

- Project ran very smoothly, among other things because of user participation and highly interested project leader, all within a relatively small company. The project leader frequently consulted (all) operator shifts on their preferences.
- The approach to user participation, and in particular the completely self-run mock-up evaluation proved to be very valuable.
- During the project it became evident that the instrumentation system itself (not part of the project scope) was less suitable for process supervision of the larger/extended plant. Also, for future extensions, operator workload issues became clear. New ergonomics projects are initiated to improve this (in 2007).
- Related to the previous remark, the results of the limited task analysis are not sufficient for further ergonomics work. Unfortunately this means, the analysis has to be redone.

CASE STUDY 7 – WASTE INCINERATOR – NEW CONTROL ROOM FOR AN EXTENDED PLANT

Project scope and organization

- organization: company, owned by the City of Amsterdam.
- scope: extension of the plant; within this project a new control room for the four existing and two new incinerator lines is needed
- investment: not known; €50 000 on ergonomic engineering
- period: 2004–2006
- management: engineering contractor
- project team: ergonomist contracted by the instrumentation engineer (taking responsibility for an integrated design of operator workplaces, which would otherwise not be covered)
- hours: 400 hours, (senior) ergonomist
- users: not formally organized.

Ergonomic topics

- main topics: control room layout, operator workplace design including videowall, graphics design
- workplaces: five operator workplaces, two supervisory desks, permit handling and social area
- role: consultant.

Procedure – ergonomic steps

Start: late (implementation phase of the project)

(1/2) feasibility: no
(3) analysis: yes, 40 hours on site (visiting all shifts)
(4) funct. design: workplaces
(5) tools: 2D sketches and drawings (mock-up proposal was not accepted by project owner)
(6) implementation: review of drawings
(7) commissioning: review on site of furniture delivery
(8) evaluation: no.

Lessons learned

- The ergonomist was consulted because the instrumentation engineer had earlier experiences (he worked at the case 4 engineering contractor); the engineer initiated some user participation.
- Within the large project organization, the instrumentation engineer and thus the ergonomic consultant, play only a minor role. In addition, support by the companies management is limited; ergonomics was considered a necessary 'evil' in view of legislative requirements.
- Integration of instrumentation, control room layout and furniture, including an ergonomics input, was not foreseen in the project.
- User participation was organized by a representative of the company's management (a former senior operator). Involvement of end users was kept low by the company, mainly because the operator had already much to do (technical reviews) within the project. The ergonomist has not been able to change this.
- To reach ergonomic design goals, it takes more time (and costs) than in other comparable projects (cases 4, 5 and 6).

CASE STUDY 8 – 'OFF-SHORE' PRODUCTION SUPERVISION CONTROL ROOM

Some aspects of this case are presented in de Groot and Pikaar [25].

Project scope and organization

- organization: process industry (exploration and production)
- scope: removal and upgrading of existing control room to another location
- investment: €200 000 exclusive of instrumentation and communication systems; 10% ergonomics engineering
- period: 2004
- management: company engineering
- project team: ergonomist/engineering contractor (architect)/instrument engineering contractor
- hours: 200 hours ergonomic engineering
- users: end users.

Ergonomic topics

- main topics: control room layout, workplace layout, detailed design, design of an overview graphic (large screen display).
- workplaces: operator console, office desk, social area.
- role: project management, ergonomic design.

Procedure – ergonomic steps

Start: Feasibility phase

(1/2) feasibility: yes
(3) analysis: existing control centre; 24 hours on site (visit all shifts)

(4) funct. design: building layout, workplaces, environment
(5) tools: 3D drawings, prototyping of graphic display
(6) implementation: managed by architect; frequent input by ergonomist
(7) commissioning: managed by architect; review visit by ergonomist
(8) evaluation: not systematically, visits to operative control room only.

Lessons learned

- The ergonomist should be prepared to, and able to, make a turn-key offer for interior/workplace design and implementation. In this case, the company had been searching for two turn-key contractors: one for instrumentation and one for all other tasks (i.e. internal building activities, control room design, delivery of dedicated furniture and so on). He also invited the ergonomics company to offer on the latter. The ergonomist and an engineering contractor (including architect) worked together on this quotation.
- After receiving the quotation, it became clear that all project steps/phases until the functional design could easily be separated from the actual building activities, which in turn could then far better be calculated. At the end of the bidding procedure the ergonomist became consultant for the company, instead of building contractor.
- A model was developed on the relative input needed by ergonomic engineering and architect/civil engineering. During the first project phases (including analysis) the ergonomist is leading (90% of man hours) and the architect is being informed (10%); during implementation this is the other way around. In between, the balance changes in an almost linear progression.
- The graphic design of a typical overview display was highly successful. It should be noted that usually the instrumentation manufacturer does not have any knowledge regarding this topic.

FIGURE 10. New control room, including overview display panel.

CASE STUDY 9 – LOCK MASTERS CONTROL CENTRE

Project scope and organization

- organization: Province of Zeeland (Netherlands)
- scope: new lock master's control building (centralised control)
- investment: not known; €10 000
- period: 1995
- project team: no project team
- engineering: architect/civil contractor
- hours: 120 hours, ergonomist
- users: not formally organized.

Ergonomic topics

- main topics: control centre layout; lines of sight study
- workplaces: two (large commercial lock, small pleasure craft lock)
- role: consultant

Procedure – ergonomic steps

Start: during functional design by architect

(1/2) feasibility: no
(3) analysis: visits to lock (total of 16 hours)
(4) funct. design: control centre layout, sketch design of workplaces
(5) tools: mock-up evaluation, 3D lines of sight study
(6) implementation: no
(7) commissioning: no
(8) evaluation: yes, after some problems had arisen (1997); see lessons learned.

Lessons learned

- The ergonomist was contracted because the project owner (engineer) had done a short course on ergonomics (he got interested in the topic).
- At the background of the project workload issues played a major role, however were no part of the project scope. A 50% reduction of lock master staff was planned, but not fully communicated to the lock masters. This made the ergonomist's job difficult (though not impossible).
- During the project we convinced the project owner to organise a full scale mock-up. This was highly appreciated by the end users and gave valuable insight in layout options. Essential outcome: lock masters have a standing (moving) job at the control panels.
- After two years the ergonomist was invited to review the control centre, because serious complaint on the workplace layout were arisen. We found the lock masters sitting at desks designed for a standing job. In addition, several mistakes (by the architect) were noted, such as heating radiators located right at the most important positions at the windows, from the lock master's point of view.
- Lesson learned: always check detailed engineering at or before implementation.

CASE STUDY 10 – BRIDGE OF TRAILING SUCTION HOPPER DREDGER

Refer to de Groot and Pikaar [25] for some details on overview graphics design.

Project scope and organization

- organization: international dredging company
- scope: ship bridge design, as part of a new dredger
- investment: not known; €40.000 ergonomic engineering
- period: 2000–2001
- management: engineering department of project owner
- project team: ergonomist is consultant to project manager
- engineering: partially internal staff/partially ship wharf (contractor)
- hours: 300 hours, ergonomist
- users: two representatives of end users, made available for the project; note: end users are working on other dredgers, all over the world.

Ergonomic topics

- main topics: bridge workplaces design, including sight lines study; later additional topics emerged, in particular regarding controls and displays (instruments)
- workplaces: two skipper and pipes operator
- role: consultant.

Procedure – ergonomic steps

(1/2) feasibility: yes, regarding bridge design
(3) analysis: 24 hours on board of an existing hopper dredger

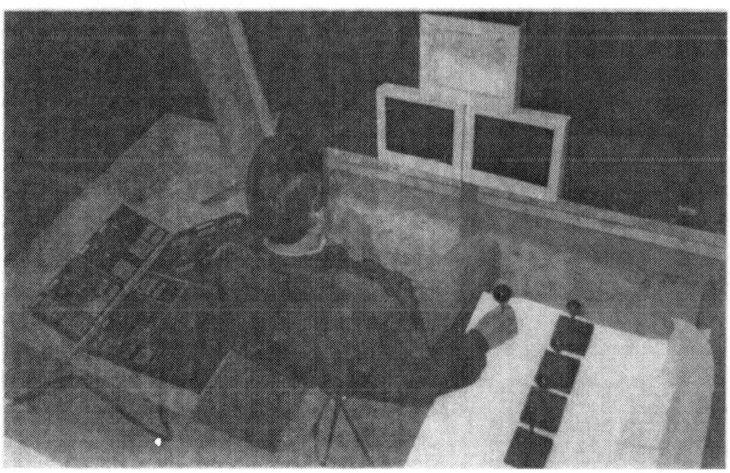

FIGURE 11. Heating radiator located exactly at the best standing position for a lock master.

FIGURE 12. Heating radiator located exactly at the best standing position for a lock master.

(4) funct. design: layout, workplaces, vision
(5) tools: mock-up evaluation, 3D drawings, prototyping displays.
(6) implementation: visits to wharf for review
(7) commissioning: no
(8) evaluation: no

Lessons learned

- A mock-up could be realized with simple materials and was in particular aimed at the lines of sight and details of displays and controls. In bridge design, a mock-up is not something new, however usually it is build as one fixed solution (to be reviewed). The ergonomic approach includes possibilities to try several alternative design solutions.
- During the project, several new ergonomic items came up and more and more engineers at the project owner became interested. Of particular interest has been the redesign of a ship's overview display.

Meeting Diversity in Ergonomics (2007)
Pikaar, Koningsveld, and Settels (Eds)

4

An Operational and Marketing Infrastructure for the Human Factors and Ergonomics Professions – a Call to Action

Glen Gallaway

Human Factors Research and Engineering Technical Advisor, Air Traffic Organization, Operation Planning Research and Development Federal Aviation Administration, Washington DC, USA
(Disclaimer: THE FINDINGS AND CONCLUSIONS IN THIS DOCUMENT HAVE NOT BEEN FORMALLY DISSEMINATED BY THE FEDERAL AVIATION ADMINISTRATION AND SHOULD NOT BE CONSTRUED TO REPRESENT ANY FAA OR DEPARTMENT OF TRANSPORTATION VIEW, DETERMINATION, OR POLICY)

Abstract. Together, all the people in and associated with our human factors professions have the knowledge, skills, information, processes, resources, and solutions that we need to make our work extremely successful. Through our efforts of enhancing the organization, application, and presentation of these resources, human factors principles can be made a central concept in our culture with all the benefits they will bring to society and our professions. This chapter is dedicated to moving us in this direction. For a number of years I have felt that our human factors profession would benefit from a thorough *enterprise analysis* to characterize how we currently do business followed by an *action program* to make improvements. This observation is based on marketplace indications that the products and services that we currently deliver do not always meet the expectations of our customers (those people who ask for human factors support and solutions). In the past, our profession may not have seen the need or benefit of such an analysis or action program, but current events such as the practitioner concerns described later, may indicate that now is the time to take action. Our professional societies, leaders, and practitioners are coming to the realization that we can offer more to society if we can better match our products and services to customer needs. In this chapter, I identify the issues, discuss the increased awareness, and then provide a specific action plan for our future direction. Although this material is initially intended for the ergonomics/human factors practitioner, it will benefit anyone interested in improvement of the ergonomics/human factors discipline.

INTRODUCTION – A REASON FOR A CALL TO ACTION

Rob Preston, VP/editor-in-chief of *InformationWeek*, in a recent article discussed the future of the information technology industry and the technical professions that support the industry [1]. As I read the article, I thought about how his words would be an appropriate introduction to my discussion of the need to change our work model and how we market the human factors profession.

'It is not the strongest of the species that survive, nor the most intelligent, but the one most responsive to change' (a quotation accredited to Charles Darwin). This notion is nonetheless a physiological truism, not just some interesting trifle of business management. Change is the

---

The concepts presented are intended to represent and apply to all of our professional dialects (ergonomics, human factors, user participation person, psychologist, etc.) and our customers and partners. Because of my human factors focus and background I have used the **human factors** OR **HF** titles throughout the text.

only constant; it is the law of life. We all nod our heads in agreement. Yet, only a relative few of us live by this code. Why? Change is uncomfortable....

Most of us spend too much time conforming to new policies and regulations rather than plowing ahead with initiatives that will really make a difference.... But when the world is moving in a new direction, you must adapt and move with it – even if you have to make concessions and mistakes along the way.... 'In times of change, learners inherit the earth', wrote novelist Eric Hoffer, 'while the learned find themselves beautifully equipped to deal with a world that no longer exists'. The history of industry is rife with examples of companies that held on to their core tenets for too long and did not survive.... Harold Wilson, a British politician of the mid twentieth century, stated, 'He who rejects change is the architect of decay. The only human institution which rejects progress is the cemetery'.

Preston's statements have many truths relative to the need for the human factors profession to embrace change. It is easy to say we need change, but harder to make change happen. I want to make it obvious that our profession needs change and then suggest one approach to making it a reality. I will start by providing a light-hearted, but serious, walk through my history in human factors. My reminiscence is just one view of how human factors science and our marketing approach started. My tale will then be juxtaposed with current discussions within the Human Factors and Ergonomics Society (HFES) in the United States and the International Ergonomics Association (IEA). The excerpts discussing the need for change provide foundation for the call to action that I will introduce.

A history of our science and application of human factors

History is always a good place to start to talk about the future. There is much to learn, especially the mistakes of the past to avoid in the future. I will give you a small history lesson about my early human factors work to provide some insight into how our profession developed its science and structure, and to show that the problems and limitations identified then are still with us.

I will start by describing a condition I experience when I have technical discussions with colleagues. I catch myself now saying things like – 'I remember when we did that 30 years ago' topics. This, for the most part, is still a topic of major concern today for the human factors professions.

In the late 1960s, our potential human factors customers would say things such as:

- They (human factors people) cannot show me the value of what they do (no business case). I think they just add cost to my work.
- I do not know what human factors is or I never heard of it, it cannot be important.
- Human factors issues are just common sense, so any human can deal with human issues.
- When we have it completed, we will ask the human factors people to review it.
- If we get human factors people involved, it will delay the schedule and cost a lot more to do. Forget it.

Human factors specialists would simply react by saying 'we need to do something to change this so customers see our value and use our services'. Over the years, there have been many projects and programs to address some aspect of those concerns. But since you have probably heard similar things said within the last year we can assume that those projects and programs did not resolve the issues. What will practitioners be saying 10 or 20 years into the future? Here are three possibilities:

- We still have the same issues. We should...
- I remembered when we used to do human factors work, but that was before it was taken over by other professions.
- I remember when we decided to address head on the issue of ensuring our knowledge and work was seen as indispensable. That was the beginning for us to become the clearinghouse for human factors issues wherever they occur.

We know that to make wise decisions about where we go in the future requires us to know where we have been. There are numerous accounts of our early technical work in human factors/machine interaction developed by the military, academia, and industry. Those early accounts show how the early work set the foundations for how we currently perform our technical work. In addition, it seems logical that this early work also set the foundations for how we define our profession, conduct business, and address the marketing of our services. I have not heard of much discussion regarding these aspects of the early work.

I believe that the high quality scientific approach we developed early on for doing our technical work, although effective at the time, may have been a process that moved us away from the practical implementation of our science. That is why it is important to understand the structure we established, the impact it has had on our current way of doing business, and to question whether it is still appropriate.

In the US, one of the most influential forces that shaped much of formal human factors history began in the 1940s as support for the military. Academic institutions supported this work because of military need and the dearth of scientific knowledge about human–machine interactions. These early customers generally focused on technical performance issues, safety, and training, and had little interest in business issues, and so cost and return-on-investment did not need to be addressed. Because of urgent war and post-war needs, during this stage of the human factors discipline development, practitioners learned to associate success with criteria such as:

- emphasize that for solutions to be valid, they must be based on research tailored to the problem.
- follow the scientific method.
- state that information only applies to the studied area and to generalize will require more research.
- assume that customers understand the scientific language.
- do not address the nonscientific aspects of a work problem (cost, politics, societal issues) because they are not valid issues in scientific work.

Using this approach, researchers developed work products of high quality, but narrow in scope, not generalizable, and not tied to documented customer value.

Because this approach to doing human factors work and dealing with customers proved successful in this specialized field, practitioners internalized and formalized this method and applied it to new work. Within this minimalist business model a number of technical requirements were established as foundational in human factors technical work: Over time, it became the basis for our human factors business culture.

Contrast this business culture with some of the basic human factors principles that human factors professionals now require:

- Know your user (customer) so your product meets their needs;
- Deliver a product that the user can understand and use; and
- Ensure the work product supports human performance, safety, strengths/limitations, and minimize training through good design.

Curiously these supposedly technical requirements were not recognized as requirements that could also support our profession meet customer's business requirements (via "know thy customer in all respects") when focused on business issues. One might wonder why our technical work should incorporate these practices and question why we have not uniformly applied them to the way we currently do business. This unfortunate dichotomy probably exists because: First we are too close to our work and cannot see its potential application; and Second that there was not a need to do so.

When our profession began to expand into commercial work, customers wanted to know what we did and what value we brought to them. Unfortunately, we could not always answer their questions. Our approach, based on our early model of success, proved counter productive to producing improved, long-term approaches to delivering products and services that meet customer need.

From the Human Factors and Ergonomics Society (HFES in USA)

The Society realizes that the world of human factors is constantly changing. In 2006, the Executive Council restructured itself with one of the objectives to help the organization focus on developing and continuously adapting strategies in response to dynamic changes occurring in the outside world. Karwowski [2], the 2007 President of the HFES, talks about the future of the profession. In his article, he asks:

- 'Where is the human factors/ergonomics discipline heading?
- How do we develop the HF/E profession to its fullest?
- How can HFES act as a leader in HF/E development?
- How can HFES add value to the global society at large?'

He states that 'we need to empower HFES members to do what they do best. To empower means to equip or supply with ability, to enable. How we empower... becomes a critical point for the success of our organization and its future'. Under Karwowski's leadership the Society will be working on developing a unified vision for the discipline in 2020 and beyond. Karwowski's focus is on how the society can support its members and the profession in the future with goals and objectives.

I applaud the Society's effort to establish its future vision and direction. My observation, however, is that traditionally the Society helped set direction and encouraged action, but is limited in what it can do to carry out implementation.

Dainoff [3], the 2006 president of HFES, made a practical suggestion on how we can move into the future. He listed the known future changes that will affect human factors professionals and discussed the need to create a different profession for the future. He made it clear that action to engage in a change process needs to begin now. Dainoff also addressed the importance of being a human factors resource of knowledge and providing the information to those who need it.

Dainoff's remarks are supported by a statement from Raymaker [4] an information technology specialist. He stated, 'As the global economy becomes stronger, the competitive edge will be data. The ~~company~~ [profession] that can collect and analyze market, product, and service data faster and better will have a competitive edge over those that can't. The data will enable a ~~company~~ [profession] to deliver products and services to the market faster at a price that the global consumer can afford'. I have substituted the word profession for company because Raymaker's advice is equally applicable to the human factors profession.

International ergonomics association 2006 world congress

In addition to all the traditional activities at the International Ergonomics Association 2006 World Congress, attendees discussed the future of human factors/ergonomics as a profession, its value to the customer, and some of the things that could be improved on in the future. For example, one discussion centered on the missed benefits of our current research processes. Another focused on the common consequence of excellent research that should be beneficial to practitioner/implementer, but ends with a published paper or a conference presentation. The discussion continued with suggestions on how to change the research process to include objectives for use, intended application, limitations of data, assumptions, and other Meta Data that would help practitioners find and apply results. Attendees also explored the lack of growth of our professional societies. Many worry that our potential customers are now employing people that they perceive to be more practical and application-oriented and who suggest that they are providing more cost- and performance-effective human factors solutions. Some of the formal sessions also addressed profession issues [5, 6]. In preparation for my closing day plenary session speech, that would cover a summary of the macroergonomics presentations, I attended many paper sessions and panel discussions. I had a preconceived idea that macroergonomics meant something like multiple ergonomics issues would be dealt within a systems context of a real world environment. I found it amazing, however, how others defined macroergonomics and that the definition often did not focus on end product delivery for real world application. My concluding remarks attempted to inspire practitioners to evaluate their current products and services to see if they matched customer and society needs.

I knew that everyone at the congress has the ability to contribute to the improvement of our profession and I wanted to encourage them to use that ability. The presentation ended with a motivational cheer 'We can do it', which I truly believe. After that presentation, a colleague told me that they had not seen someone that evangelistic about human factors in many years (I interpreted the comment to mean over the top). Maybe our profession needs more evangelists for action to occur.

In a recent article, Kolasinski [7] discusses human factors evangelism. Kolasinski suggests that we need to expand beyond talking to ourselves at professional meetings and communicate to others in all sectors of life. She believes all Society members should do their part of evangelizing by publishing common language human factors related articles on topics that interest us and promote human factors to others in our interest fields.

The most important motivator for change – the practitioners and the customers

Over the years you probably have heard the thoughts of countless numbers of human factors practitioners who have expressed opinions on topics such as: our profession's strengths and weaknesses; what is being done right and what is not and suggestions for improvement. Our external customers (and potential customers) also have their own thoughts on these topics. Unfortunately, those thoughts have collected for years in the information space of quantum ether and, for the most part, have not been put to practical use.

Many in the profession have made a case for change. But change how? Knowing some of our basic weaknesses and strengths will provide insight into our needs and requirements.

In my own field, for example, it is understood that aviation safety improvements are dependent on developing an aviation system that is not only technically sophisticated, but also human performance-based and human-centered. One of the lessons the industry has learned the hard way is that ignoring human factors can cost both in the expense of re-engineering and in schedule delays. Technology and human factors simply cannot be separated from one another if safety and efficiency are to improve. It is essential that human factors specialists remain full partners in the development and deployment of advanced aviation technologies. That is why the aviation industry is working to systematically integrate human factors at each critical step in the design, testing, and acquisition of new technology introduced into the national aviation system. Through research in areas such as selection, training, automation, workload, and communication, the industry is identifying the most effective procedures to be used in combination with new technology applications and a more capable workforce to make the global air transportation system of the future more safe and efficient.

This, as you know from your experience, is not an easy task. To succeed as a profession in tackling issues like these we need to take a close look at how we do business, how we can improve our processes, and how we can re-engineer the discipline to ensure our future success.

PROFESSION WEAKNESSES

Weaknesses provide opportunities to do better. Strengths help us accomplish our objectives. Having an awareness of some of the profession's primary weaknesses will help later in understanding why some of the specific project tasks are recommended for early action. It is obvious that other professions also have these weaknesses so why single them out for our profession? By our recognition of them, we can be proactive in eliminating them.

Cost, time, and project limited evangelism

We all know that most of what we do in life has a start, a middle, and an end. But when you apply an end date to establishing acceptance of human factors by customers, educating people about and how to apply human factors, defining the profession, and other such important issues, working with end dates may not be the best approach. The following sections provide specific examples of how our current approach would be better served by using our efforts to cause a cultural embedding of human factors.

Project funding dependence

We have all read about the successful rollout of an awareness campaign or educational effort to improve people's understanding of human factors. Over time, however, the effort fades and then disappears. Often, these good intentioned human factors promotional projects are dependent on finitely funded/limited time projects. When we base our promotion efforts on projects, brochures, and training programs that terminate when the funding ends and/or time runs out, the benefits of that campaign generally also end. It is clear, that much of our effort to promote our profession is based on business practices that are not conducive to a sustained, long-term promotion of the benefits of human factors.

Lesson. Establish a central professional infrastructure or organization to synthesize the individual results into an ongoing effort.

Personality dependence

As is the case in many areas of our society, things get done because someone champions an idea, is passionate about a cause, or wants to make a difference. Usually these people accomplish a great deal, but when those accomplishments are dependent on the presence of a single person, the efforts die when the individual fulfills his/her personal goal. In human factors, many great people have put a huge amount of effort into promoting human factors, but lost their momentum because of a lack of time, money, support, and competing priorities.

Lesson. Focus the work products of these exceptional people into strengthening the cultural base for human factors.

Our promotions focus on projects, programs, & training

We desire that human factors products and services be recognized as a normal way of doing business. When you look at most of our efforts in the past to promote human factors you will see very scholarly communications projects, brochures, and training programs that often have a limited life, narrow focus, and limited audience. This approach produces incremental value to promoting human factors but not the desired 'I cannot live without discipline' that we hope for. If we approach our promotion from the viewpoint of creating a cultural acceptance of human factors as the authoritative representative of human/system interaction, we will then have a sustainable demonstration of our value to society. Embedding our contributions and services in the culture that would look to us to facilitate human/systems solutions is possible, practical, and appropriate. (Note: Getting what we ask for will also be a heavy responsibility for human factors practitioners to live up to the high standards that would be required.)

Lesson. Focus our effort of promoting human factors in work that will have long-term positive impacts and cultural change in addition to short-term results.

Focus on academic purity vs practical application

We, as human factors practitioners, often feel that we must pass on to our potential customers our knowledge about sample size, statistic limits, control groups, the technical nuances of our work, and the fine distinctions between flavors of human factors. These distinctions and discussions probably should be reserved for our professional get-togethers and paper presentations. If you are an engineer or a manager with a set project schedule, you simply do not have the time or resources to figure out the importance of this type of knowledge. As a manager, you are focused on the bottomline and want help in solving your human problems through whatever means necessary.

Lesson

Stop talking and listen to the customer. Find out what they need and address those issues.

The silo effect

Much of our work is focused on a small technical, geographic, influence, business, or academic area. The customers, practitioners, users, funding sources, and application and impacts of human factors are often limited to these closely linked communities of work. These communities often have limited interaction with other-focus communities of work. Products developed in such vertical structures (known as silos) often focus on issues within the silo with the result that system components outside the silo may not be accounted for and solutions stay within the silo.

Lesson

Our efforts to develop work products and promote human factors will benefit from solutions developed jointly by human factors professionals in concert with other disciplines and our customers, to address the complete issue at hand.

We don't have to do it all

We have often been less than happy when others try to do our work and not use our services. However we sometimes try to do work that other specialists should be doing. Most work that has human factors issues has many other issues that are related to work that other specialists should be asked to support. Because of our better understanding of human systems issues, we probably should be the facilitating leader for many interdisciplinary projects.

Lesson

We have to realize that it is our responsibility to accomplish the best solution called for in our work and that will often involve asking others for help when we are not the best resource.

Our value has limited distribution

We do an excellent job of talking amongst ourselves about the value of human factors. It is comfortable and no one asks questions. Unfortunately, we are not the paying customer. Talking to customers is often difficult because we have not learned to present our work and ourselves in a manner that is understandable to a non-technical audience. The result of these encounters is not always beneficial or pleasant and so contact may be avoided.

Lesson

Work with other human factors professionals to develop a positive message, but work with the customers to learn how to present it so that the customer will understand it and wants to take action on it.

CHARACTERIZING OUR PROFESSION'S STRENGTHS

Human factors people are truly unique in the professional world. Our outstanding talents in systems integration (human-physical world) are a foundation upon which our profession can build. Since we may not always realize the value of our talents, it is important to explore a few of the reasons why we are unique and valuable.

System thinking

Because of the complexity of human thinking and actions, the human factors practitioner must always attempt to account for a broad spectrum of issues relative to the human. This requires thinking on a broad, systems level.

Education

Human factors professionals are very diverse because they come from different technical areas, such as electrical engineering, farming, psychology and history. The common thread for all, though, is that they have extensive human factors education in addition to their other specialty.

Human – world connection

Besides the concern for humans, the human factors specialist makes the connection between the human and the physical world (although there is are a great number of cognitive issues in the connection). Other professions also deal with human issues (architects design houses for their human customer for example), but they usually work on specific topics using standard practices, such as wiring a house by applying handbook information.

Human life cycle

The human factors discipline deals with humans across all phases of a system or organization. This continuity of human factors work makes it possible for solutions in one phase to take into account impacts from, and to, other phases of the life cycle. This continuity sets us apart from many disciplines that only work in one phase of a life cycle. (see Fig. 1)

Broad discipline and training

Our system thinking and broad work area facilitates our ability to work with professionals from many other disciplines. We have the skills to serve as integrators between multiple disciplines. These skills position us to work on interdisciplinary programs and to serve as the

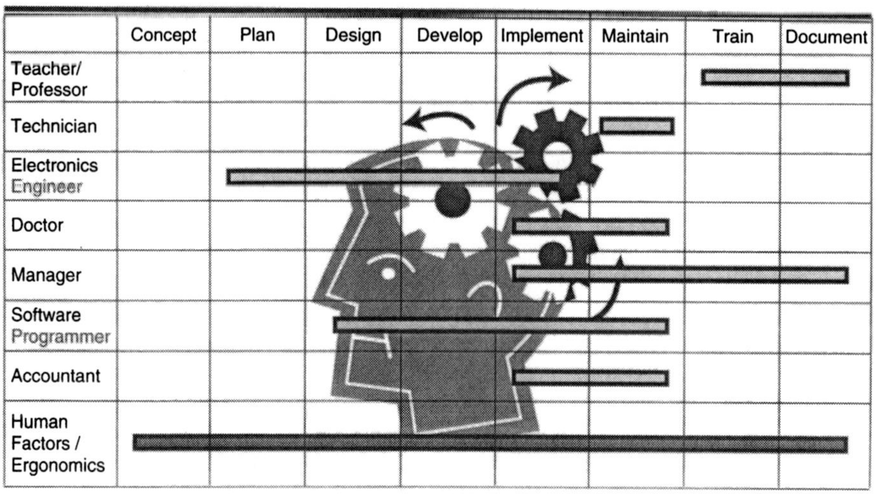

FIGURE 1. Human Factors cuts across many parts of all life cycle work.

integrators of various technologies in a human systems context. The systems attributes that greatly contribute to our integration abilities include:

- Broad systems thinking.
- Core human factors with one or more other major areas.
- Human factors make the connection between the human and the physical world.
- Deal with humans' complete experience with the physical world.
- Work across life-cycle phases providing continuity.

TAKING ACTION – ROADMAP TO THE FUTURE

We must take action now to reform our discipline so that human factors practitioners will be the ones who plan, implement, and control our future in human factors work. Traditionally, in a purely academic discussion, I would proceed with the discussion by saying *we should do this or that or whatever.* But, it is too late for shoulds. I am therefore going to describe what we WILL do to advance the human factors profession. I have taken this approach because I believe that a new human factors infrastructure will change our culture, give us a more viable discipline, and ensure our place in the future. Success depends on a unified effort to develop an action plan, apply a long-term human resource strategy, and obtain sustained funding so that action can begin. I sincerely hope that this preliminary planning can motivate you to take action to make our profession more viable in the future.

The remainder of this chapter is based on an assumption that we are going to proceed with a long-term program to influence and support the future direction of the human factors profession. The 'we' is us the practitioners, customers and partners needed to develop and execute our new beginning.

The roadmap to change

I am going to provide one roadmap for establishing a new organization, a business infrastructure support organization, to move our profession into the future. This roadmap is intended to kick start the change effort. Over time, the new organization will be matured into an ongoing service unit for the profession to provide us with information, tools, processes, communications, and other resources. To succeed in the long-term we will also concentrate, as individual human factors practitioners, our efforts to use new business practices and promote the human factors profession. Such an effort will reinforce the business support side of our profession and will build on (and support) the current professional organizations and certification groups.

In keeping with good scientific and business practices, the roadmap requirements will meet the following purpose and objectives:

Purpose

Make our profession the leader for human factors knowledge and work.

Objectives

- Learn what our customers need in terms of our services and establish a mechanism where we will continue to be in tune with their needs
- Establish a business-based human factors support infrastructure that has the objective of supporting all human factors professionals

- Provide our services in a timely manner to meet needs
- Have the tools, processes, and information to support our work so that we will become and stay leaders in human factors work
- Make human factors a cultural requirement (taking the place of programs and training as a means of promotion of human factors)
- Learn to deal with human factors issues in a life cycle and systems context
- Learn to apply human factors solutions in indirect systems (automated and systems that don't have direct human/system interfaces) that impact direct human/system interaction

Likelihood of success

Changing how a profession views itself and changing its operational culture would seem to be a daunting task, but this is the type of business we do as human factors specialists so it should not be difficult. It has been said that the most important condition to be met for re-engineering success is that the person or organization recognizes the need for change and wants to change. I believe we meet these conditions. The initial uncertainty relates to who will initiate the work to create a new support organization. Once that is overcome, success is ours.

ESTABLISHING THE INFRASTRUCTURE SUPPORT ORGANIZATION

Identifying and implementing an effective architecture for the support organization that will be our dedicated long-term resource will be crucial to the organization's success. A wise first approach will be to determine if there is a current organization that would be appropriate to become a 'business' support function to human factors practitioners, partners, and customers. Because this organization will be involved in day-to-day business operations of practitioners and those they support, my guess is that no current professional organization would be appropriate. Likewise, I don't believe any other organization currently fits the role because if it did, it would have already addressed many of the issues identified below and that is not the case.

I am therefore proposing a tested and effective architecture for a distributed work environment (one used at one of my early employers). This architecture consists of two main work function areas. The first area would be the core, centralized group of human factors and business professionals that support the practitioners. This would be the infrastructure support organization. This core group would manage system integration issues, develop tools and procedures that could be used by the direct support people, identify appropriate training and education for individual support personnel, identify applicable project standards, ensure funding was appropriated, and marketed for their participation in each organization. They would also provide administrative support and other services as identified.

The second work function area is the real world of the decentralized practitioners who are in the field supporting the customer. The practitioners identify the support that they and their customers need, and provide market intelligence to help build the profession's knowledge. They collectively direct the core group's activities.

Our new support organization can be tasked by individual and groups of practitioners to address all manner of issues as they arise. It will provide support for customers and business partners who need human factors work done. It will act, in a sense, as a broker, a knowledge sharing network, a resource guide, a policy and education center, and as an advocate.

Naturally, there will be numerous organizational issues to overcome (who is in charge, competition with other organizations, focus, return on investment, types of resources) and professional issues to be resolved. None are insurmountable in light of the value that can be achieved.

Business infrastructure support organization development issues

Once the basic organizational structure question is answered, we have a number of administrative issues to deal with that will establish the operational guidelines for the new organization. It is important to address each of these issues up front so that we can develop an enterprise approach. To create a viable organization, we must think about ongoing management, coordination and facilitation. I suggest practitioners in the ODAM/Macroergonomics Technical Interest groups of IEA and HFES might be excellent leads for these early dialogs.

Core support organization personnel

The people who will manage and work in the Infrastructure organization will be experienced human factors professionals who have a strong business, marketing, and communications background and record of accomplishment.

Paid management

Management will focus on the objectives that will eventually accomplish our cultural change. Although volunteer lead functions can do good work, they are often can only provide a part-time focus for the people concerned. A paid management staff is critical to sustained progress.

Oversight board

An elected volunteer oversight board will direct the support organization management to ensure objectives are met.

Link to all other disciplines

The organization will concentrate efforts on 'Human Factors' issues and people. The organization will develop strong links with all other disciplines that human factors people interact with. A systems approach to teaming will be followed.

Volunteer support and guidance

We want and need many professional volunteers from the human factors discipline and many other disciplines to participate in all support organization activities. As characterization of the organization begins, emphasis will be on tapping the extensive knowledge and skills of working and retired volunteers to lead and participate in many of the work tasks.

Virtual work location

The organization will be grounded in the concept that we will have virtual work relationships and locations in much of our future work. Keeping in tune with this concept, the functioning and management of the organization will be from a virtual location. The management will be directly accessible by all as a part of the facilitation process.

Figure 2 describes the organization of the Business Infrastructure Support Group.

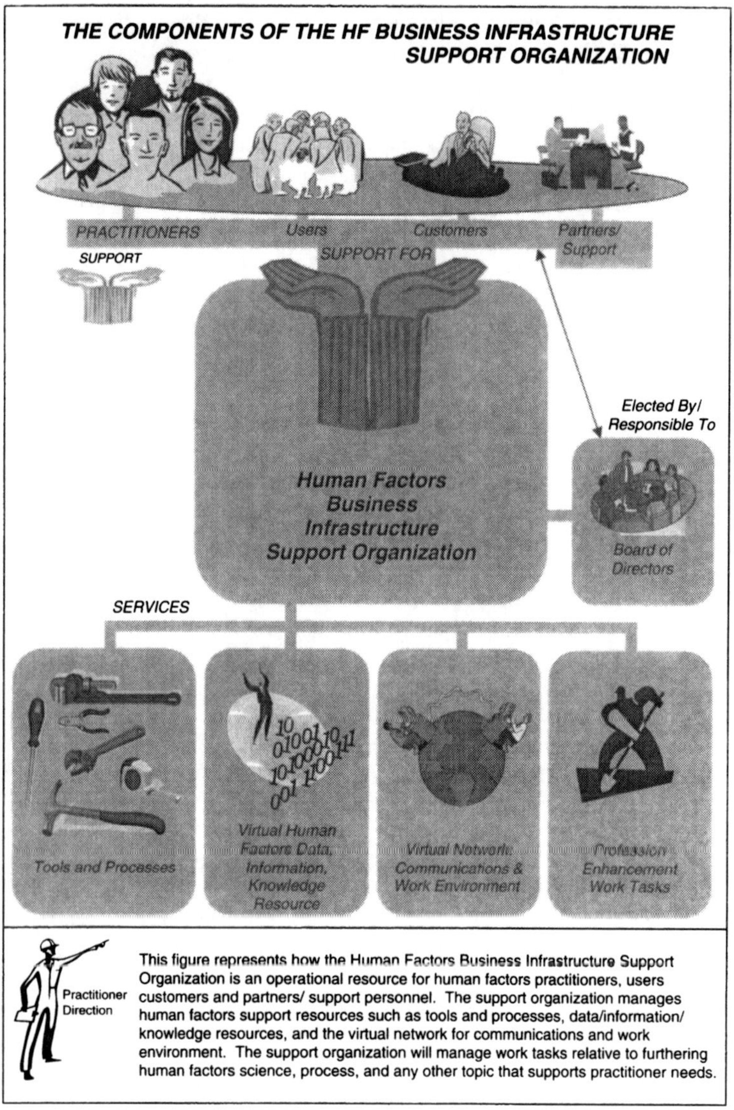

FIGURE 2. The relationship of the components of the HF Business Infrastructure Support Organization.

Funding

Startup sponsor – government, academic or industry

We can and will accomplish a great deal of our startup work on a volunteer basis lead by the paid management person/team. Progress will depend on an initial funding source – preferable a sponsor(s) who can provide money and/or resources to initiate the organization, support initial operation, and help build an ongoing infrastructure support organization. A special technical area of human factors (the insurance industry) already uses a sponsor to support their work and this relationship has worked well.

Baseline cost/benefit of current way of doing business

Proposing improvements to the organization's future financial model will be dependent on understanding the current financial reality of doing business. Revenue streams from sponsors, business partners, fees for service arrangements, and other models will be employed to make the program financially viable to all partners, customers, and users. The organization can explore whether a practitioner/customer fee could eventually be requested to fund the organization.

Virtual network of human factors practitioners, customers, and partners

As individual human factors practitioners, we know that interpersonal and organizational networking and communications with co-workers, partners, customers, management and others in our work environment is extremely important. One of the reasons we have created professional societies like HFES is to share information about capabilities, availability, status, schedules, resources, availability and myriad technical topics. Such formal and informal communication networks prove invaluable to our work. In today's work environment, communications technologies allow us faster communications at reduced cost.

The infrastructure support organization will benefit from minimizing its facilities requirements and administrative staff. The organization will benefit from having immediate, low cost access to professionals, tools and information that can be accomplished with communications technologies. The support infrastructure organization will base its operations on the technologies that permit operations and management in a virtual form (from any location to any location). The operation will network human factors people, information resources, customers, tools etc. An example networking tool that might be used is the Easy Ergonomist Finder, the professional website launched at the IEA 2006 Congress. The IEA 2006 Foundation is sponsoring this site and awaiting acceptance by the IEA Executive Council.

The virtual network will also provide an effective means to gather, process and distribute human factors business related data/information/knowledge. Information technology will be used to establish an information resource for all to use to support their human factors work. Currency, accuracy, availability and control of the data, information and knowledge will be maintained in a real time environment. The infrastructure organization will facilitate the linking of the data, as necessary, both via computer and via interpersonal interaction.

IN PURSUIT OF ACTION

The creation of a human factors infrastructure support organization is just one step in developing a new human factors culture, establishing our profession and us as the go-to people for projects in a wide variety of human impacted disciplines. While we ponder how to create the organization and who will lead it, it is time for real action. With or without the new startup group in place, we will begin work on a number of things critical to changing our culture that will prepare us for future endeavors.

As a community, we will establish a set of goals and tasks that will secure our place as future leaders in our ever changing and dynamic world. It goes without saying that somebody needs to do this work. I am asking you, the reader, to take action by establishing a task-related

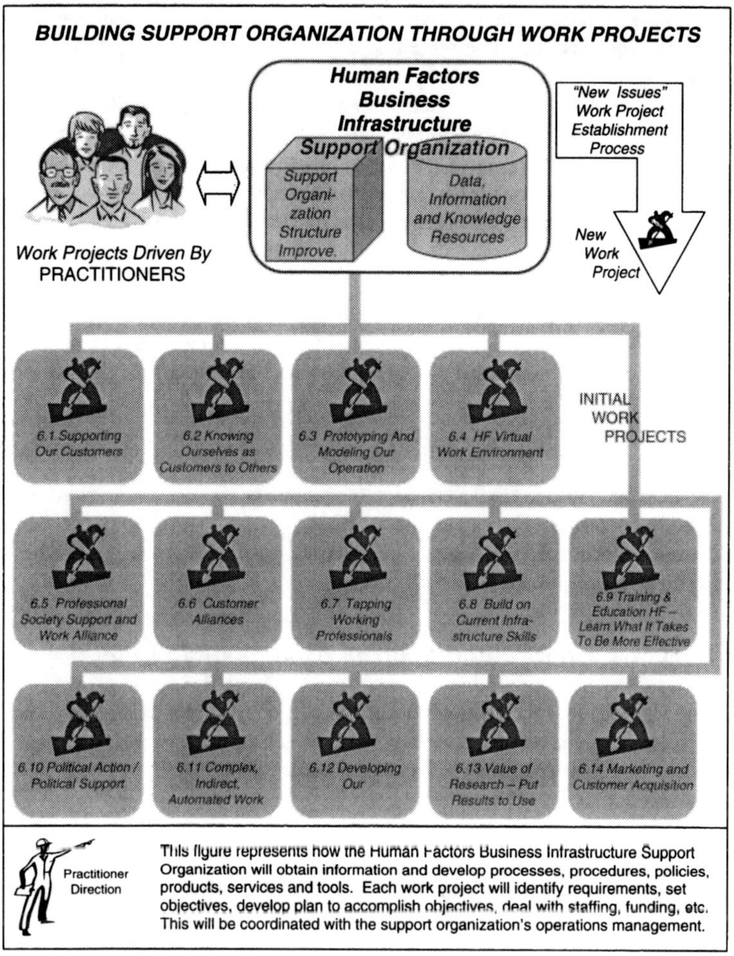

FIGURE 3. Process for building support organization through work projects.

startup group. I believe that if even a small number of practitioners take action, we will have begun a process to address the needs of many and improve the future of human factors.

At this point I could step away and wait for someone else to volunteer to get things started. Then I thought that if I wasn't willing to help get this work started, then I should not be proposing the support organization and change process.

Therefore to get the task process going, I have identified 14 work projects which are presented in Fig. 3. Using a standard human factors approach, for each work project I have identified for each work item its objective and tasks.

Work project 1 – supporting our customers

We know that survival as a community and as a profession requires that we pay attention to the needs and requirements of our customers/users/operators/participants. Those who request the work, pay the bills, sign contracts, manage the work, set policy, administer the programs and pay for the work may often be more important to the success of our technical work than good technical results.

Yes, we can say that we already interact with these people, but usually this is from a procedures viewpoint where we merely deliver administrative products, such as monthly reports or technical specifications, on request. If we are to go beyond the contractual relationship to support our customers in a manner that keeps them coming back, we will realize that knowing our user/customer and their needs in depth is the foundation of all work that follows.

Success depends on:

Objective

Improving our ability to identify quickly and effectively customer problem space, work with the customers to identify the needed product results, and develop action plans to deliver those results.

Task/Subtasks

- Develop an initial generic characterization of customers
- Refine the characterization on a continuous basis
- Conduct ongoing focused work area characterization so that knowledge can be tailored to the specific needs of a work project.

Work project 2 – knowing ourselves as customers to others

We have often heard that the 'Clothes make the person' and 'Image is everything' when it comes to projecting an image of ourselves. Well, it is time we establish an image (knowing how others see us) appropriate to our customers needs.

Success depends on:

Objective

Providing guidance on developing an effective image of ourselves and of our profession.

Task

Developing processes/tools and other resources that will help human factors practitioners learn about their image as seen by customers, and meet their needs and requirements in order to help us improve the image we present to customers.

Work project 3 – prototyping and modeling our operation

Having a means to perform evaluations of tactical and strategic decisions before they are implemented has become an important business tool for organizational success. Since our efforts require the establishment of some form of business structure we have an opportunity to incorporate a modeling/prototyping process into that structure. Each task performed to build our infrastructure can provide content for a model that will continuously improve our ability to meet customer needs, minimize cost, and improve performance.

Success depends on:

Objective

Employ a distributed management/parctitioner environment modeling support tool.

Task

Obtain and customize a modeling package that will support our business. Use it in a continuous improvement mode.

Work project 4 – creating a human factors virtual work environment

Many small organizations have found that they no longer can afford 'Brick-and-mortar' facilities to do business. They have to use communications and computer technology to work with distributed resources that need not be co-located. Our human factors infrastructure will be based on a virtual work environment model. There will be three components to the virtual model as follows:

The virtual communications and work network of human factors practitioners, customers, and partners

I introduced the practitioner communications virtual network as a foundational requirement for our infrastructure organization. This virtual network of people will provide the means to manage all the communications and work throughout the world. The network will be directly accessible to all partners as part of the facilitation process.

Virtual repository of data/information/knowledge

You have probably heard the old saying, 'Information is Power'. Here is a recent quote that indicates the importance of our management of information (words in parentheses are my changes). 'As the global economy becomes stronger, the competitive edge will be data. The ~~company~~ (practitioner/organization) that can collect and analyze market, product, and service data faster and better will have a competitive edge over those that can't. The data will enable ~~companies~~ (practitioners/organizations) to deliver products and services to the market faster at a price that the global consumer can afford' [4].

This basic underlying concept is already having a profound impact on our current work and on our future viability, because a large component of our work is in the development, processing, and application of data, information and knowledge. Since data, information and knowledge are the primary commodities that we develop and use, a great deal of attention needs to be paid to how we manage this area. I am not advocating the creation of an all-encompassing human factors database or document library – that would be too labor intensive, time consuming, and costly. With the data management technologies available we need only to be able to find, process and deliver data from current sources.

Success depends on:

Objective. Providing human factors communications, data, information and knowledge resources when, where and how the practitioners need it.

Task/Subtasks

- Develop an effective means to gather, process, and distribute human factors business related data/information/knowledge, in a form that is appropriate, usable and economical.
- Establish procedures to maintain currency, accuracy, availability and control of the data, information and knowledge that will be maintained in a real time environment.

- Implement a process to link human factors Metadata (data about data) such as source, validity, purpose and meaning to our information to increase its value and validity for implementation.
- Human factor the network so that special user skills, computer hardware or software will not be needed.
- Open information resource to customers, researchers, and others who will benefit.

Work project 5 – developing professional society support and work alliances

Our professional societies and certification organizations are a great resource for promoting our profession. We will work with these professional groups to establish effective working relationships so that everyone can accomplish their goals and objectives.

Success depends on:

Objective

Becoming supporting partners in the human factors profession.

Task

Develop cross-organizational working relationships.

Work project 6 – establishing customer alliances

Who better to identify customer needs and requirements than those we serve, wish to serve, and those who do not want our services, or who have had bad experiences with us.

We know that customers are not always experienced in identifying and defining their human factors problems and needs. The closer we are to our customers the easier it will be for us to clarify those needs. Working directly with customers will also help us to recognize the nonhuman factors issues that we will have to deal with. Since our objective is success for ourselves and for our customers we will encourage partnering and alliances with customers to satisfy their need to address human factors issues.

Success depends on:

Objective

Establishing our profession as the primary human factors support organization and showing that we are primarily concerned about customer well being

Task/Subtasks

- Develop ethical and business rules to work under
- Develop a marketing strategy for working with our customers as work partners
- Develop a business process that includes continuous improvement

Work project 7 – tapping working professionals' knowledge

We are an extremely important resource for one another. We can each provide strategic and tactical direction to identify and supply content for improving our community. Currently our professional society meetings are the only venue (time and place) used to deal with business, support infrastructure and professional topics such as 'What is our science' type issues. At our annual meetings, we rarely discuss these types of issues. We will create new venues to develop business methods and discuss sales, customer issues, trends, finances, training, legal issues, tools, procedures training and new resources like tools and processes.

Success depends on:

Objective

Establishing and maintaining a participatory work program venue for the practitioners.

Task/Subtasks

- Establish work processes and standards.
- Identify funding approaches for practitioner events
- Tie into virtual network communications.

Work project 8 – building on current skills

Some human factors practitioners might feel that they will never be able to understand how to develop and use business case information and consider socio-technical issues. Some prefer to do work that does not involve dealing with these issues. Others who do not often get to see how their work is actually used in a real world application may wonder how they would account for applications (i.e. integration into a system) in their work.

Fortunately, you know a lot more than you think you do. Whenever you do a research project, develop a product or participate in project work, you actually deal with many business case issues, such as finance, project management, project organization, record keeping (admin istrative) and many other issues. You might also address issues such as educational requirements and cultural impacts, which are socio-technical issues. The fact is, you already used some of the skills and knowledge needed to build business and socio-technical cases for your work.

The difference between what we already do and know, and what customers are expecting from us, involves a change in how we approach our work. We will make building a business and the socio-technical case an integral part of our work. Additionally, we will also consider the integration of the product, service or process into the world environment [8].

Success depends on:

Objective

Increasing the comprehension and proficiency of our practitioners to address customer infrastructure information needs

Task/Subtasks

- Identify proficiency of practitioners' current basic skills in infrastructure
- Use infrastructure information requirements from the customer analysis task to determine what proficiency enhancements are needed by our practitioners
- Establish a process whereby practitioners can identify their specific needs and broker solutions and make this a continuous improvement process

Work project 9 – training and education human factors

There are numerous education and training courses available for almost any technical aspect of human factors, management and administration. What often is missing is guidance (education, training, books) on how to bring all the pieces together and apply them. A more formal plan for learning and sharpening the business, socio-technical and system application case skills will be established. As a community, we will work to broker the education and application experience that will meet all practitioner and customers' requirements.

Success depends on:

Objective

Providing the ability to assess what we need to know and identifyi the training and education that can be used to fulfill our needs

Task/Subtasks

- Develop/provide educational assessment tools for organizations and individuals
- Increase the awareness of school offerings in applying the business case issues to technology
- Ensure that human factors and ergonomics societies sponsor educational programs, workshops and seminars
- Sharing human factors lessons learnt

Work project 10 – taking political action/achieving political support

If we look at politics as the process of negotiating with others to agree on the importance of one's point of view and support that view under given circumstances, then we will need to pursue politics. Technical people often find the concept of dealing with political issues as too challenging and so they do not readily pursue them. Well, the practicality of the way our world works is that many decisions are made in a political setting. Human factors practitioners will increasingly find that it is to their benefit to know how and when to be politically active.

Success depends on:

Objective

Taking advantage of political opportunities to promote the value of our profession

Task/Subtasks

- Define the political practices that will be supported by our effort, while providing rules to minimize any negative effects
- Identify training and education opportunities for practitioners to learn to be effective in politics
- Develop an information package that can be used in political activity and keep it current

Work project 11 – creating our own opportunities

Four changes in human/system interaction are being brought about by technology: indirect human factors work; human factors in automation; component and system complexity and socio-technical issues. These changes are providing new opportunities and challenges for the human factors discipline. All four are complementary in that they all indicate our need to pursue work that may not have direct user/system interfaces or they may change the form and function of those interfaces.

Indirect human factors work

Some human factors practitioners think that their work requires an identified direct human interface. Even in systems design work, we often think of groupings of things with direct user interfaces working together. If there is not a direct user interface, then we may think that there is no work there for us. On the contrary, much of what we do is to develop work systems that support humans and the environment they are in. Human factors practitioners are in the enviable position of being a human factors expert and the system expert who can understand and help resolve these complex multilevel issues.

Human factors practitioners have a tremendous opportunity to take on a leadership role in complex systems and automation work by broadening our view of what constitutes human factors work and expanding our skills so we can work on complex systems.

Human factors in automation

Most of us are experiencing an ever increasingly automated world. In the past, we primarily performed manual processes (communications, transportation, cooking, entertainment, washing, etc.). Now many of us use automated devices to do such work. Initially we might think that there are fewer or no direct user interfaces, but in reality, those user interfaces are still there, just in a different form and function.

Automated systems are still designed to accomplish their task in a manner that supports the user. To do that, automation development will deal with the human factors issues that are inherent in these systems. Human factors professionals need to be an integral part of automation design to ensure that work-design, process design performance specification and many other conditions we normally attribute to human work are appropriately designed in automated systems.

Component and system complexity

When products, services, work processes and system components become abstract, have many interacting parts, nondeterministic results and require extensive training/education to be understood (to name a few attributes), they can be considered complex. Much can and needs to be done to insulate humans from the complexity. The next level of increased complexity occurs when the products, services and work processes are combined in systems where they are interactive and inter-dependent, and produce new products and services that are a result of those interactions.

Socio-technical issues

Socio-technical systems issues (management, organizations, administration, finance, employee pool, population shifts) become a mandatory design consideration in complex system development. Socio-technical systems (that include the people and organizations that use, operate, support and maintain these new technologies) will be considered as an integral part of the technology development and deployment process in the life cycle of a system.

Success depends on [9]:

Objective. Expanding our involvement in complex indirect systems level human factors work

Task/Subtasks

- Develop a simple description of indirect and automated human factors work (include examples)

- Characterize human factors roles in indirect, automated and complex human factors work
- Identify and brokering on-demand support tools, services and training needed to work effectively in this area and marketing work with customers and partners
- Incorporate socio-technical requirements into the basic structure of all work

Work project 12 – developing our science of human factors

We need to be clear to all of our customers, partners and any others we work with about what we are. When someone asks what we do we should be able to provide a general description of our profession, a tailored answer for our specific individual focus, and the value this all brings to the customer. The answer should be short and concise.

During the IEA 2006 World Congress, attendees had a number of discussions about the science of human factors. Human factors practitioners at the congress agreed on the need to address and provide definitive answers to the following questions:

- Do we have a unique science or not?
- Do we need our own science or just use the appropriate tools, processes, methods, equipment etc. for all the other work areas? If this is our approach, how do we represent it?
- What are the components of our profession that make up our profession?
- How would we proceed to develop our science if we thought we needed to?
- Do our customers, our partners and us see value in a science of human factors?
- Do we already have a science, but it needs to be described?
- We have a profession that brings many skills (different topic areas vs. different skill sets). The variations of our profession (ergonomics, human factors, industrial psychology, operations research etc) all can work together to present a common face to the customer. How can the special skills of each be applied to our work in a way that would benefit all?
- Would a business model and structure be beneficial to human factors for promoting and facilitating human factors activity?

Success depends on:

Objective

Describing the science and application of our profession

Task/Subtasks

- Develop a means that will allow the science issue to be discussed and defined
- Task a group to define our profession for business purposes, including all major human factors professions and associated resources
- Update this description as a continuous improvement requirement
- Use the results of this task in market development resource information

Work project 13 – putting our results to use

It is no longer enough to conduct research for research's sake. Since we expend a great deal of resources to produce a research product, we must use the products extensively to receive their value. Unfortunately, many research results are never used and their application value cannot be easily identified. In addition, we rarely think about potential users as a part of the research, and have no intent to market the results [10].

Without doing this, the value of the research is not obvious. To prove the value of our research, we will:

- Always identify all the customers of our work (technical and non-technical)
- Always identify all customer requirements (technical and non-technical)
- Deal with business case requirements
- Deal with socio-technical case requirements
- Deal with application and systems integration requirements

Success depends on:

Objective

Making research results more valuable as a source of information to support human factors work and fill customer needs.

Task/Subtasks

- Identify the additional information needed with traditional research findings to make it more valuable
- Influence the research process to include additional information about the research
- Index research (with Meta Data) and providing a means for matching research findings with customer need

Work project 14 – marketing and customer acquisition

You obtain customers because you have a product or service that will help them achieve their goals. We will develop an effective process of making ourselves known to potential customers and convincing them that we are the ones to help them. This marketing process will be an integral part of the infrastructure support organization that will have the ongoing task of defining and supporting the marketing process with leads, information and tools. The process will be continually upgraded to improve the effectiveness of the process.

Success depends on:

Objective

Establish a high quality marketing process that can be tailored to demand and be continuously updated.

Tasks/Subtasks

- Develop the process that will inform potential customers about how human factors can support them:
  - Search for people/organizations who know they have human factors problems
  - Search for people/organizations who don't know they have human factors problems
  - Advertise
  - Network with other disciplines
  - Demonstration work showcasing human factors
- Develop tools for problem identification, cost estimating and value received. Develop tools that will help potential customers and practitioners identify human factors problems and quickly and effectively build a technical value and business case projection for a problem solution.

- Establish procedures requiring all work performed by the support organization to provide a business case, ROI and value created information. This information will be used as a marketing tool as a basis for making estimates based on historical results.

As we start this work, it is critical to remember that to accomplish the goals in each work project, we must do more than identify needs. To succeed, we will also:

(1) Set up a work organization that will pursue requirements development, planning, scheduling, resource acquisition and pursue specific technical tasks.
(2) Put together a needs/requirements and task plan document that is based on customer discussions and consultation with human factors professionals.
(3) Obtain funding to accomplish the work.
(4) Solicit and consolidate input from the wide customer base that is supported by our professions. Include people who presently do not support human factors as a basis for knowing the problems that people perceive with current human factors support.

FOCUS ON VALUE TODAY – NOT TOMORROW

Work that is put off to tomorrow may never get started or when it does its' contribution may be diminished. We have an opportunity to begin our pursuit of making our human factors profession more competitive starting today. We will take action not tomorrow, but today.

I have detailed my suggestions on what we will do (our roadmap) to get started to ensure our future value. I have provided a starting point – and together we can do it – we can create a more viable, sustainable human factors discipline and community.

It is clear that we need to change our culture, the way we do business and how we view ourselves if we are to succeed. We have to think about our economic future, our credibility and our utility as a professional community. We will define our destiny by shaping our profession – we need to do it now.

How can we get started? I suggest that we begin discussing this issue with our colleagues, customers and partners. Remember the effectiveness of one person calling and talking to two people and asking them to call and talk to two others and asking each thereafter to call and talk to two more people. It won't take long to start managing how our profession serves its customers and society. I am excited about what we can accomplish. I want to take an active role in helping us develop that enhanced future.

Remember:

WE NEED TO DO IT

WE CAN DO IT

WE WILL DO IT NOT TOMORROW, BUT TODAY

REFERENCES

[1] Preston, R. (2006). Are you a change agent? Easier said than done. *InformationWeek*, 6 November, 1, 113.

[2] Karwowski, W. (2006). From past to future: Building a collective vision for HFES 2020+. *Human Factors and Ergonomics Society Bulletin*, 49, 11.

[3] Dainoff, M. (2006). Will HFES be here in 50 years? The need for provocative discussion. *Human Factors and Ergonomics Society Bulletin*, 49, 9.
[4] Raymaker, R. (2006) Speed – or lack of it – kills. *InformationWeek*, 25 September.
[5] Gallaway, G. (2006). Linking organizational, managerial, administrative, and financial business issues to human factors science. In: Pikaar, R.N., Koningsveld, E.A.P., Settels, P.J.M., (eds), *Meeting Diversity in Ergonomics*. Proceedings IEA 2006 Congress. ISSN 0003-6870, Elsevier Ltd.
[6] Boff, K.R. (2006). Revolutions and shifting paradigms in human factors and ergonomics. *Applied Ergonomics*, 37 (4), 391-400.
[7] Kolasinski, E. (2006). Guidance, young grasshopper, for your mission as an HF/E evangelist. *Human Factors and Ergonomics Society Bulletin*, 49, 5.
[8] Kortum, P. and Motowidlo, S. (2006). It takes more than technical knowledge to be an effective human factors professional. *Proceedings of the Human Factors and Ergonomics Society*. 50th Annual Meeting.
[9] Chabrow, E. (2006). 'Metrics, The science of gauging IT's contribution to business effectiveness must go beyond measures like ROI'. *InformationWeek*, 23 October, 1, 111.
[10] Gallaway, G. (2007). *Aviation Psychology Research – Extending Scientific Method to Incorporate Value and Application.* International Symposium on Aviation Psychology. Ohio: Dayton.

APPENDIX

TERMS AND CONCEPTS DESCRIBED

The concepts presented in the paper may not be new to you, but for purposes of having a common basis for understanding, the following definitions are provided. When you think about your own application of the concepts, use terms and definitions that are pertinent to your application.

Business case

Financial, administrative, managerial, and/or organizational information about a product.

Complexity

Things not obvious, take numerous steps, have interdependence, abstract, indirectly involved, many components.

Components

Necessary parts of a whole system.

Concept

Representation of an idea, product, service, process, or whatever content and focus of our work

Cultural embedding

An idea or concept that has been accepted by a culture or society as normal and customary.

Holistic view

Used in the broad sense of looking at all human factors aspects of a product's technical and business case requirements from a life cycle view in the context of the project and end use application.

Infrastructure

The basic facilities, services, and installations needed for the functioning of a community or society.

Mind-set

The way someone normally views something.

Product

Any/all type(s) of physical or cognitive concept, thing, process, or service that is the focus of development, implementation, and integration in a system or application.

Project

An organized endeavor to perform work to develop/implement requirements, and/or products.

Sociotechnical case

Technical and business case implementations of a product that affect social (environment, work, and personnel) systems that the product operates. Those external systems impacted are the sociotechnical case issues. Examples: education, transport, finance, family, etc.

System integration

The incorporation of a product in the context that it will be used.

Technical case

The performance, acceptance, cognitive, attitude and other information that is associated with a human, a product, or the human–product service interaction.

Meeting Diversity in Ergonomics (2007)
Pikaar, Koningsveld, and Settels (Eds)

5

The Value of Participation in Ergonomics

Andrew S. Imada
A. S. Imada & Associates

Abstract. Ergonomics has evolved beyond its original intent of changing work. Finding solutions beyond better products, recreation, learning and living applications are challenging. We are confronted with the paradox of finding limitless applications, users and emergent technologies with finite tools and knowledge base. A macroergonomic approach that considers organizational and technological contexts is a way of meeting this challenge. Engaging people in implementing ergonomic change is one macroergonomic method. Successes in participatory ergonomics are attributed to several keys: focussing on local solutions; considering the organizational context; and addressing human needs. Finding ways to access the emotional context is an important component in the successful implementation of ergonomics. Participation may be a necessary condition for ergonomics to be applied in new arenas where behaviour patterns are already well established.

Keywords: participatory ergonomics, requisite variety, change strategies, appreciative inquiry, emotional context

THE CONTEXT

Consistent with this book's theme – Meeting Diversity in Ergonomics – ergonomics has, and will continue to address diverse emergent issues. This diversity will manifest itself in the applications of the sciences, the people who benefit from ergonomics, industries and locations around the globe. All of this is expected and consistent with an increasingly technological society with far-reaching interconnectivity. What initially started as increasing efficiency has now moved to product design, work processes, environments and organizations. Our ability to continue meeting these needs depends on our ability to adapt our technologies to these different cultures and settings.

THE CHALLENGE

As with any technology or organization, ergonomics' ability to satisfy a wider diversity of people, settings, outcomes, and organizations is based on its willingness to change and develop with the world around it. We have often argued that the robustness of our principles and science-based findings make our profession generalizable to other situations. Therefore, it does not matter whether the workspace is a factory, a control room or a living space; the basic ergonomic principles can be used to satisfy these needs. At the same time, we have used anthropometry and measurement of individual differences to accommodate a range of people to things and places. This confluence of principles, measurement and our research has led to effective practice and application of ergonomics.

However, if we extend this argument, we would have an infinite number of people and situations served by a finite set of ideas and principles. Clearly, as we enter new arenas (e.g., habitation of outer space, nanotechnology), ergonomics will need to develop new paradigms, principles and methodologies. Boff [1] describes four transitions we are traveling through en route to these new arenas. We will be confronted by increasingly complex questions as we change simple tools to adapt to human characteristics, harmonize humans and technology for appropriate cognitive fit, optimize human physiological and cognitive capabilities in harmonized systems, and possibly biologically alter humans to enhance human and system performance. These will not be seamless transitions. In the interim, we are confronted with using what we know from our profession to what we apply. Herein lies a potential contradiction. At one point the diversity of applications will exceed our knowledge and capacities to deliver solutions. Stated another way, can we solve an infinite number of problems with a finite set of principles?

An example of this became evident to the author when working as part of a team commissioned by the International Ergonomics Association (IEA) and the International Labour Organization (ILO). Our goal was to take the existing body of knowledge and generate a checklist that people in the industrially developing world could use to improve safety and efficiency. For our first task we asked a team of experts to develop an initial set of principles. Next, we identified the key principles that could be verified by our science and data. Finally, we limited the principles to those that would be relevant to the specific user population (managers and stakeholders of production facilities in industrially developing countries). The outcome from this effort is a publication entitled Ergonomic Checkpoints [2]. This publication lists ergonomic checkpoints or principles that can be used to improve safety, health and working conditions. These same 128 principles have been used in many training programs affecting thousands of people across a wide range of facilities and countries [3]. Can this finite set of ideas satisfy a global diversity of needs, differences and applications?

The ability of ergonomics to solve an ever-widening diversity may depend upon an idea first proposed by Weick [4] called requisite variety. All organisms or organizations need the ability to respond to environmental demands. Moreover, the amplitude of this response must be at least as great as the environmental changes. Imada [5, 6] cites technology and globalization as two such changes that will require people and organizations to respond in a way that is commensurate with these changes. The ability or capacity to respond or provide a response potential is essential for the entity to survive. As an example, Imada [6] cites an article that points to the potential need for every producer in the market place to reduce prices by 30% to meet the 'China Price'. Failing to respond to this environmental change with the requisite intensity (in this case, reducing prices by 30%) may mean the end of an enterprise.

Returning to our example of the ergonomic checkpoints, how is it that these same 128 principles can remain relevant to a wide range of people, workplaces, problems, personalities, and cultures? Perhaps it is because these ergonomic principles have a solid foundation. Perhaps it is because they are so general that they can be applied across this wide chasm of diversity. Is there sufficient elasticity in these ergonomic checkpoints to continually expand into new nonproduction, out of the ordinary settings? How can we continue to develop capability (variety) to respond with a finite set of ideas? These are the challenging questions that ergonomics must face if we embrace diversity.

A WAY TO MEET THE CHALLENGE

There has been a natural progression for introducing ergonomics from the purely physical, to cognitive and macroergonomic levels. This last approach emphasizes the organizational (contextual) interface [7]. As opposed to focusing on individual or subsystems (microergonomics), macroergonomics considers the context (social, organizational, work system) in which the ergonomics will exist. Since the 1980s, Participatory Ergonomics has been developing momentum and support as a macroergonomics methodology for introducing ergonomic change. Imada [8] provides the rationale for involving end-users and stakeholders in the process. This is useful because it enables people to understand ergonomics, take ownership of change and adapt solutions for future change.

Perhaps one way of thinking about Participatory Ergonomics is to conceive it as 'middleware' that allows technology to be useful to a wide range of users. Information technologists have identified machines (hardware) that can execute the demands that the programs (software) instruct them to do easily when there is a single, integrated user, program and machine. However, when there is a wide range of users across disciplines, time, space, applications, media and hardware, the system requires 'middleware'. Middleware is the interface that translates across these differences to meet either computing or delivery functions. In one sense, middleware lubricates the connections among these differences to reduce friction in each transaction.

Participation from end-users, designers, stakeholders, owners, managers or supervisors translates needs and ergonomic principles to usable solutions. We argue here that as with the single user, task, application and outcome, ergonomic principles (such as the checkpoints) may be able to solve problems. However, to continue being effective over a much wider range of users, the principle alone is not enough. It is, in fact, the participation that allows the ergonomic principle to accommodate the diversity.

During the 2006 IEA congress, numerous presentations highlighted the ways in which participation allows ergonomic technology to increase response to a diverse set of situations.

Global ergonomics, local solutions

Kogi's [3] presentation summarized the valuable lessons learned from his many years of working with industrially developing countries. Starting in the mid 1980s with small enterprises, Kogi and his colleagues have made thousands of improvements to workplaces in countries throughout Asia and Africa. In one example, there were more than 4000 improvements made by farmers in Vietnam in one year. In the 12-year history of POSITIVE program, they have engaged more than 26,000 participants in nearly 900 workshops with a network exceeding 1200 trainers. Participatory workshops using photographs to illustrate good examples, checklists to analyze and implement changes and illustrated how-to manuals are keys to these programs.

There are three keys to Kogi and his colleagues' successful participatory strategy:

(1) Target a stepwise process that builds on local good practices. Finding examples of what people are doing well or what they are capable of achieving is a strong basis for making future change. Build progressively on these small positive changes.
(2) Focus on low-cost improvements that implement basic ergonomic principles. The simplicity of locally developed, readily available solutions reduces risks for trying new ideas and is likely to increase voluntary participative behaviour.
(3) Develop a network of trainers to compound the growth of the participatory action-oriented tools. This creates synergies for further increased capability to respond.

Organizational context

Successful participatory strategies depend on the infrastructure, support and organizational context in which the ergonomics technology is introduced. This point was made clearly in two important papers at the 2006 IEA Congress. First, Vink, Koningsveld and Molenbroek [9] identified factors that their review uncovered as important to participatory strategies for increasing comfort and productivity. These factors involve two general categories.

(1) Involvement
  (a) Direct worker involvement
  (b) Strong management support
  (c) Involvement of key stakeholders
(2) Process
  (a) Inventory of what is going well, and the problem
  (b) Step-by-step approach
  (c) Review and accountability
  (d) Steering group to guide the process

Vink et al. cite works in the literature that support the basic assumptions of this successful model for improving health and productivity. There is an important parallel between this paper and one by Dul and Neumann [10] who argue that ergonomics has value to organizations and can only be appreciated by those organizations if it is built into the business strategy. That is, ergonomics must be built into the context to achieve organizations' and peoples' objectives rather than being an end in itself. The parallel with the Vink et al. article occurs in the way that they define strategies. Corporate strategies define major objectives that require co-ordination over larger regions of time and space. This requires involvement from participants much removed from the work or processes that will be affected by the ergonomics. This speaks directly to Vink et al.'s first involvement category. Levels of involvement include top management, unions, key stakeholders, staff functions whose expertise will be required, and the works themselves. Coordination and communication of this larger vision requires a broad range of skills. Business process strategies are ways to achieve the corporate strategies and usually involve more concrete actions and a shorter time horizon and decision-making. This includes Vink et al.'s identification of a step-by-step approach, review and checking, and steering groups to guide and lead the process. This strategy is more tactical, measurable, definable and absolutely necessary to meet the corporate or strategic vision.

Dul and Neumann assert that omitting ergonomics from corporate or tactical strategies is a mistake. They also rightly point out that ergonomics outside the context of the enterprise causes it to be perceived as being associated with health and safety legislation rather than as part of the enterprise's success. Employing the wrong strategy with a given group is equally devastating. Imada [11, 12] has argued that each of these groups has a vested interest, language, rituals, and cultures that must be addressed through their participation. Violating these normative behaviours is what leads to unsatisfactory change interventions. Vink et al. and Dul and Neumann have identified participatory requirements for successfully building ergonomics into the context. This is the essence of the macroergonomic approach [13].

Addressing human needs

Participatory ergonomics and other involvement strategies have another feature that is critical to the success of ergonomic intervention – addressing individuals and their needs. Participatory ergonomics can be defined in several ways: a) an array of tactical tools for engaging owners, users, customers and stakeholders as a means to implement ergonomics; b) a macroergonomic method for creating change; or c) a philosophy about creating continuous, effective and sustainable change in people and organizations using ergonomics technology. We can arrive at our own definition by answering another question: What is it that we give people when we make ergonomic changes? Possible deliverables include: 1) our technical know-how and science-based solutions, 2) our experiences and good practices that we have found work elsewhere, 3) processes to get the desired outcomes without personal agendas, 4) positive attention to people who suffer from bad ergonomics by validating their concerns, fears, or pain, and 5) hope that changes can indeed be made.

Practical utility might cause us to choose one benefit over another. Our priorities might change if we assume a different perspective. Which of these five deliverables does the person in the photo below most benefit from? What needs can be satisfied through ergonomics? Which needs can be met through participation? What does this person think about all day and dream about at night?

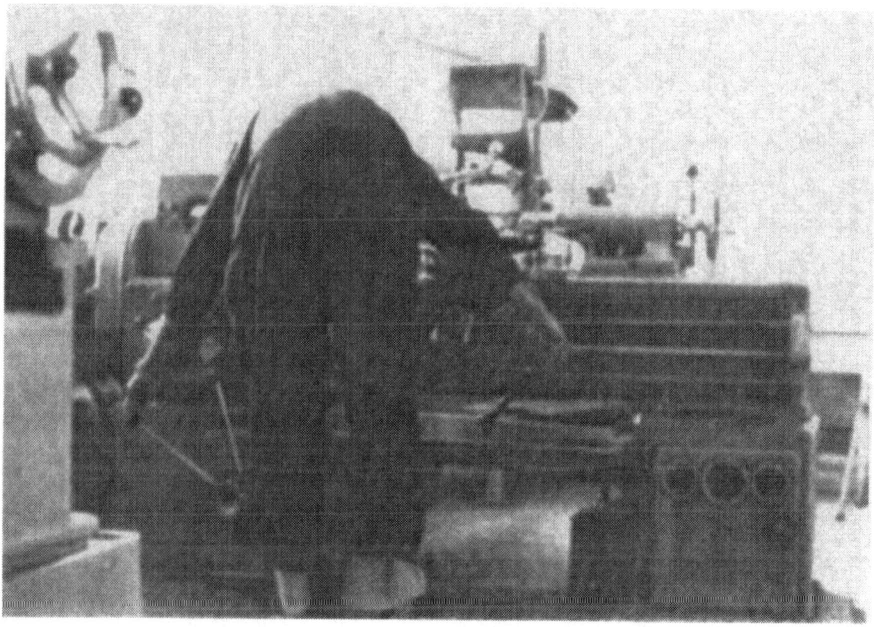

Technology Transfer

Worker operating a lathe in an industrially developing country.

Clearly, deliverables 1 and 2 speak to participatory ergonomics as a method for delivering technical ergonomic solutions (definition a). Delivering 1, 2 and 3 speak to more than the tactics but address a broader macroergonomic and social change strategy. When we deliver all of these outcomes we begin to address real sustainable change in people and organizations. We then begin to recognize the need to understand organizations as living, complex entities made of diverse and necessary parts existing in time and space. It also recognizes the very human qualities that are often the reason why change strategies are unsuccessful.

Imada [14] cited macroergonomic successes and asks what are the reasons that make us successful. The research and technology our discipline provides is indeed useful. Imada believes that participatory ergonomics' success extends far beyond the technical reasons. First, citing work on a human-centered approach for improving safety and health. [15, 16] Imada points to the importance of understanding the emotional and affective human components.

This work reveals that many accidents and risky behaviours occur because of negative feelings or responses to other aspects of work and life (e.g., supervision, family) rather than to the design of the work itself. Second, the work on affective neuroscience (e.g., Davidson et al. [17]) shows how negative emotions can cause a pattern of brain activity that predisposes people to greater vulnerability and to more negative states. Fortunately, it appears that these are alterable states and positive steps can be taken to reverse these trends [18]. Therefore, people exposed to poorly designed environments may experience negative states, which in turn cause them to process incoming information more negatively (e.g., perceiving pain more severely). Through participatory ergonomics, people may be able to convert these negative states into constructive and positive actions.

Accessing the emotional context is one of the keys. Imada [14] believes that these can be captured through stories and vignettes that clearly illustrate the design problems and the emotional components that accompany such conditions. In participating in ergonomic solutions people can validate their concerns, allay fears, reveal unexpressed anxiety, or simply be heard. Imada cites a clinical psychologist [19] who concluded that his role was to actively listen to 'help people tell their story well'. It is the telling of the story that sets the teller free of anger, shame, pain or fear. He believes that participatory practices work because it allows people to access, address and solve their problems in line with their own interests. In so doing, we create commitment and a sustainable change. Helping people to tell their stories well is one of the keys. It requires a deep understanding of the problem, not only in its physical, visible, or work presentation and a keen engagement with the whole person. This requires listening and asking questions that go well beyond the obvious. This leads to a deeper understanding of what people say and do. The teller also comes to a deeper understanding of their situation and a position from which they can positively impact their condition. Any change becomes easier to implement when it is taken from this personal context.

This approach to creating change is consistent with work done in an area called Appreciative Inquiry [20]. This four-step process has been helpful in implementing change to industrially developing and developed parts of the world. Steps include: 1) Discovery. What gives this organization or situation life? 2) Dream. What might or could be? 3) Design. What should be the ideal? 4) Destiny. How to empower, learn, adjust and improvise to meet the dream? Traditional ergonomic interventions begin at step 3 – design, preceded by analysis. Appreciative Inquiry however, begins similarly to Kogi's [3] keys; understand what is going well, what people do well, why they do what they do, or how well they are performing given the current situation. This can be an affirming first step. It validates what people bring to the situation rather than focusing on them as part of the problem. Allowing people to dream beyond the problem to higher goals and objectives is an important step that removes the limits of traditional problems. When confronted with a 'problem', people simply seek to make it 'not a problem' rather than a higher order objective. Designing an ideal against the dream is also a liberating process. It keeps the higher order goal in mind and invariably allows people to achieve more than they would have through traditional design strategies. Finally, Appreciative Inquiry focuses on the participatory mechanism that empowers people to do the work and realize their potential to implement changes on their own.

CONCLUSION

The core body of knowledge that makes up the ergonomic profession is a powerful basis for improving how humans live, work and play. Our ability to make this basic knowledge available to an ever-increasing variety of situations and audiences is the challenge that our diverse world poses. The three themes identified at the 2006 IEA Congress – global ergonomics with local solutions, organizational context, and addressing human needs is one way to increase the ability of ergonomics technology's relevance to more situations. In this sense, participatory ergonomics acts as the middleware by enabling an interface between the same basic technology across a wide range of situations, applications, users, time and space.

Recent technological introductions suggest that user input may not be necessary when there is no context, expectations, habits or conceptions about how the product may be used. In fact, it is true that technology has uses that users don't even know they need. However, once behaviour patterns and schemas are developed, the change that is required demands participatory strategies for introducing change. In this sense, participatory ergonomics may be a sine qua non for effective ergonomics in these situations.

Participatory strategies enhance ergonomics technology implementation. Because the participatory approach addresses social processes, it provides a necessary and unique contribution as a macroergonomic method. Guided by a consistent philosophy about ergonomists' role, we can make meaningful changes using participatory ergonomics.

REFERENCES

[1] Boff, K.R. (2006). Revolutions and shifting paradigms in human factors and ergonomics. *Applied Ergonomics*, 37 (4), 391–400.

[2] ILO. (1996). Ergonomic checkpoints. *Practical and Easy-to-implement Solutions for Improving Safety, Health and Working Conditions*. Geneva: International Labour Office.

[3] Kogi, K. (2006). Participatory methods effective for ergonomic workplace improvement. *Applied Ergonomics*, 37 (4), 547–54.

[4] Weick, K.E. (1988). Enacted sensemaking in crisis situations. *Journal of Management Studies*, 25, 305–17.

[5] Imada, A.S. (2003). Technology in organizations. In *Human Factors in Organizational Design and Management* – VII. (H. Luczak and K.J. Zink, eds) Santa Monica, CA: IEA Press, pp. 55–62.

[6] Imada, A.S. (2005a). Meeting future work and workforce needs with participatory ergonomics. In *Human Factors in Organizational Design and Management* – VIII. (P. Carayon, M. Robertson, B. Kleiner, P.L.T. Hoonakker, eds) Santa Monica, CA: IEA Press, pp. 397–400.

[7] Hendrick, H.W. and Kleiner, B.M. (2002). *Macroergonomics*. Mahwah, NJ: Lawrence Erlbaum.

[8] Imada, A.S. (1991). The rationale and tools of participatory ergonomics. In *Participatory Ergonomics* (K. Noro and A.S. Imada, eds) London: Taylor & Francis, pp. 30–49.

[9] Vink, P., Koningsveld, E., and Molenbroek, J. (2006). Positive outcomes of participatory ergonomics in terms of higher comfort and productivity. *Applied Ergonomics*, 37 (4), 537–46.

[10] Dul, J. and Neumann, W.P. (2006). The strategic business value of ergonomics. This book.

[11] Imada, A.S. (1994). Overcoming cultural barriers within organizations. In *Human Factors in Organizational Design and Management – IV* (G.E. Bradley and H.W. Hendrick, eds) North Holland: Amsterdam, pp. 625–30.

[12] Imada, A.S. (1990). Ergonomics: Influencing management behaviour. *Ergonomics*, 33, 621–8.

[13] Hendrick, H.W. (1984). Cognitive complexity, conceptual systems and organizational design and management: Review and ergonomic implications. In *Human Factors in Organizational Design and Management* (H.W. Hendrick and O. Brown Jr., eds) The Netherlands, North Holland: Amsterdam, pp. 15–31.

[14] Imada, A.S. (2005b). Macroergonomic contributions: Understanding the causes of our success. In *Human Factors in Organizational Design and Management – VIII* (P. Carayon, M. Robertson, B. Kleiner, P.L.T. Hoonakker, eds) Santa Monica, CA: IEA Press, pp. 35–42.

[15] Nagamachi, M. (1984). *Knowledge About Safety Management*. Tokyo: Daily Engineering Newspaper.

[16] Nagamachi, M. and Imada, A.S. (1994). Human-centered safety: A macroergonomic approach to safety management. In *Human Factors in Organizational Design and Management – IV* (G.E. Bradley and H.W. Hendrick, eds) North Holland: Amsterdam, pp. 769–74.

[17] Davidson, R.J., Jackson, D.C., and Kalin, N.H. (2000). Emotion, plasticity, context and regulation: Perspectives from affective neuroscience. *Psychological Bulletin*, 126, 890–906.
[18] Davidson, R.J., Kabat-Zinn, J., Schumacher, J., et al. (2003). Alterations in brain and immune function produced by mindfulness meditation. *Psychosom Med.*, 65 (4), 564–70.
[19] Kopp, S.B. (1972). *If you Meet Buddha on the Road, Kill him*! New York: Bantam.
[20] Cooperrider, D.L. and Whitney, D. (2005). *Appreciative Inquiry*. San Francisco: Berrett–Koehler.

Part II
State-of-the Art Papers

Meeting Diversity in Ergonomics (2007)
Pikaar, Koningsveld, and Settels (Eds)

6

Physical Ergonomics and Musculoskeletal Disorders: What's hot? What's cool?

Allard J. van der Beek

Department of Public and Occupational Health, EMGO Institute, VU University Medical Centre, Amsterdam, The Netherlands
Body@Work TNO VUmc, Research Centre Physical Activity, Work and Health, The Netherlands

Stefan IJmker

Department of Public and Occupational Health, EMGO Institute, VU University Medical Centre, Amsterdam, The Netherlands
Body@Work TNO VUmc, Research Centre Physical Activity, Work and Health, The Netherlands
TNO Quality of Life, The Netherlands

Abstract. This chapter's aim is to summarise the Triennial IEA World Congress 2006 – IEA2006 – highlights on physical ergonomics and work-related MusculoSkeletal Disorders (MSDs). Two general trends were observed. Firstly, physical ergonomics has been positively influenced by other disciplines, such as occupational epidemiology. Interventions were more often evaluated in well-designed epidemiological studies, when compared with earlier IEA congresses. However, there were fewer presentations of biomechanical and physiological studies on underlying mechanisms in IEA2006 than before. Secondly, modern work has changed, and so has physical ergonomics, focussing on MSDs. Many contributions involved computer use.
Several trends as to diversity were observed. Unfortunately, there are striking differences between countries in ergonomic state-of-the-art. We also observed too much disparity between scientific research from academics and case studies from ergonomic consultants. Finally, ergonomists seem to be increasingly aware of the diversity in worker populations. Attention was paid to topics such as gender and work, ageing workers and people with disabilities, as well as diversity and variation of physical exposures, described in an excellent keynote lecture by Mathiassen. For the future, we expect further collaboration between researchers performing experimental laboratory studies and those performing epidemiological field studies. We also foresee a further increase in well-designed, high quality evaluation studies with a cost-benefit analysis from the company perspective.

Keywords: Physical ergonomics, musculoskeletal disorders, IEA2006, major topics, trends, developments, future directions

INTRODUCTION

The objective of this chapter is to give a summary of the highlights of the Triennial IEA World Congress 2006 – IEA2006 – on the topic of physical ergonomics and work-related Musculo-Skeletal Disorders (MSDs). What have been major topics? What have been the trends towards new developments? What have been the most important advances in physical ergonomics? What have been the significant results of panel sessions or round table discussions? In other words, Physical Ergonomics and MSDs: What's hot? What's cool? This text is largely based on a summary presentation given by this chapter's first author at the closing plenary session of IEA2006.

We will use the model 'sequence of ergonomic prevention of MSDs' as a framework to describe the highlights of IEA2006. The sequence of ergonomic prevention of MSDs is

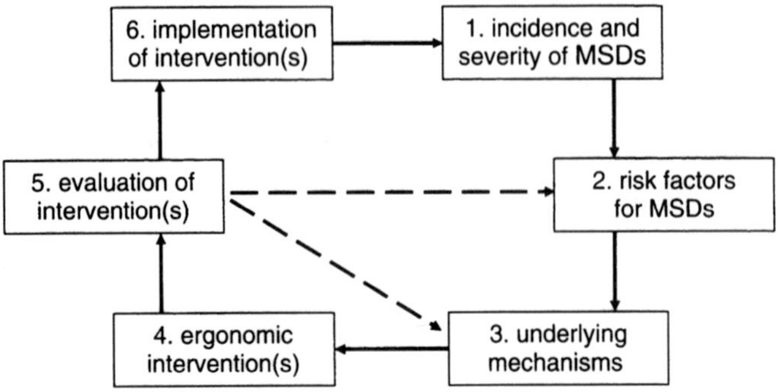

FIGURE 1. The sequence of ergonomic of musculoskeletal disorders (MSDs); largely based on Van Mechelen (1997).

largely based on a comparable model, which has been proposed by Van Mechelen [1] for the prevention of sports injuries. Figure 1 shows that the sequence of ergonomic prevention of MSDs consists of six steps. Firstly, the problem itself needs to be examined. In this initial step basic epidemiological data is needed. What is the extent of the musculoskeletal problem in the (working) population of interest? Most often, incidence or prevalence figures are given for particular MSDs. Apart from location, information on the severity of these MSDs can be gathered in more detail. For instance, for companies and society, in-depth information about disabling MSDs is valuable, since costs related to sick leave and disability for work, respectively, are the principal drivers of the financial burden resulting from MSDs. The second step in the sequence identifies the factors playing a role in the occurrence of these MSDs. Ergonomic epidemiological studies are needed to gain insight into the aetiology of MSDs. In general, cross-sectional studies give indications as to risk factors, whereas case-referent studies, or preferably prospective cohort studies, are required to make a proper distinction between causes and effects. However, it should be noted that various interacting risk factors are involved in the occurrence of work-related MSDs. The third step in the sequence unravels the underlying mechanism of the occurrence of MSDs. For instance, duration of computer-mouse use appears to be a risk factor for hand-arm symptoms among office workers [2]. However, this is not the end of the story; we need to know why this is the case in order to generate possible ergonomic solutions. If exposure to non-neutral hand postures while holding the mouse is the underlying cause, then interventions might aim at improved design of the workstation. If exposure to static postures is the underlying cause, then interventions might aim at an increased number of micro pauses induced by, for instance, computer software. Ergonomic studies designed to shed light on the underlying mechanism can take place either at the workplace or in the laboratory. The fourth step in the sequence is to introduce an ergonomic intervention which is likely to reduce the risk and/or severity of MSDs. This intervention is based on the aetiology and underlying mechanism of MSDs as identified in the second and third steps. The fourth step starts with risk assessment to find out whether interventions are needed in a specific situation for a specific population of workers. Moreover, risk assessment provides indications as to which preventive ergonomic actions should be prioritised. This fourth step is the most important playground of Ergonomics/Human Factors, since it is the ultimate example of 'fitting the job to the worker'. The workplace or organisation of the work should be (re)designed such that the risk of occurrence of MSDs is minimised. It is beyond debate that this step requires ample ergonomic expertise. Most of these ergonomic interventions aim at optimisation of mechanical load of (parts of) the worker's locomotor system, thus reducing the risk of developing MSDs due to overload. It is postulated that exposure to mechanical load is one of the most important risk factors for the occurrence of MSDs. The fifth step in the sequence is to evaluate the effectiveness of the ergonomic intervention. A large number of studies have been performed on ergonomic interventions, but there has been little evaluation of whether or not this intervention indeed resulted in positive effects on MSDs. In other words, in many studies it was shown that a relatively simple ergonomic intervention, such as re-design of the workstation, established a decrease in exposure to risk factors. Unfortunately, the use of outcome measures regarding MSDs has been sparse in the ergonomic intervention literature. Moreover, some of the few well-designed studies offered a negative conclusion: ergonomic interventions did not have an effect on the risk or severity of MSDs [3, 4]. If the fifth step produces unsatisfactory results one might be forced to study once more the risk factors – short-cut to the second step – and/or the underlying mechanisms – short-cut to the third step – after which the fourth and fifth steps are repeated. Finally, the sixth step in the sequence is the large-scale implementation of ergonomic interventions proven to be effective. In an optimal situation, this implementation would result in a positive effect on the incidence and/or severity of MSDs, as monitored in a repetition of the first step among population(s) in which the ergonomic intervention has been implemented. Hence, the sixth step closes the circle towards the first step.

During IEA2006 a large number of sessions were dedicated to MSDs. We mention several sessions, categorised according to the steps in the earlier described model. 'Risks for complaints of arm, neck and shoulder (CANS)' focussed mostly on the second step. 'Physical aspects and manual materials handling', and particularly 'Biomechanics' and 'EMG' were typical examples of sessions paying attention to the third step of the model. The two-session symposium 'Unravelling the causes of upper extremity disorders among computer users' aimed to bring researchers dealing with the second, third, fourth and the fifth steps together into one forum. The multiple-session symposium 'Prevention of Work-Related MSDs' also covered the second, third, fourth and the fifth steps. Some of the interactive sessions were perfect illustrations of the fourth step, in which the ergonomic solutions are to be developed. The session 'Production systems and Manufacturing' touched on aspects related to the third, fourth and fifth steps. The sessions within 'Effectiveness of Interventions on Musculoskeletal Disorders', 'Hands on: The ergonomics of hand (tools)', and partly 'Economics' focussed on the fifth step of the model. Furthermore, sessions on target groups, such as health care workers or construction workers, were of significant interest for MSDs ('Health care ergonomics – patient handling' and 'Ergonomics in building and construction', respectively). Finally, in the design sessions 'Applied ergonomics in design' and 'Methods/design methods' several papers on MSDs were presented. Concluding, the second, third, fourth and fifth steps of the sequence of ergonomic prevention of MSDs were well represented in IEA2006, whereas the first and the sixth steps received far less attention.

RISK FACTORS AND UNDERLYING MECHANISMS

Ergonomists need knowledge on the causation of MSDs to be able to optimally target ergonomic interventions in the workplace. Information on risk factors and underlying mechanisms can be gathered in both field and laboratory studies. Two general trends were observed. Firstly, fewer biomechanical and physiological studies on the underlying mechanisms were presented in IEA2006 than in earlier IEA congresses. There appears to be a tendency towards more applied ergonomic research on the one hand, and ergonomic epidemiological research on the other. However, it was also noticed that researchers performing epidemiological field studies and those performing experimental laboratory studies were in closer contact than they were before. The two-session symposium 'Unravelling the causes of upper extremity disorders among computer users' can be mentioned as a successful example of this reciprocal influence. We will further describe these sessions later on in the text.

Secondly, the development of instruments for direct measurements of mechanical load in large-scale field projects has been less fast than hoped for. It has been postulated that the lack of measurement precision, by using questionnaires or observations to quantify mechanical load, likely leads to underestimation of the effect of mechanical load on MSDs [5, 6]. As a result, valid quantitative information on the risks of mechanical load related to the occurrence of MSDs is lacking. This information is needed, however, for functional development of preventive policies in order to reduce the burden of MSDs in companies. Direct measurements may fill this gap in knowledge, but these types of measurements require extensive resources when applied in large field studies. Fortunately, there were some promising advances for the near future. For example, Trask and co-workers (session: EMG) presented a study in which workers were measured during a whole working day simultaneously with EMG and inclinometry. In addition, Lutmann et al. (session: EMG) measured EMG during a whole working day among office workers. This type of extended direct measurement of mechanical load might provide more insight into the development of MSDs over time.

The majority of the contributions to the IEA2006 congress focussed on office work, with special emphasis on computer use. De Kraker and colleagues (session: Risks for complaints of arm, neck and shoulder (CANS)) took an innovative approach to studying risk factors among computer users by looking at software characteristics. They found that high precision demands for mouse- clicking, insufficient opportunities for the use of shortcut keys, and having difficulty reading information on the screen due to a small font size, were the main software-related risks for experiencing musculoskeletal symptoms of the upper extremity. Longitudinal studies are needed to confirm these findings. In a laboratory study, Visser et al. (symposium: Unravelling the causes of upper extremity disorders among computer users) showed that shoulder load had negative consequences for blood flow in the wrist area. This finding sheds new light on the mechanisms leading to MSDs among office workers.

A number of contributions challenged general ergonomic knowledge and practice. Lindegard et al. (symposium: Prevention of Work-Related Musculoskeletal Disorders) reported that work postures and working technique were not longitudinally related to the development of MSDs in computer users. IJmker et al. (symposium: Unravelling the causes of upper extremity disorders among computer users) systematically reviewed longitudinal studies and presented evidence that a longer duration of computer use is related to hand-arm symptoms, but not to neck-shoulder symptoms.

A hot topic in risk assessment is exposure variability, which can be considered diversity in itself. In his excellent keynote lecture, Mathiassen [7] pointed out that ergonomic studies should take into account the variability of exposure to mechanical load over time and design measurement strategies accordingly, in order to arrive at valid conclusions. This was confirmed by Paquet (session: Effectiveness of Interventions on Musculoskeletal Disorders), who discussed Observational Ergonomic Job Analysis. He stressed that variation between and within workers should be taken into account to best estimate the level of postural exposure, since hierarchical bootstrapping showed a great deal of exposure variability. However, Mathiassen [7] also mentioned that the lack of variation in mechanical load in working life and leisure time might be an important risk factor for the development of MSDs, but this hypothesis still awaits a thorough evaluation in longitudinal ergonomic studies.

ERGONOMIC INTERVENTIONS

The development of ergonomic interventions is based on the know-how and skills of ergonomists combined with scientific evidence as to the aetiology and underlying mechanism of MSDs. This topic was not well represented in the normal sessions consisting of the presentation of papers. However, original and innovative sessions were organised in which risk assessment tools and ergonomic solutions were interactively presented. In general, these so-called 'interactive sessions', in which a diversity of scientists and practitioners intermingled in hands-on type of workshops, were successful.

Firstly, risk assessment should take place in order to find out whether interventions are needed. In general, hardly any improvement in the scientific quality of ergonomic checklists, instruments and methodologies was observed when compared with those developed and presented before. In ergonomic practice, a large number of risk assessment tools can be used, e.g. RULA, HAL TLVs, and OCRA Index. Two contributions (Bao and co-workers, symposium: Prevention of Work-Related Musculoskeletal Disorders; Hoehne-Hückstadt et al., symposium: Prevention of Work-Related Musculoskeletal Disorders) found that different risk assessment tools lead to variable outcomes as to the quantification of exposure to (combinations of) risk factors. Inevitably, this disagreement between tools leads to unwanted differences in proposed ergonomic interventions to prevent MSDs. This trend was confirmed in the interactive session 'Musculoskeletal disorder risk assessment - shall we agree to disagree?'. The purpose of this interactive session was to understand whether there are significant differences between MSD risk assessment tools and the application of them by different assessors. It is, for instance, suggested that there are cultural, professional and individual differences in the perception of risk, which may need to be addressed within a global organisation. Participants had an opportunity to review a task and assess the MSD risk in this interactive session, using one of the tools described earlier. In another interactive session, 'A different approach for the evaluation of pushing and pulling in practice', participants learnt a fairly simple, hands-on method for ergonomic risk assessment of pushing and pulling carts and four-wheeled containers, especially regarding shoulder load.

Secondly, ergonomists should (re)design the workplace or organisation of the work such that the risk of MSDs is minimised. Surprisingly, this topic was hardly taken into account in the normal sessions consisting of the presentation of papers. Again, an interactive session was one of the positive exceptions to the rule. The interactive session 'A revolution in media distribution logistics: The case of the case', which was organised by Kees (a Dutch name, pronounced as 'Case') Peereboom, re-examined an ergonomic intervention for a distribution centre, in which 300 types of magazine are distributed to 9000 selling points throughout the Netherlands. The interesting process of developing the best ergonomic and logistic solution was simulated by actively determining which case out of ten would be the best one to be used in a logistic system in the distribution centre. This resulted in stimulating discussions among the participants with different backgrounds.

EVALUATION AND IMPLEMENTATION OF ERGONOMIC INTERVENTIONS

More studies evaluating the effectiveness of ergonomic interventions in the workplace on musculoskeletal health were available. In general, the quality of these evaluation studies has increased, when compared with earlier IEA congresses. More ergonomic evaluation studies used experimental study designs with a control group and with proper randomisation of subjects to intervention arms. This is good news, since high-quality studies such as these are urgently needed to convince policy makers and company management. In general, however, the results of the evaluation studies showed mixed findings regarding the effectiveness of ergonomic interventions. Amick III et al. (session: Effectiveness of interventions on Musculoskeletal Disorders) showed that the increase in musculoskeletal symptom intensity over a year was lower in subjects who received an adjustable chair in combination with an ergonomic training in comparison with subjects who only received an ergonomic training. Interestingly, a replication study confirmed the finding of this study that ergonomic training alone was not effective in preventing increases in musculoskeletal symptom intensity over time. Rempel et al. (symposium: Unravelling the causes of upper extremity disorders among computer users) summarised the available literature in a systematic review on the effects of workstation and behavioural interventions on musculoskeletal symptoms among computer users. They used methodological quality criteria and consistency of results to weigh the available evidence. Their results showed that there was moderate evidence that workstation adjustment and breaks combined with exercise had no effect. In addition, moderate evidence for a positive effect of alternative pointing devices was available. For all other interventions either mixed or insufficient evidence of effect was concluded. Haukka et al. (session: Effectiveness of interventions on Musculoskeletal Disorders) investigated the effectiveness of a participatory ergonomics intervention in a randomised controlled trial among kitchen workers. The intervention did not have an effect on the prevalence of musculoskeletal symptoms.

Ergonomic interventions targeted at reducing musculoskeletal symptom severity or consequences of musculoskeletal symptoms – e.g., sick leave – were also evaluated. Voerman and colleagues (session: Effectiveness of interventions on Musculoskeletal Disorders) were not able to show in a randomised controlled trial that myofeedback training in combination with ergonomic counselling over four weeks was more effective in reducing symptom severity than ergonomic counselling alone. However, in both groups symptom severity decreased significantly during the four-week period. Anema and co-workers (session: Effectiveness of interventions on Musculoskeletal Disorders) investigated a participatory ergonomic intervention among workers sick-listed due to low back pain. They found that subjects receiving the participatory ergonomic intervention returned to work faster than subjects receiving usual care provided by their occupational physician. Some laboratory studies showed that the effects of ergonomic interventions might be highly individual. Burgess-Limerick and Cook (symposium: Unravelling the causes of upper extremity disorders among computer users) showed that providing a trackball mouse for computer users displaying non-neutral wrist posture while using the computer mouse resulted in neutral wrist posture for one subject, but not for another. In a similar vein, Graf and Kothiyal (symposium: Prevention of Work Related Musculoskeletal Disorders) showed that the effect of sitting on spinal loading, measured as spinal shrinkage, was highly individual, regardless of providing a chair with a fixed or freely moveable backrest. More attention to individual differences, i.e., diversity, might help in the evaluation of ergonomic interventions in laboratory and field studies, as well as in ergonomic practice.

GENERAL OBSERVATIONS AND FUTURE DIRECTIONS

1. Physical ergonomics has been positively influenced by other disciplines, such as occupational epidemiology. Two examples of this trend can be cited. Firstly, ergonomic interventions were more often evaluated in well-designed epidemiological studies. Secondly, in systematic reviews of studies on ergonomic risk factors or ergonomic interventions, more

often a proper quality assessment of the included studies was involved and conclusions were given with levels of available scientific evidence. To illustrate both examples at once, in their systematic reviews of the literature on computer workers IJmker et al. and Rempel et al. simply excluded all non-longitudinal studies on risk factors and all ergonomic intervention studies without a proper control group, respectively.

For the future, we expect that ergonomic interventions will be evaluated more often using well-designed epidemiological studies of high methodological quality – i.e., controlled trials or randomised controlled trials – to make sure that the study results are free of bias. These studies are also likely to far more often include a cost-benefit analysis using a company perspective to facilitate the communication of results with the companies. In the end, this will drastically improve the chances of implementation of the most cost-efficient intervention(s).

2. Modern work has changed – see Fig. 2 –, and so has physical ergonomics focussing on the prevention of MSDs. In IEA2006, by far the most attention was paid to office work, with special emphasis for computer use and its association with upper extremity symptoms. We expect that this attention will continue to grow in the near future. However, other topics will also become more important. Sedentary work is associated with overweight and obesity. Given the expected epidemic of overweight and obesity with the resulting adverse health effects, we foresee that there will be more attention paid to the promotion of physical activity and physical exercise among office workers and other workers with low intensity jobs. The session 'Recommendations on sufficient physical activity in low intensity static jobs' can be cited as an example of this tendency.
3. Ergonomists seem to be increasingly aware of the diversity in target populations. The fact that the IEA Board decided to approve the proposed new IEA Technical Committee 'Gender and Work' can be cited as an illustrative example. Sluiter [8] illustrated in her keynote presentation that population pyramids show an increasing labour participation of ageing workers, and even people aged over 65 years. This means that there will be much more diversity in the work force, which will inevitably also lead to ergonomic attention for ageing workers. Similarly, the symposium 'Rehabilitation Ergonomics' can be cited as an example in which diversity as to people with disabilities is explicitly taken into account.

Future directions of research may include investigation of what works for whom and when. Subgroup analysis and individual variability in effects of interventions will be a focus of future intervention studies. Starting from these results, more attention will be paid to the implementation of ergonomic interventions, from simple recipes to theory-based specific

FIGURE 2. Modern working life has changed, and so does physical ergonomics focussing on prevention of musculoskeletal disorders.

solutions. Implementation theory is needed to guide ergonomists in attaining maximum results with their (proposed) interventions. In this way better research can lead to better practice.

4. There is much diversity in ergonomic state-of-the-art. First of all, the differences between countries are striking. Unfortunately, studies from industrialised countries are generally of higher quality than those from developing countries. We sincerely believe that lots of effort should be put into this unequal partition. Secondly, we observed disparity between scientific research from academics and the case studies from ergonomic consultants. In our opinion, the gap between both 'worlds' can and should be bridged.

   More in-depth knowledge in the field of physical ergonomics and MSDs can help ergonomic practice to further refine its methods in the future. A prerequisite is closer collaboration between ergonomists in practice and ergonomic researchers. Collaboration between researchers performing experimental laboratory studies and those performing epidemiological field studies is warranted to increase knowledge on the causation of MSDs in the workplace. It is recommended that more precise 'laboratory' measures of mechanical load be included in epidemiological field studies. The challenge to take into account diversity and variation in mechanical exposures within and between workers also fits with this recommendation.

REFERENCES

[1] Van Mechelen, W. (1997). Sports injury surveillance systems: 'One size fits all'? *Sports Med.*, 24 (3), 164–8.

[2] IJmker, S., Huysmans, M.A., Blatter, B.M., et al. (2007). Should office workers spend fewer hours at their computer? A systematic review of the literature. *Occup. Environ. Med.*, 64 (4), 211–22.

[3] Westgaard, R.H. and Winkel, J. (1997). Ergonomic intervention research for improved musculoskeletal health: A critical review. *Int. J. Ind. Ergon.*, 20, 463–500.

[4] Linton, S.J. and Van Tulder, M.W. (2001). Preventive interventions for back and neck pain problems What is the evidence? *Spine*, 26 (7), 778–87.

[5] Van der Beek, A.J. and Frings-Dresen, M.H. (1998). Assessment of mechanical exposure in ergonomic epidemiology. *Occup. Environ. Med.*, 55 (5), 291–9.

[6] Winkel, J. and Mathiassen, S.E. (1994). Assessment of physical work load in epidemiologic studies: concepts, issues and operational considerations. *Ergonomics*, 37 (6), 979–88.

[7] Mathiassen, S.E. (2006). Diversity and variation in biomechanical exposure: What is it, and why would we like to know? *Appl. Ergon.*, 37 (4), 419–27.

[8] Sluiter, J.K. (2006). High-demand jobs: Age-related diversity in work ability? *Appl. Ergon.*, 37 (4), 429–40.

LIST OF CITED ABSTRACTS PUBLISHED IN THE CONGRESS PROCEEDINGS

Trask and co-workers, 2006, session, EMG. In: Pikaar, R.N., et al., (Eds), *Proceedings Meeting Diversity in Ergonomics*, Elsevier.

Lutmann, et al., 2006, session, EMG. In: Pikaar, R.N., et al., (Eds), *Proceedings Meeting Diversity in Ergonomics*, Elsevier.

De Kraker and colleagues, 2006, session, Risks for complaints of arm, neck and shoulder (CANS). In: Pikaar, R.N., et al., (Eds), *Proceedings Meeting Diversity in Ergonomics*, Elsevier.

Visser, et al., 2006, symposium, Unravelling the causes of upper extremity disorders among computer users. In: Pikaar, R.N., et al., (Eds), *Proceedings Meeting Diversity in Ergonomics*, Elsevier.

Lindegard, et al., 2006, symposium, Prevention of work related musculoskeletal disorders, In: Pikaar, R.N., et al., (Eds), *Proceedings Meeting Diversity in Ergonomics*, Elsevier.

IJmker, et al., 2006, symposium, Unravelling the causes of upper extremity disorders among computer users. In: Pikaar, R.N., et al., (Eds), *Proceedings Meeting Diversity in Ergonomics*, Elsevier.

Paquet, 2006, session, Effectiveness of interventions on musculoskeletal disorders. In: Pikaar, R.N., et al., (Eds), *Proceedings Meeting Diversity in Ergonomics*, Elsevier.

Bao, and co-workers, 2006, symposium, Prevention of work-related musculoskeletal disorders. In: Pikaar, R.N., et al., (Eds), *Proceedings Meeting Diversity in Ergonomics*, Elsevier.

Hoehne-Hückstadt, et al., 2006, symposium, Prevention of work-related musculoskeletal disorders. In: Pikaar, R.N., et al., (Eds), *Proceedings Meeting Diversity in Ergonomics*, Elsevier.

Amick III, et al., 2006, session, Effectiveness of interventions on musculoskeletal disorders. In: Pikaar, R.N., et al., (Eds), *Proceedings Meeting Diversity in Ergonomics*, Elsevier.

Rempel, et al., 2006, symposium, Unravelling the causes of upper extremity disorders among computer users. In: Pikaar, R.N., et al., (Eds), *Proceedings Meeting Diversity in Ergonomics*, Elsevier.

Haukka, et al., 2006, session, Effectiveness of interventions on musculoskeletal disorders. In: Pikaar, R.N., et al., (Eds), *Proceedings Meeting Diversity in Ergonomics*, Elsevier.

Voerman, and colleagues, 2006, session, Effectiveness of interventions on musculoskeletal disorders. In: Pikaar, R.N., et al., (Eds), *Proceedings Meeting Diversity in Ergonomics*, Elsevier.

Anema, and co-workers, 2006, session, Effectiveness of interventions on musculoskeletal disorders. In: Pikaar, R.N., et al., (Eds), *Proceedings Meeting Diversity in Ergonomics*, Elsevier.

Burgess-Limerick, and Cook, 2006, symposium, Unravelling the causes of upper extremity disorders among computer users. In: Pikaar, R.N., et al., (Eds), *Proceedings Meeting Diversity in Ergonomics*, Elsevier.

Graf, and Kothiyal, 2006, symposium, Prevention of work related musculoskeletal disorders In: Pikaar, R.N., et al., (Eds), *Proceedings Meeting Diversity in Ergonomics*, Elsevier.

Meeting Diversity in Ergonomics (2007)
Pikaar, Koningsveld, and Settels (Eds)

7

Ergonomics and Work – Different Approaches and Challenges for the Future

Laerte Idal Sznelwar

Departamento de Engenharia de Produção, Escola Politécnica da Universidade de São Paulo, São Paulo, Brazil

Abstract. This chapter presents some aspects of ergonomics and its diversity, based on studies published at the triennial IEA Congress 2006, particularly in the fields of ODAM, user participation, activity theories and human error. The text is an extended version of the summary presentation given by the author at the closing session of the congress.

Keywords: ODAM, user participation, activity theories

INTRODUCTION

Since the IEA proposes to be an umbrella organization for ergonomics and human factor societies all around the world, it will be important to accept and put in evidence its diversity. Articulating and reflecting on our individual and collective experiences in ergonomics are not easy; therefore we continually need to improve discussions about ergonomics; its schools, theories and practices. Ergonomics is a relatively new field, and experiences are being gained in geographically and historically distinct countries and regions. Concepts and words to make these experiences explicit result from these processes, so it will be important to ask whether ergonomists using the same words are referring to the same phenomena. Different readers may interpret some of the themes in this chapter differently.

de Looze and Pikaar [1] emphasize that by definition ergonomics is linked to people and diversity. The aim of creating designs to serve everyone includes an implicit challenge related to variety, which intensifies if we propose to link ergonomics to cultural diversity in its worldwide implementation. The authors also point to diversity due to differences brought about by information technology, automation and mechanization, and diversity in goals.

About diversity

Diversity can be considered both as a strength and a weakness of ergonomics. It is a strength because the principal aim in ergonomics is to contribute to human well-being, maintaining a constant dialogue with productivity and quality. Work content, organizational structure, timing, tools, norms, rules, procedures, production systems and training are examples of objects of research and fields for intervention. One of the main challenges in ergonomics should be to reconnect engineering with the biological and human sciences, by considering other points of view, other epistemologies and divergence from a mechanistic and functional approach to science and life, where the disjuncture of phenomena is an important basis. Based on a complex approach, it's possible to state that in ergonomics we are building bridges. Ergonomics is spread around the world, and adopted by people and companies in so many situations, that it is very hard to trace a complete typology of approaches, methods and links in different cultures and societies.

Weakness can be explained by the difficulties of communicating within the field of ergonomics, due to different ways of reasoning and different approaches to research. If the epistemological affiliations are made explicit, results can be understood better and researchers, professionals and students will thus have more information about their meaning and about the purposes of ergonomics actions. Some actions are developed from the point of view of occupational health, some are based on work results in terms of quality, productivity and systems reliability; others are centred on human tasks and thus based in physiology and psychology. Still others are centred on work activities and on individual and collective worker actions.

It is possible to distinguish different ways of doing ergonomics based on surveys, interviews, observations, images and audio recordings. Consequently, the data collected and results are dissimilar. One of the aims of ergonomics should be to provide clarifications and promote debates, including dialogue both inside ergonomics and with others' fields of research.

It should also be possible to provide arguments for reflective actions concerning the future of ergonomics, especially if it is to be consolidated as a scientific discipline, a technology or, as proposed by Wisner [2], as an art, like the arts of engineering and medicine. These reflections include understanding the future in terms of designing objects (mainly artefacts) and designing work. In this text, the main questions are related to work and the relationship between artefacts and human activities.

Future development

According to Boff [3], beginning in the middle of the twentieth century, 'futurists predicted a new era, in which technology would relieve people from tasks that are difficult, time-consuming and subject to error'. However, technology has not evolved exactly like that. An important reflection related to the field of information technology is that while it has provided enormous benefits, there have also been a lot of costs, for example, in terms of health problems and the difficulty or even impossibility of performing some tasks. Reflecting on the possibility of people to have interesting jobs, no significant changes in the division of labour can be observed. There are still too many work situations where people perform tasks with very poor content (repetitive tasks, using information technology in real time). Another challenge for ergonomics is to help designers, managers and others responsible for production systems to provide more interesting and challenging jobs, thereby affording conditions for people to develop themselves individually and collectively.

Understanding diversity in ergonomics also means taking a look at its evolution. Boff [3] divided evolution in ergonomics into four generations. The first aimed to adapt equipment, the workplace and tasks to human capabilities and limits. The second attempted to harmoniously integrate humans, technology and work to enable effective systems. The third generation could be seen as enhancing human capabilities to amplify human physical and cognitive work performance capabilities through a symbiotic coupling with technology. The fourth and last stage is still embryonic and should provide modifications in human capabilities (physical and cognitive) to maximize human effectiveness. Boff also discusses some ethical implications, mainly for the third and fourth generations. It is clear that these proposals come from a specific stance and focus on different aspects of human actions. Collective work, social relations at work and the psychological effects of enhanced capabilities, especially cognitive abilities, are not considered. Many controversial points should be considered especially concerning divisions of labour and the ethical limits of changing human kind.

MACROERGONOMICS AND ANTHROPOTECHNOLOGY

Macroergonomics

Macroergonomics is defined by Hendrick [4] as a top-down socio-technical approach to work system design, and its carry-through to jobs and related interfaces between humans and machines as well as humans and software. The ultimate purpose of macroergonomics is to ensure that work systems are fully harmonized and compatible with their socio-technical characteristics. The point of view is based on systems theory. Hendrick affirms that the main goal is to harmonize the parts of the system with themselves and with the environment.

Hendrick [5] also proposes a discussion based on a macroergonomic approach to environmental uncertainty. Rather than environmental stability, uncertainty is more often the case in real actual situations, where conditions are unstable and changes are unpredictable. When dealing with environmental uncertainty, organizations must have a low degree of formalization, a high degree of professionalism and relatively few hierarchical levels. It will be important to provide favourable conditions for employees to acquire high levels of education and to consolidate their experience.

This point of view contradicts a more classical approach where stability and predictability are considered to be characteristic of production systems and are also seen as a goal. Routines, standardized coordination, control procedures, a high degree of formalization, centralization and moderate-to-high vertical differentiation were main premises for achieving the objective. One could question whether stability was ever possible, or just a non-achieved perspective. Stability could also be considered desirable, but not a realistic goal. Signs of uncertainty were always present, but perhaps they went unrecognized by researchers, since they were considered to be noise or something that had escaped from the predicted scenario.

According to Kleiner [6] macroergonomics can be considered as an extension of ergonomics where the integration of human systems is the main proposal. This approach is helpful in the analysis and redesign of systems. This analysis involves collecting and systematizing data from the outside environment, technology, personnel, the internal environment and the organization of the subsystems of the work system. The results of this analysis are important to a human centred design process that could support operators as they attempt to prevent or control key variances in the work process.

Considering the dynamics of production systems and technological and social development, Carayon [7] proposes six principles for continuous macroergonomic system adaptation and improvement. She states that it is important:

(1) to guarantee active participation by customers and end users in system design activities;
(2) to provide continuous interactions between customers and product/service design and redesign;
(3) to incorporate and adapt the product or service continuously;
(4) to provide an adaptable system in the long run;
(5) to provide learning opportunities for the individual and the organization; and
(6) to have a change process (and results) that makes sense.

The design process should be continuous and incorporate an adaptive product or service. The system should be adaptable in a long run. Another principle concerns the learning possibilities for the individuals and for the organization. Making sense should be also one of the attributes of the changing processes and their results.

These questions are significant to the challenges Carayon identifies when working across organizational, geographical, cultural and temporal boundaries. Members of design teams working in different parts of the world, those designing systems for cross-cultural work and ergonomists developing participatory studies and interventions in various cultural environments are all working in domains where these principles can be applied. Although Carayon doesn't propose an ergonomic work analysis (EWA), there are interesting links with anthropotechnology, such as the concern with different cultural scenarios and their importance to work results in terms of health and performance.

According to Imada [8], it is important to use methodologies that engage people's 'emotions' when they are working on what they perceive as real problems. When their needs are being fulfilled, when they find validation of their concerns and finally, when they are working on problems that affect them directly, success should be guaranteed.

From a similar perspective, but at another level of discussion, Karsh [9] refers to macroergonomics as distinct from other areas of human factors and ergonomics, since it focuses on the relationship between macro-level variables such as organizational design, structure and culture, and individual outcomes such as performance, stress and injury. However, Karsh thinks that it still misses giving a precise explanation of how variables at different levels, such as individuals, groups and organizations, or variables at different echelons, such as levels of hierarchy, causally influence each other. It is important to understand that employee behaviour may be influenced by:

- group norms (group level at the lowest echelon);
- a single opinion leader among the employees (individual at the lowest echelon);
- a single supervisor (individual at a higher echelon);
- the safety committee (group level at a higher echelon); or perhaps
- organizational culture (organization level).

It will be a challenge to propose a method where data could be collected from those different levels. That's why he proposes a *mesoergonomics* approach in order to provide results that could help to fill this gap, and to develop theories that bridge the individual, group and organization levels or bridge levels of organizational hierarchy.

Kogi [10] proposes an interesting idea for discussions on macroergonomics. He describes an inventory of participatory approaches used for improving workplaces in a programme developed for small enterprises of home workers, construction workers and farmers in southern Asia. For Kogi it is very important to respect local possibilities and needs, and to involve people in planning and controlling a significant portion of their own work activities. To a large extent, the results obtained depend on the implementation of an inter-country network to provide adequate action-oriented participatory methods. Kogi reinforces some tenets of macroergonomics, and also shares some of Wisner's concerns related to the implications of ergonomics for improving quality of work and life in industrially developing countries.

Anthropotechnology

Anthropotechnology is a useful approach when transferring technology to culturally different regions. It is important to highlight the lack of fit between what designers expect and what workers actually do, in order to create better designs. Themes discussed by Wisner [11, 12] can be useful for ergonomists, macroergonomists and anthropotechnologists:

- Intellectual knowledge is not sufficient for performing practical activity.
- Mental models are formed in diverse situations, such as in family living, school, playing games and in professional and social experiences.
- Knowledge is not just a cognitive field and many body techniques are acquired from childhood.
- Experience is embodied and many skills are latent even when not formalized.
- Worker involvement is also important in designing new work systems, which should be analysed and discussed together with workers using their previous experience as reference.

Geslin [13] discusses the impact of anthropotechnology in the debate within ergonomics and also on social scientists such as anthropologists. The contributions of Wisner and others researchers in the domain have made it possible to take into account the relationship between the microscopic characteristics of human activity and a broader point of view, including the functioning of society and culture. There are other points that can be shared with macroergonomics. One of the greatest interests of anthropotechnology is to understand cultural differences and their influence on the success or failure of technology transfers, as well as the damage and weak repercussions from attempts to improve local economies and mitigate social disturbances and bad working conditions. Geslin considered another aspect of Wisner's proposals: its declared intention to 'break the usual compartmentalization of the disciplines' and thus propose to anthropologists the inclusion of material dimensions and the role of technical objects in the process of constructing knowledge.

ACTIVITY AS A PERSPECTIVE IN ERGONOMICS AND DIVERSITY IN APPROACHING IT

Activity is the centre of at least two approaches in ergonomics. The French-speaking approach is focused on how work really occurs in confrontation with expectations and prescriptions. For the Nordic schools, activity conveys the notion of tension and contradiction. Some concepts that are considered as a basis of these approaches:

- Actions are not simple, linear and respecting a causal determination.
- Actions are linked to meanings that are often not evident.
- Events are causes for action and there is need for knowledge and collaboration to cope with events.
- Work is an instrumental activity mediated by society and culture.
- Safety, efficiency or health viewed from an operative perspective.
- Activity can be understood in its categories of dynamism, complexity, uncertainty and controllability.
- Activity is also concerned with an analysis of meaningful signs in the environment.

French school

Daniellou [14] discusses the evolution of concepts from the difference between real and prescribed work to activity. The concept of activity emerges as a result of the congruence of three phenomena:

(1) mode of thinking since the beginnings of ergonomics in France, the relationship between what needs to be done and how it is done;
(2) the traditional use of the term 'activity' in psychology; and
(3) the notion of activity developed in Soviet psychology.

For Daniellou, ergonomists' interventions in response to requests generally cannot deal with issues of occupational health or production effectiveness simply by understanding them through a formal description of work in terms used by management or the formal procedures. To understand work activity, it is important to gather evidence of the difficulties encountered by workers and the strategies they develop to achieve goals. This is possible if there is collaboration between analysts and workers. Workers do not perform tasks just to achieve the greatest efficiency, but also to 'safeguard their normality, partake in social relationships and resolve ethical dilemmas'. Some of the questions raised by Daniellou are related to:

- the collective dimensions of the activity;
- the production of rules, debate of norms, antecedent norms;
- the fact that future activity is not predictable – the paradox of designing – a non-direct link between the existing situation and future potential activity.

The concept of activity appears as a general framework of thought whose boundaries have evolved according to the types of problem that ergonomists have encountered. This is due to technological and social changes, the new fields that the methodological developments have opened up for inter-disciplinary discussions [14].

Cunha and Lacomblez [15] discuss the relevance of the strategies and renormalizations that workers do every day during their work activity, to establish commitments that make it possible to manage the activity's objectives and the real needs of the customers. They have shown how workers build collectively negotiated compromises to achieve everyone's objectives and regulate a shared 'way of living'. Adopting an interpretation derived from an ergologic approach [16], they suggest that visible deviations of the real in relation to the prescribed could arise from the application of another type of values in response to the challenges that face the sector (bus transport), thus transgressing the antecedent norms, or that the results should be interpreted as a sign that there should be public debate on the values associated with providing a service.

It is also possible that evaluations in companies are derived from a theoretical point of view about work, related to what is expected and prescribed. Evaluating work performance and understanding the practices of the operators based on what is actually done would contribute to existing situations and the design of new ones. Regarding this theme, Daniellou [17] discusses a design perspective for the future activity, and its importance in preventing problems in efficiency and safeguarding workers' health. Work activity is not merely the carrying out of the prescribed tasks. It includes coping with unpredicted variability, mobilizing personal and collective resources, experiencing contradictions and debating values. It implies personal costs and social contradictions. Béguin [18] emphasizes that this approach locates the person as an 'intelligent' agent (and not as a component in the human–machine system), with a set of skills and shared practices based on work experience with others.

De la Garza and Fadier [19] also propose that analysis of the work activity could be a real tool for better design, because this approach affords data to enrich design models and anticipate the majority of critical events. This process involves choices of design safety, health risks and real difficulties.

Collective work and opportunities to develop it are also part of activity approaches. Caroly and Weill-Fassina [20] discuss the relevance of an approach that locates the network of interactions among members of a work group (rather than the individual operator) at the heart of the analysis. They propose to analyse the organization of such groups as an emergence from these interactions. Studying health services, they propose a discussion in which major dysfunctions might be caused by the lack of collective organization of work, as well as the difficulties involved in delivering quality of service to patients. For them, 'collective work' does not pre-exist; it is constructed, and it is important to provide opportunities to develop individual styles that encourage caution and work-saving strategies are developed by the workers in order to stay healthy.

For Rabardel and Béguin [21] activity is instrument-mediated, which is the result of the characteristics of the individuals, the instruments and contexts. Subjects' activities are oriented towards an object, others and themselves. Various types of constraints and various shaping modalities define an 'open space of possibilities' within which subjects develop their actions according to their objectives and motivations in line with their own characteristics. An activity analysis should be considered as an intrinsic approach that seeks and identifies a subject's subjective grasp of reality, aiming to account for the modalities of engendering the activity and the use of artefacts by people, as well as the modifications they potentially impose on them to adjust them to the needs of their actions. This is an anthropocentric view, the technical device being examined in terms of the person's activity and the problems he/she faces when using the artefact in daily life or a work situation.

Instruments are mediators of the users' finalized action and activity; they allow the transformation of an object in order to reach a goal. An instrument is a composite entity made up by the artefact and a scheme component, which are associated and have a relative independence.

According to these concepts, a new artefact solves old problems and creates new ones. Thus, a new process to transform it into an appropriate instrument is necessary for individual and collective uses. An artefact must allow a dynamic space for functional action possibilities, allowing users the opportunity to finish the design [21].

For Hubault [22] working conditions are contexts of activity, experienced by subjects in a double sense – the self-realization in the work and the realization of the work by a self-realization. So, in terms of technical systems, the human is the centre of decisions, the one who provides an interpretation of the events. With this paradigm the concept of activity is in opposition with the one of functioning. Enterprises are considered from the point of view of complexity, adopting the hypothesis of irreducibility of different points of view and the need to manage tensions between them. The transversal concept is the one of compromise.

Nordic school

Virkkunen [23] proposes concepts such as:

- Activity should be understood as a balancing of contradictory demands.
- The possibility of applying the principle of collective invention in creating a new operating concept.
- Approaches to activities break with linear sequential logic and functional specialization.
- It is possible to understand work processes in a dynamic network of co-operation.

Nosulenko et al. [24] consider activity as the centre of an anthropocentric approach. The human is not simply an element of a human–technology system, but the core that organizes the functioning of the system and also provides flexibility to achieve goals. They propose two kinds of research for designing work. One is centred in the conception of the technical system and another is to analyse activity within this system. It is important to 'describe the psychological constituents [in order] to anticipate the different variants of realization of actions and to evaluate the way the proposed interfaces ensure that the operator can anticipate the system's functioning.' The results of this design method include a definition of zones of activity that need 'creative decisions'. They state that it is important for the system reliability to propose a mixed control (automatic and manual) in order to give operators an opportunity to observe and also to influence the conditions of the objects under their control.

The concept of *affordance* is treated by Norros and Savioja [25] as a part of a discussion on how behaviour is structured. They propose the concept of *prehensility,* which is the subject's ability to grasp the environment, relating it to the specific dynamism, complexity and uncertainty. The concept of *habit* is also used as a pragmatic notion that helps to understand meaning and style in worker activities. They focused on some aspects that could result from activity analyses:

- Reveal the course of action.
- Construct a timeline of the process phases that take place during the operational situation.
- Facilitate operators' recognition of the state of the process and their control over it.
- Reveal operators' habits of action (working practices).

The concept of habit is also discussed by Norros [26] who proposes to include it in the analysis of situated actions. The principal aim of the method is to promote understanding of the core content of a particular work task and the dynamics of constructing situated actions. Based on a socio-historical approach, this proposition is used to put in evidence the possibilities of action and the demands that must be met with respect to the present and future task. Other points of interest are the meaning of the activity, the laws and rules taken into account by workers and their insertion in the society.

The concept 'habits of action' is proposed. First of all, the aspects of repetition and regularity of behaviour are related. Habits of action reflect the 'personal sense of actions that are constrained by the situational features of operations'. This also refers to thinking, communicating and co-operating. The method focuses on the identification of the semiotic structures of behaviour and includes observations, recordings of actor's behaviour, process training, and interviews, all used to understand the meaning of actions, which are determined by the environment and the actor. 'The actor's subjective point of view denotes the environment as a personally meaningful object of action and the investigator must make an analytical effort to become knowledgeable of the personal sense of action, it's the reason-based analysis of actions'.

Norros proposes a reflective process for organizations in order to facilitate the development of adaptive mediated actions to obtain insights into the nature of phenomena and in learning, focused in, for example, a problematic event (object-oriented in a subjective sense). Reflection is a more or less intuitive acknowledgment of the object as a source of new knowledge, not just as a target to be controlled. The construction of actual actions in work – the dynamics of the courses of action, the development of expertise, learning from experience – constructing knowledge of the uncertain world is a continuous social process within the communities of practice.

Owen [27] proposes that context enables and constrains opportunities for formal and informal learning in workplace practice. In addition the implementation of technological changes should be taken into account. Examples of some concepts used by this author based on activity theory (see Engström, Leontiev and others) are:

- Activity is never fully achieved, since it transforms and represents a never-ending horizon.
- Actions are afforded by socially created artefacts, which have practice embodied in them.
- Tensions and contradictions are inherent in work organization and its developmental trajectory.

An application of the theoretical approach to activity is used to propose a new concept for work related to well-being in situations of rapid change. For Launisa and Pihlajab [28] recurrent changes at the workplace seem to increase the haste, stress, health complaints and safety problems of workers. Asynchronies cause frustration, confusion, and lower employees' work motivation and perceived well-being. The transition process is not as smooth and linear as management experts and consultants usually claim. The work units and individual workers often experience the changes as challenging, but also as exhausting. In order to cope with redesigning production/services without undue extra health and safety problems for employees, new types of encounters and alliances are needed between management, production designers, health and safety specialists and local work communities. A common language is not the only aspect required for a new kind of collaboration. New ways of conceptualizing, models and tools are needed to analyse and interpret transition processes and solve the ever-increasing asynchronics and collapses in production and service concepts. In rigorous transformation processes, easing of production and service disciplines improves both the health and safety of personnel and their productivity.

Conclusion

This section provides only a narrow glimpse of the concepts, applications and results obtained by different schools of ergonomics. Issues relating to opportunities for collaboration, to placing points of view in evidence and to building dialogue and comparisons are of major interest. It seems that schools of activity theory can communicate rather easily, since they have similar, and partially the same, concepts. They also propose bottom-up approaches and an intense dialogue with strategic perspectives and organizational constraints on companies.

Macroergonomics is proposed as a top-down approach, and peoples' activity is not proposed as an axis of analysis and a guide for design. Moreover, it is possible to identify points of dialogue and perspectives for comparison. A discussion concerning participation and human error is proposed to try to establish links and themes for this purpose.

PARTICIPATION: WHAT IS ITS TRUE MEANING?

Many field experiences deal with participation. Social actors claim to be really involved in ergonomic interventions. When starting research, ergonomists do not know what workers are doing, how they are using things, how they are acting and experiencing their work. Using different techniques and bringing experiences from other interventions, ergonomists can help to catalyse transformations of work. For Daniellou [17] participation is very important but this is not a shared point of view among work designers. In a more traditional approach users do not participate; 'those are modelled'. Others propose the participation of users or 'user-similar' individuals as a controlled experiment, to analyse their behaviour and to ask subjects about their feelings regarding the usability of the simulated system.

Design and management are processes where people have significant involvement in the planning and control of their own work activities, and sufficient knowledge and power to influence both the process and outcome in order to achieve the desired goals [29]. Ergonomics by definition involves people. The question is, to understand whether the approaches proposed in ergonomics are really participatory. Some examples: does participation lead to better solutions, more effective design processes, an easier implementation phase, better commitment to change, an improved organizational climate, learning experiences for designers and users and spreading ergonomic interests and expertise throughout an organization? Difficulties and potential problems may be:

- lack of confidence among people due to autocratic cultures in organizations;
- lack of time to be involved effectively; or
- need for investment of time and resources.

It is also important for the design process to consider Béguin's proposal: 'every professional category not only perceives the work situation from its distinct perspective, but will also act in order to maintain the process dynamics within its "world"'. During the design process there is a disparity in the proportional attention given, on the one hand to the specifications of the machines or organization, and on the other hand, to those who, through their activity, ensure the function [18].

Hartmann and Ryom [30] discuss the results of adopting a method of analysis and design defined as a double process, which is simultaneously 'top-down' and 'bottom-up'. This results in a change of attitude, working on employees' self-esteem, leading workers to take responsibility for job satisfaction and creating responsible, motivated production personnel. The fact that unskilled employees often do not respond when offered greater influence over their jobs, could be the result of a long-term culture of lack of influence.

Vink et al. [31] discuss the negative connotations linked to coupling ergonomics to illness, complaints or guidelines. They propose another view, focused on the 'positive side'. Ergonomics should be linked to health and innovation, to comfort and productivity, without denying the importance of studying risks. Their model is based on three keywords: goals, involvement and process. Involvement is central in their model; goals are better achieved when end-users participate in the design process and empowerment is possible.

Blewett [32] defines strategies used by those she calls 'workers of influence' to influence management decision-making in the process of organizational change. The positions of these workers are central, because of their leadership role in the change process. They use identifiable strategies, to contribute to and shape organizational change. However, this category of leadership goes unrecognized in management literature, and thus these leaders are lost. She also postulates that 'finding' and identifying this important group and their role in organizational change opens up new avenues for research that may refine management theory.

Zink [33] discusses the ergonomic benefits for companies and not ergonomics as an application of worker protection laws bearing costs. Different experiences show that when ergonomics is included in business strategy, it is possible to reduce health problems, work related injuries and increase customers' satisfaction with better products. Cost reductions and contributions to create a positive company image are also possible and desirable results deriving from this inclusion. In that sense, management should adopt ergonomics and restructure its product development processes, adopt integrative management systems, create win-win situations and allow people in companies to reflect. This last statement is very interesting, since one of the justifications proposed is that involving workers is a recognition that 'employees know much more about their work situation than management experts'. It's also important to consider that when workers are involved in creating solutions, they are more easily accepted than when a top-down approach is adopted.

Darses and Wolf [34] discuss some of the difficulties in achieving a participatory design and having future users participate in design meetings. The challenge for designers is to enrich and expand the vision of operator needs. Designers should adopt a user-centred approach, including techniques such as prototyping and user trials. Even if participatory design is a goal proposed by many ergonomists, it is not the prevalent approach. Designers often make assumptions about user needs. Some difficulties one encounters adopting a participatory approach are [34]:

- designers' professional training;
- human factors analysis generally does not lead to design solutions;
- knowledge and competences are not well represented in guidelines; and
- incomplete and out-of-date ergonomic guidelines.

The results of this study are divided into three categories of operator-references:

(1) Operators are thought of in general terms regarding human–machine interaction.
(2) Operators are represented as elements of an imagined scenario; the designers figure out how the future users are supposed to perform their task.
(3) Operators are considered as subsystems of the human–machine system; their role is evaluated according to the interactions with the technical system; they are considered as the non-automated component of the device.

These results illustrate the designer's difficulties in understanding and including the operator's point of view in the design process. It will be interesting to obtain more data and adopt approaches to bring operators' activities into the design process, even though at a first glance this process could be considered as more costly.

Rabardel and Béguin [21] propose a 'distributed design process', conceived as a mutual learning process with a comparison of different forms of knowledge between designers and workers/users. Achieving this goal means creating social and cognitive conditions where users and designers can share references in a process of adjusting heterogeneous views. This process can be characterized as constructing a 'common world', which takes usage and activity into account. Involvement also means making work activities readable for project participants, which will facilitate worker participation in defining tasks. This action should be maintained by the ergonomists who will put to use techniques and tools to analyse activity.

Conclusion

It is interesting to observe that all quoted authors propose participation, and define it as fundamental for ergonomic interventions. Certainly the approaches and techniques are not the same and there is unequal worker involvement, but when workers are involved they use their experience and their knowledge; they give evidence of what they do; and what the difficulties are. They can tell about their strategies to cope with variability, deficiencies in productions systems, failures in information technologies, problems related to servicing others, etc. Perhaps here we can find themes to create a dialogue and build comparisons with respect to diversity in ergonomics.

HUMAN ERROR: IS IT A SUSTAINABLE CONCEPT?

People adopt different points of view to analyse the actions of others, based mainly on their own expectations about what others should do. Sznelwar [35] and Bouyer and Sznelwar [36] propose the concept of unsuccessful actions. These are actions that didn't achieve their goals, or ended in an incident or accident. By avoiding a focus on guilt, it will be easier to provide tools, work content and a work organization that minimizes unsuccessful actions.

According to Re et al. [37] latent conditions can predispose a system to potential failure since errors are a constitutive part of the human condition. In an error-centred approach, designers and decision makers maintain the assumption that the operator is the weaker part of the system, or at least the most sensitive indicator of the system's vulnerability. However, it is more productive to focus on the strategies that operators activate to assure a successful outcome under uncertain and unexpected conditions. In this context, reliability can be defined as the measurable capacity of a process, procedure or service to perform its intended function in the required time, under common and uncommon circumstances.

Guimarães et al. [38] put in evidence that one prevalent characteristic of safety procedures is to prohibit some activities that have previously resulted in an incident. In the long run, these prohibitions can turn into restrictions that might reduce people's actions in such a way that it is no longer possible to carry out the task.

Kraemer et al. [39] propose a distinction between human error as a failure of planned actions to achieve a goal and *violations*, or deliberate deviations from practices believed to maintain safe or secure operations. For them, violations can only be described in their social context, where behaviour is governed by operating procedures such as codes of practice, rules and regulations. They explain the propagation of violations in the work environment as one of the consequences of time pressure or an overly large number of tasks to perform.

It is important to discuss some of the conditions that could prevent events that can potentially disturb the work process, or even result in major problems such as accidents. It is important to create conditions that don't limit the effective sharing of knowledge in complex task settings. Caldwell [40] discusses whether the significance of an event should be attributed to failures in the communication links between critical decision makers, a lack of awareness of timing and the rates of progression of events and/or inappropriate protocols for determining responses to emergency requests for disaster relief resources. The discussion proposed by Owen [41] helps us to understand some issues of human reliability and the importance of work activity in managing incidents. It is important to understand the spatiotemporal features, their complexity and the need to act interdependently with others. For Owen it is important to provide multiple forms of co-ordination and innovation within short and temporary time-spans in contexts of uncertainty that have significant consequences. In this perspective practices will support shared knowledge and the development of processes that support high reliability. His results show that practices in work activity are shared by operators involved in a related example of high-reliability work and are inherent in all activity systems. It would be important to know the processes through which activity is structured and to put into evidence the fact that work and its complexity occurs through emergent variability.

Conclusion

Discussing 'errors' is always controversial; often people are somehow involved and try to defend themselves. Incident and accident analysis are highly contaminated with emotion; people die, are injured, and material loss can be considerable. Obtaining the truth is difficult. Perhaps it would be better to reinforce the analysis of working situations, spreading a cult of reliability, and initiate action before the situation gets out of control.

To understand this question it would be interesting to propose an analysis based on paradigms related to the Theory of Complexity [42, 43]. This theory represents a huge advance from the classical theory, going beyond a systemic worldview. Complexity is based on a non-fragmented point of view; it is based on existing links between people and how systems are built. Ergonomics provides some principles that are important to the understanding of the relationship between work and the chance of incidents or accidents. Emergence, relationship between order-disorder-organization, auto-eco-regulation, and the recursive properties of phenomena are examples of concepts that could help employers to understand events, and to propose design processes in which it would be possible to take such principles into account in order to improve safety.

DISCUSSION

Proposing a dialogue among the different schools and approaches in ergonomics does not mean attempting to equate them, or to propose a single way of thinking, or seek some 'truth' that would pervade all the concepts and practices. The proposal is that it should be possible to clarify and enrich our several points of view.

There are many questions that should be discussed among ergonomists in order to clarify their positions [44]. Presently, there is no consensus among ergonomists and perhaps there never will be. One consequence will be the impossibility of establishing an epistemological basis for the discipline. How many different approaches in ergonomics exist and how many are yet to come?

It is also important to take into account decision processes among all levels of production. Ergonomics should also treat the way working processes are regulated. It is linked to regulations between subjects, agents acting according to an intentional and limited rationality [45]. It is important to put in evidence arbitrages made by workers in an infinitesimal level [46]. They refer to decisions based on values revealed on analysing sequences of operations and multiform interventions related to relations between: safety and productivity; speed and quality; care with the individual economy and care with the collective well being; health and performance. Decisions are always present since ergonomics deals with human beings and production systems, and variability is one of the main phenomena. To put in evidence those variables means to understand what happens with people individually and collectively, many of them are not explicit and are even unconscious. For Schwartz [16] the concept of activity includes:

1. dramatic uses of the self;
2. is a 'transgressive concept';
3. is a possibility to sew again (to link);
4. is related to every dimension of a human being.

In the field of ergonomics, proposals to consider what really happens in working situations are gaining growing interest. What people do in the context of production becomes more and more significant to ergonomists' approaches. In this paper some examples that confirm this are given. Boff [3] proposes an interesting challenge. He refers to a naturalistic context for understanding work: 'Information technologies enabled new "affordances" for studying and understanding how people perceive, think and act in naturalistic contexts and, in turn, revolutionizing the practice of Human Factors and Ergonomics'. Understanding human factors is important, as is understanding business strategies, the kind of production system, the products, quantities, the quality goals and so on.

Perhaps it's not the label of the school of ergonomics that matters the most, but what point of view is adopted, and from that, to build debates and links between different approaches in ergonomics.

TO CONCLUDE...

After studying people at work in many different contexts over a period of about 20 years, it seems to the writer that one of the most important points defining an acceptable environment is 'what actually makes sense to people'. In many situations, workers were suffering, complaining and even becoming ill because the content of the tasks and the way management was acting created a work context devoid of sense. In some situations workers avoided showing colleagues and members of the hierarchy what they were actually doing, because their actions could be considered as violating the rules. But acting in that way was what made sense to them.

Making sense also means to actually know the utility and the 'beauty' of what workers are doing [47]. For example, stopping the bus outside the formal bus stop boundaries is considered a serious transgression of the norms, but for the driver it is a way of serving an old woman. Speaking with a patient in a hospital room provides cleaners with a feeling of being part of the system, of helping in the clinical process. Both examples show that people's jobs are not limited to work procedures. More recent studies made by our group reinforce this point, and their relevance for re-designing work processes [48, 49].

Of course, ergonomists should take into account working conditions, tools, machines, architectural aspects of the plants, information technology, norms, rules, etc. The question should be how all those aspects that are constituents of the tasks afford or create impediments for people to develop their work. Perhaps a starting point should be about what we are obtaining as results of our interventions. Our results should not only be analysed by the ergonomics staff, but must also be validated by the end users. Their point of view is fundamental to our profession.

It is important to understand that in ergonomics we are dealing with system dynamics, such as continuous change, order and disorder and the emerging properties of phenomena. Being active as ergonomists means working with people. It is always interesting to understand and to deal with different points of view.

Ergonomists face many different challenges in order to achieve interesting results. One of them should be to reinforce links between theory and practice and between practice and theory. Many questions have to be faced in a transformation/innovation process; some of which have been discussed in this paper. Certainly there are many things that have not been discussed here and there are many controversial points. The main conclusion is that it is important to continue the processes of enriching our discussions with result from experiences in the field and from reflections made in academic and non-academic situations. There is a lot of work to be done!

REFERENCES

[1] de Looze, M. and Pikaar, R. (2006). Meeting diversity in ergonomics. *Appl. Ergon.*, 37, 389–90.

[2] Wisner, A. (1993). A metodologia da ergonomia: ontem e hoje. In *Alain Wisner – A inteligência no trabalho: textos selecionados de ergonomia* (L.L. Ferreira, org.) São Paulo: Fundacentro, pp. 87–107.

[3] Boff, K.F. (2006). Revolutions and shifting paradigms in human factors and ergonomics. *Appl. Ergon.*, 37, 391–9.

[4] Hendrick, H.W. (2005). ODAM and macroergonomics 20 years later: You've come a long way baby! In *Human Factors in Organizational Design and Management – VIII* (P. Carayon, M. Robertson, B. Kleiner, P.L.T. Hoonakker, eds) Santa Monica: IEA Press, pp. 25–34.

[5] Hendrick, H.W. (2006). Macroergonomic analysis of structure, anthropotechnology, and modification of three classical research methods for macroergonomic application. In: Pikaar, R.N., Koningsveld, E.A.P., Settels, P.J.M. (eds), *Meeting Diversity in Ergonomics*. Proceedings IEA 2006 Congress. ISSN 0003-6870, Elsevier.

[6] Kleiner, B.M. (2006). MacroErgonomics analysis and design. In: Pikaar, R.N., Koningsveld, E.A.P., Settels, P.J.M. (eds), *Meeting Diversity in Ergonomics*. Proceedings IEA 2006 Congress. ISSN 0003-6870, Elsevier.

[7] Carayon, P. (2006). Human factors of complex sociotechnical systems. *Appl. Ergon.*, 37, 525–35.

[8] Imada, A.S. (2005). Macroergonomic contributions: Understanding the causes of our success. In *Human Factors in Organizational Design and Management – VIII* (P. Carayon, M. Robertson, B. Kleiner, P.L.T. Hoonakker, eds) Santa Monica: IEA Press, pp. 35–42.

[9] Karsh, B. (2006). Meso-ergonomics: A new paradigm for macroergonomic research. In: Pikaar, R.N., Koningsveld, E.A.P., Settels, P.J.M. (eds), *Meeting Diversity in Ergonomics*. Proceedings IEA 2006 Congress. ISSN 0003-6870, Elsevier.

[10] Kogi, K. (2006). Participatory methods effective for ergonomic workplace improvement. *Appl. Ergon.*, 37, 547–54.

[11] Wisner, A. (1991). New technologies and old thoughts: Relations between ergonomics and cognitive anthropology. In *Collected Papers – Ergonomics, Cognition, Anthropotechnology* (A. Wisner, ed.) Paris: CNAM.

[12] Wisner, A. (1995). Situated cognition and action: Implications for ergonomic work analysis and anthropotechnology. *Ergonomics*, 38, 1542–57.

[13] Geslin, P. (2005). The development of anthropotechnology in the social and human sciences. Its applications on fieldworks. In *Human Factors in Organizational Design and Management – VIII* (P. Carayon, M. Robertson, B. Kleiner, P.L.T. Hoonakker, eds) Santa Monica: IEA Press, pp. 455–60.

[14] Daniellou, F. (2005). The French ergonomists' approach to work activity: Cross-influences of field intervention and conceptual models. *Theor. Issues Ergon. Sci.*, 6(5), 409–27.

[15] Cunha, L. and Lacomblez, M. (2006). Driver activity in an ergologic sense: New territories of knowledge and intervention. In: Pikaar, R.N., Koningsveld, E.A.P., Settels, P.J.M. (eds), *Meeting Diversity in Ergonomics*. Proceedings IEA 2006 Congress. ISSN 0003-6870, Elsevier.

[16] Schwartz, Y. (2006). Insight on cultural history of the concept of activity. In: Pikaar, R.N., Koningsveld, E.A.P., Settels, P.J.M. (eds), *Meeting Diversity in Ergonomics*. Proceedings IEA 2006 Congress. ISSN 0003-6870, Elsevier.

[17] Daniellou, F. (2006). Simulating future work activity is not only a way of improving workstation design. In: Pikaar, R.N., Koningsveld, E.A.P., Settels, P.J.M. (eds), *Meeting Diversity in Ergonomics*. Proceedings IEA 2006 Congress. ISSN 0003-6870, Elsevier.

[18] Béguin, P. (2006). Taking activity into account during the design process. In: Pikaar, R.N., Koningsveld, E.A.P., Settels, P.J.M. (eds), *Meeting Diversity in Ergonomics*. Proceedings IEA 2006 Congress. ISSN 0003-6870, Elsevier.

[19] De la Garza, C. and Fadier, F. (2006). Learning from experience: A theoretical framework for the work activity analysis and safe design. In: Pikaar, R.N., Koningsveld, E.A.P., Settels, P.J.M. (eds), *Meeting Diversity in Ergonomics*. Proceedings IEA 2006 Congress. ISSN 0003-6870, Elsevier.

Dudziak, E., Sznelwar L.I., and Plonski, G.A. (2005). Learning organizations and the paradigm of complexity: Work design approach. In *Human Factors in Organisational Design and Management –*

VIII (P. Carayon, M. Robertson, B. Kleiner, P.L.T. Hoonakker, eds) Santa Monica: IEA Press, pp. 335–40.

[20] Caroly, A. and Weill-Fassina, A. (2006). How do different approaches to collective activity in service relations call into question the plurality of ergonomic activity models? In: Pikaar, R.N., Koningsveld, E.A.P., Settels, P.J.M. (eds) *Meeting Diversity in Ergonomics*. Proceedings IEA 2006 Congress. ISSN 0003-6870, Elsevier.

[21] Rabardel, P. and Béguin, P. (2005). Instrumented mediated activity: From subject development to anthropocentric design. *Theor. Issues Ergon. Sci.*, 7(5), 429–61.

[22] Hubault, F. (2004). Do que a ergonomia pode fazer a análise? In *A ergonomia em busca de seus princípios: debates epistemológicos* (Daniellou, F. coord.) São Paulo: Edgard Blücher, pp. 105–40.

[23] Virkkunen, J. (2006). Collaborative development of a new concept for an activity. In: Pikaar, R.N., Koningsveld, E.A.P., Settels, P.J.M. (eds), *Meeting Diversity in Ergonomics*. Proceedings IEA 2006 Congress. ISSN 0003-6870, Elsevier.

[24] Nosulenko, V.N., Barabanshikov, V.A., Brushlinsky, A.V., and Rabardel, P. (2005). Man–technology interaction: Some of the Russian approaches. *Theor. Issues Ergon. Sci.*, 6(5), 359–83.

[25] Norros, L.L. and Savioja, P.J. (2006). Towards a theory and method for usability evaluation of complex human–technology systems. In: Pikaar, R.N., Koningsveld, E.A.P., Settels, P.J.M. (eds), *Meeting Diversity in Ergonomics*. Proceedings IEA 2006 Congress. ISSN 0003-6870, Elsevier.

[26] Norros, L. (2005). The concept of habit in the analysis of situated actions. *Theor. Issues Ergon. Sci.*, 6(5), 385–407.

[27] Owen, C.A. (2006a). Analysing the activity of work in emergency incident management. In: Pikaar, R.N., Koningsveld, E.A.P., Settels, P.J.M. (eds), *Meeting Diversity in Ergonomics*. Proceedings IEA 2006 Congress. ISSN 0003-6870, Elsevier.

[28] Launisa, K. and Pihlajab, J. (2006). Changes in production concepts emphasize problems in workrelated wellbeing. In: Pikaar, R.N., Koningsveld, E.A.P., Settels, P.J.M. (eds), *Meeting Diversity in Ergonomics*. Proceedings IEA 2006 Congress. ISSN 0003-6870, Elsevier.

[29] Wilson, J. (2005). Participation and its role in human factors. In *Human Factors in Organizational Design and Management – VIII* (P. Carayon, M. Robertson, B. Kleiner, P.L.T. Hoonakker, eds) Santa Monica: IEA Press, pp. 53–62.

[30] Hartmann, M. and Ryom, P. (2006). Reducing repetitive work by organizational changes and job development. In: Pikaar, R.N., Koningsveld, E.A.P., Settels, P.J.M. (eds), *Meeting Diversity in Ergonomics*. Proceedings IEA 2006 Congress. ISSN 0003-6870, Elsevier.

[31] Vink, P., Koningsveld, E.A.P., and Molenbroek, J.F. (2006). Positive outcomes of participatory ergonomics in terms of greater comfort and higher productivity. *Appl. Ergon.*, 37, 537–46.

[32] Blewett, V. (2006). Navigating the industrial turf: Lost leaders in organisational change. In: Pikaar, R.N., Koningsveld, E.A.P., Settels, P.J.M. (eds), *Meeting Diversity in Ergonomics*. Proceedings IEA 2006 Congress. ISSN 0003-6870, Elsevier.

[33] Zink, A. (2006). Human factor, management and society. *Theor. Issues Ergon. Sci.*, 4, 437–45.

[34] Darses, F. and Wolf, M. (2006). How do designers represent to themselves the users' needs? *Appl. Ergon.*, 37, 757–64.

[35] Sznelwar, L.I. (2005). Aspectos psíquicos: métodos e técnicas para avaliação. In *1º Curso Internacional de Confiabilidade Humana*, Rio de Janeiro, 6–8 dezembro 2005 (oral presentation).

[36] Bouyer, G.C. and Sznelwar, L.I. (2006). Actuationism and action in parceled work: An ontological rupture. In: Pikaar, R.N., Koningsveld, E.A.P., Settels, P.J.M. (eds), *Meeting Diversity in Ergonomics*. Proceedings IEA 2006 Congress. ISSN 0003-6870, Elsevier.

[37] Re, A., Montagna, S., and Ferrari, E. (2006). Improving risk management in complex systems: From reliability to collective competence. In: Pikaar, R.N., Koningsveld, E.A.P., Settels, P.J.M. (eds), *Meeting Diversity in Ergonomics*. Proceedings IEA 2006 Congress. ISSN 0003-6870, Elsevier.

[38] Guimarães, L.B.M., Costella, M.F., and Saurin, T.A. (2006). Demything "human error" by reanalyzing incidents in a heavy machinery manufacturer. In: Pikaar, R.N., Koningsveld, E.A.P., Settels, P.J.M. (eds), *Meeting Diversity in Ergonomics*. Proceedings IEA 2006 Congress. ISSN 0003-6870, Elsevier.

[39] Kraemer, S., Carayon, P., and Clem, J.F. (2006). Characterizing violations in computer and information security systems. In: Pikaar, R.N., Koningsveld, E.A.P., Settels, P.J.M. (eds), *Meeting Diversity in Ergonomics*. Proceedings IEA 2006 Congress. ISSN 0003-6870, Elsevier.

[40] Caldwell, B.S. (2006). Issues of task and temporal coordination in distributed expert teams. In: Pikaar, R.N., Koningsveld, E.A.P., Settels, P.J.M. (eds), *Meeting Diversity in Ergonomics*. Proceedings IEA 2006 Congress. ISSN 0003-6870, Elsevier.

[41] Owen, C.A. (2006b). Using activity theory to analyze development of work practice and learning in High-3 work environments. In: Pikaar, R.N., Koningsveld, E.A.P., Settels, P.J.M. (eds), *Meeting Diversity in Ergonomics*. Proceedings IEA 2006 Congress. ISSN 0003-6870, Elsevier.

[42] Morin, E. (1998). *Ciência com consciência*, 2nd ed. Rio de Janeiro, Brasil: Bertrand.

[43] Morin, E. and LeMoigne, J.L. (2000). *A inteligência da complexidade*, 2nd ed. São Paulo: Cortez.

[44] Dejours, C. (2004). Epistemologia concreta e ergonomia. In *A ergonomia em busca de seus princípios: debates epistemológicos* (Daniellou, F. coord.) São Paulo: Edgard Blücher, pp. 199–215.

[45] Maggi, B. (2006). *Do agir organizacional: um ponto de vista sobre o trabalho, o bem-estar e a aprendizagem*. São Paulo: Edgard Blücher.

[46] Schwartz, Y. (2004). Ergonomia, filosofia e exterritorialidade. In *A ergonomia em busca de seus princípios: debates epistemológicos* (Daniellou, F. coord.) São Paulo: Edgard Blücher, pp. 199–215.

[47] Dejours, C. (2003). *L'évalutaion du travail à l'épreuve du reel*. Paris: INRA Editions.

[48] Sznelwar, L.I., Lacman, S., and Uchida, S. (2006). Working at streets as public agents – how to learn and create a profession. In: Pikaar, R.N., Koningsveld, E.A.P., Settels, P.J.M. (eds), *Meeting Diversity in Ergonomics*. Proceedings IEA 2006 Congress. ISSN 0003-6870, Elsevier.

[49] Quaggio, F.M., Silva, M.T., Sznelwar, L.I., and Mascia, F.M. (2006). Relating service operations and work in professional services. In: Pikaar, R.N., Koningsveld, E.A.P., Settels, P.J.M. (eds), *Meeting Diversity in Ergonomics*. Proceedings IEA 2006 Congress. ISSN 0003-6870, Elsevier.

Meeting Diversity in Ergonomics (2007)
Pikaar, Koningsveld, and Settels (Eds)

8

Ergonomic Analysis of Work Activity and Training: Basic Paradigm, Evolutions and Challenges

Marianne Lacomblez

Faculdade de Psicologia e de Ciências da Educação da Universidade do Porto, Rua do Dr. Manuel Pereira da Silva, Porto, Portugal

Marie Bellemare

Département des Relations industrielles et Chaire en gestion de la santé et de la sécurité du travail Université Laval, Québec, Canada

Céline Chatigny

Université de Sherbrooke, CRFP and CINBIOSE, Canada

Catherine Delgoulet

Ergonomie, Comportement et Interactions, Université de Paris 5 René Descartes, EA 4070, Paris, France and CREAPT – CÉE 29 promenade M. Simon, Noisy le Grand, France

Alessandra Re

Dipartimento di Psicologia, Università degli Studi di Torino, via Verdi, Italy

Louis Trudel

Département de réadaptation, Faculté de médecine, Pavillon Vandry, Université Laval, Québec, Canada

Ricardo Vasconcelos

Faculdade de Psicologia e de Ciências da Educação da Universidade do Porto, Rua do Dr. Manuel Pereira da Silva, Porto, Portugal

Abstract. The use of ergonomic analysis of work activity (EAWA) has constantly evolved during the last 15 years. For researchers adopting the ergonomic approach, 'object training' has also greatly evolved, raising epistemological, ethical and pragmatic questions. The reflections concluding the symposium *Ergonomic Analysis of Work Activity and Training* of the IEA2006 Congress were prompted, in most cases, by the mid-term and long-term effects of ergonomics-related interventions in the context of technical and organizational changes. The focus is on how to accomplish transforming actions, how to merge description with intervention.

From this perspective, training is a constitutive element of the ergonomic analysis of work activity, both directly and indirectly, mainly thanks to workers' participation in the research process. EAWA is one of the few methods that never leave the research subject anonymous; on the contrary, it postulates an active and controlled relationship, on a methodological level, between the 'research object' and the subject. These evolutions justify a collective reflection about the theoretical frameworks and methodological perspectives assumed in the papers presented at the symposium. Moreover, some questions concerning evaluation should not be underestimated.

Keywords: Ergonomic analysis of work activity, training, technical and organizational changes, methodology, evaluation

INTRODUCTION

Training techniques in the workplace have progressed in terms of their impact upon health and safety issues and upon professional skills development. When integrated with the ergonomic approach, training has been increasingly transformed, accompanying the evolution of referential frameworks, the objectives of the intervention, and the target people. Work analysis has now moved from a status of 'technical tool' used in the detection of work demands and required competence to a status of 'object' of 'training for action' addressed to different work actors – aiming at increasing their professional mastery and their awareness both of working conditions and of the transformations needed. Nowadays, training often acquires the dual status of a programme aimed at the development of technical and non-technical skills, and at health preservation. But 'object training' itself has also greatly evolved, raising epistemological, ethical and pragmatic questions that cut across different experiences. The high number of studies available on the subject permits now to accomplish a state-of-the-art assessment of this 'tradition'.

Despite the great diversity of the studies presented and some noticeable divergences, the papers presented in the symposium *Ergonomic Analysis of Work Activity and Training: Assessment of 15 Years of Intervention and Research* of the IEA2006 Congress, reinforce the need to assess the state and the development of some important questions prompted, in most cases, by the mid-term and long-term effects of the referred interventions in the context of (often profound) technical and organizational changes.

Preventionists' training

Nowadays we are more ready to admit that health, in a broad sense, is not independent from 'professional mastery' – with its developmental, dynamic and collective dimensions. In this sense, the actors whose mission is related with health and working conditions recognise training as a useful resource in the exercise of their professional role or of their mandate.

Nevertheless, it is important to remember that the experiences of training preventionists to undertake an ergonomic approach to work are part of an evolution of conceptualizations that, for a long time, favoured a tradition inspired by the worlds of sports and the military. Founded on holding the individual responsible, on the image of strictly physical work and of an ideal worker – the industrial athlete – these conceptualizations valued prescriptions of the use of more or less uncomfortable protection equipment or of acquisition of safety behaviours, in terms of correct gestures and postures, with no reference to real work activity nor to the characteristics of the individuals [1].

A synergy developed between prevention and ergonomics media that led an increasing number of preventionists to consider training in Ergonomic Analysis of Work Activity (EAWA) as an instrument for improving prevention services to workplaces. As a result of this training, they are expected to adopt a different point of view on risk prevention; a more positive vision of the role played by workers themselves in the prevention of occupational accidents and injuries; a broader point of view on the way to decrease risks in work situations.

For example, in the Portuguese construction and building industry [2], which is characterized by a high rate of occupational injury and fatal accidents, current practices of 'safety coordinators' are based on a pejorative vision of the worker, considered to be a source of error and bad initiatives rather than an active prevention actor facing ever-changing and unpredictable situations. There is also a tendency to focus on the immediate causes of risks on site rather than to look for the roots of the risk back in the planning of the project, where determinants of the risk are put into place. Including training in EAWA in the basic training of these practitioners would help them achieve a more 'contextualized prevention'. It would also help in countering the tendency to limit prevention to procedures and prescriptions merging with the spirit of EU directives, which is supposed to guide 'safety co-ordinator' practices.

But it may be asked if, after the introduction of such training, the trainees effectively modify their work practices? For Duwelz et al. [3], who assessed long-term (3–5 years after training) effects of such a training programme, it seems to produce interesting results. From interviews conducted with the former trainees, the authors found that CRAMs'[1] inspectors transmit their ergonomic approach to companies they deal with. They transmit especially a broader view of prevention where non-technical responses are often required to solve technical problems. Their analysis reveals also the following changes in trainees' work practices: change in the way they organize company visits and in the type of information taken into account when making their diagnosis. The time spent with workers to understand their real activity increases, while the time spent on technical aspects decreases; they also enlarge the scope of prevention acting as adviser when companies, especially the small ones, are buying new equipment, giving them criteria related to health and safety; trainees also say that they transfer a new vocabulary to the companies; changing, for instance, the expression 'break time' by 'recovery time'. Despite this demonstration of the interest of training health and safety practitioners to EAWA, the authors suggest that the integration of training content into the practice 'depends on many factors, extending far beyond the preventionists' straightforward intention to implement it' [3].

Another interesting point to discuss is: How such training should be designed, what kind of activities should it include? The trainer can be a certified ergonomist as in CRAM, or a multidisciplinary team as suggested by Valverde and Lacomblez [2], who stress the possibility of an alternative process between classroom and field activities. Adopting a common vocabulary allows those professionals to create, beyond the training process itself, forums where they can discuss the problems they are facing in their practice. Some researchers [4] suggest that looking directly at their own activity might be a good way for trainees to improve their 'professional mastery'. Using the ergological perspective of Schwartz [5], the training given to health surveillance workers (VST) in Brazil is based on a co-analysis (by researcher-trainer and trainees) of work situations. The process starts with the trainees discussing which particular work situations should be analysed. Relying on filmed situations where one is performing an intervention at a workplace, the co-analysis is first made by the protagonist and a researcher and their discussion is video-recorded. These 'self-confrontation' scenes are then analysed by a peer who is also filmed. Finally, from this 'crossed-self-confrontation' [6], audiovisual material is created for use within the group of trainees. This process opens a 'permanent debate between values and norms' [5] and contributes to empowerment of the VST by the formalization of their knowledge. This new methodological device appears to be promising in the training of health and safety practitioners, especially to address the question of values. This conceptual and methodological framework, that relies on constituted knowledge of researchers and invested knowledge of workers, has also been used for internal practitioners (members of Health and Safety Committees) by Prévot-Carpentier and Wild [7] in order to support debates of value between employees and management.

The debate of values now occupies the forefront of issues concerning preventionists, namely institutional preventionists, who intend to pass from a 'normative and prescriptive prevention to a formative and participative prevention' [8]. The conditions of feasibility of action are the main focus of the questions raised in the sequence of training programmes in ergonomics, by

[1] Caisses Régionales d'Assurance Maladie.

the awareness of the contradictions between what is demanded of the preventionists and what they aspire to develop in their practices.

This concern crosses also the analyses that directly or indirectly deal with the evolution of the representations of the participants in training for action, aiming at an increase of their awareness of working conditions and of the transformations needed.

Representations transformation

Past symposia, concomitant with the different IEA Congresses [9–18], have enhanced the interest on training actions based on the following principles [19–21]:

- the acknowledgment of the experiential knowledge of the operators (workers' technical and non-technical skills);
- the need to start from the initial representations and knowledge of the trainees and to consider their point of view;
- the appropriation of the concepts and methods of work analysis, facilitated by the 'opportune' use of basic knowledge (conceptual, methodological and strategic);
- the concern with working the language in order to facilitate the sharing and the confrontation of knowledge, and to gradually pass from the formulation of the problems to its formalization and generalization, opening possibilities for collective action;
- training through action and reflection about action; the construction of knowledge with the help of reflection about practical experience of analysis in real situation; and
- the training situation is conceived as an opportunity of reciprocal learning, valuing the collective dimension of the process.

This tradition, herein synthesized, differentiates itself from two others. There is one that tends to consider EAWA essentially as a 'tool' used in the detection of required competences. That's why only a few papers assumed this perspective in the 2006 symposium. Nevertheless, it is clearly present in the text of Santos et al. [22], as they refer to work that leads to the redesign of an industrial production process in the rubber sector, using training to alert operators to the need for a change in their attitude with respect to production and market requirements.

A second tradition has left its marks in some contemporary projects; one that considers the adult as a 'blank slate', without taking their professional experience and background into account. They were explicitly or implicitly asked to forget things they already knew [23]. This concept is still encountered today in workplace training [24]. Besides these questions, some of the papers presented in the symposium also refer to processes in which ergonomics training is interrelated with organizational transformations. This is the case, for instance, of an ergonomics training procedure, evaluated by Montreuil et al. [25] at a service company. However, it is not the case among the small businesses that took part in an ergonomics assistance programme presented by Ulin et al. [26], several of which made significant changes in workstations and equipment, without, however, developing a framework for incorporating ergonomic job assessment into their organizational structure (one of the programme's objectives).

In most of the companies that participated in the studies presented in the symposium, the hierarchy and workers had a more active role to play in developing occupational health and safety than was the case prior to the ergonomic intervention. Notably, in the framework of EAWA in learning situations (formal or informal), importance is given to the adult's work experience in his or her role as a learning individual, both actively and reflectively in the learning situation [27, 28]. This alternative concept of adult learning leads, among other things, to a greater emphasis on the collective dimension of the learning and situational transformation process, based on such concepts as construction of a common operative referential system [29, 30] or of shared skills [31, 32], which are essential elements in the collective action of co-ordination. The learner is no longer alone, without any reference points, in relation to the new knowledge to be acquired. On the contrary, he or she may draw on internal resources (previous knowledge, expertise), external resources, such as human resources (for instance, the trainer, other learners, colleagues, clients), material resources (objects, rules, instructions) or organizational resources (training or professional mechanisms) that are, more or less, facilitating individual and collective learning and training time [33, 34]. However, this conception of learning, which seems so complex to implement and maintain at companies once the ergonomist has left, becomes relevant and indispensable in analysing informal training sessions, even in work situations in which there is very little manoeuvering room for learning. There is the example of female shelter workers, who collectively build their expertise on a daily basis and through formal training during the development process, by sharing and transforming their representations concerning the work and the organizational methods for supporting the development of their competencies [35]. There are the truck drivers who must learn, but this time on their own, how to regulate their state of fatigue along with production requirements [36]. Changes in workers' representations concerning the demands of their work must be accompanied by transformations in organizational constraints [37].

These studies reveal the different approaches and expectations of the actors, including ergonomists, concerning work, learning and the roles of the actors involved. Barros Duarte and Lacomblez [38] clearly illustrate these issues and their complexity. They refer to encountering occupational health models based on antagonistic logics that do not offer much room for the individual. On one hand there are individuals and their needs, and, on the other, there are policies and standards that seek to control human behaviours and do not allow significant room for workers to make an effective contribution. Individual and group interviews with workers have led to the emergence of a new vision of the health–work relationship based on a collective approach that fosters the creation and integration of a 'reflective assistance network'. The authors discuss the importance of developing organizational structures in companies – but also outside the companies – to permit the individual and collective construction of competencies and structures necessary for development of health.

It is clear that the issue of actors' representations and their roles is central to the transformation of their health, their competencies and their working conditions. Whether involving work analysis training or training situation analysis, the subsequent transformations of the workplace often seem to depend on the state of knowledge and representations of actors concerning the changes to be made and the means to be used. Even when the request comes from the company, and workers contribute to the process, the training intervention does not necessarily result in concrete transformations.

The real political will of companies may be questioned. We may also wonder if participants being trained in ergonomic principles have sufficient preparation and organizational support to carry out other case studies after the ergonomists have gone. Isn't the necessary ergonomic support essential for lasting, concrete transformation of work and training, based on alternative approaches and procedures concerning adult learners in a learning situation?

It is therefore important to reflect on the social, institutional and organizational conditions that may help or hinder each party's efforts to verbalize and to lead co-elaboration analyses of shared knowledge about work and health. An important issue is the impact of this learning process in terms of prescription or development in the work and training situations examined. The training of company actors in EAWA gives them a new power to act, but also new responsibilities to improve their working conditions. What are the ramifications of these new responsibilities for the hierarchy and for the workers themselves? Given that time is sometimes limited for appropriating the ergonomic approach and its criteria, how can workers reinvest the process in their work? Moreover, in certain studies in which the hierarchy and professional ergonomic resources are not involved in EAWA training, what are the indirect effects of changes made to one workstation on activities, other workstations and workplace colleagues?

What we are dealing with here is a model of a worker defined as competent, but that is at the risk of finding himself isolated in a project of work transformation. In this context, a set of studies presented in the symposium provides a framework that enlarges the scope of this reflection.

The representation of the 'model of worker' and of his/her competence

These studies, related to individual and collective processes of experience elaboration, contribute to fostering the notion of a worker as a 'competent individual', and producer of a personal project and expertise. These studies cut across different sectors of application, ranging from high-tech environments such as research reactors and offshore platforms, which are subject to strict safety standards, to sectors less strictly regulated, and where technology plays a very limited role, as is the case with, for instance, installers of party structures employed by a town administration, public and traffic agents.

The difference between prescribed and actual work has characterized the whole history of EAWA and still plays a fundamental role. However, the analysis of this difference is now conducted based on a notion of actual work as producer of context-embedded competencies, from which the following corollaries are derived: the focus on the working subject, the description of forms of expanded competence and of processes of reformulation of actual-work rules aimed at an adjustment to actual needs.

In order to get closer to actual work, ergonomic work analysis never studies work in general. The focus is less on the process (leadership, climate, communication) than on the subjects at work, who take part in the research and who may either be reluctant to comply or, on the contrary, as in the studies presented, may ask researchers to delve into the matter more deeply in order to challenge the current situation and define 'opportunities for improvement'.

EAWA is one of the few methods that never leave the research subject anonymous; on the contrary, it postulates an active and controlled relationship, at a methodological level, between the 'research object' and the subject, mainly thanks to the participation of workers in the research process. In this light, an extended research community comprising researchers, the commissioning party and a group of workers is mobilized in order to encourage comparing and developing the actors' viewpoints.

In the studies presented, EAWA broadens its original setting of work analysis by exploring the meanings attributed by subjects to their activities in order to understand and fully describe their operational and social intelligence. In the case of less structured jobs, e.g. party-structure installers [39], results highlight the development of an expanded professional competence, where the operational skills are embedded with the managing skills of a wider system. For instance, observation confirms that experienced workers not only perform gestures, but they also develop a reflexive activity on the team's work, by co-ordinating some stages of the job, supervising the group and the novices, informally structuring roles within the team, and by protecting themselves and others as well as the equipment. The team recognizes their competence and their informal contribution to the group, therefore accepting an even distribution of efforts between workers and avoidance strategies aimed at preserving senior workers from potentially dangerous tasks. Another study [40] explores how public agents working on the streets enhance the value of their job, building day by day a relationship with the residents and improving the democratic use of public space. Thanks to the ergonomics of the activity, which highlights the informal strategies and instruments used by the operators for the work activity and risk management, operative work proves to be producer of knowledge and experience with an original value.

The divergence between prescribed and actual work plays a major role also in more regulated sectors such as the oil industry, where it is described as intrinsic to the complexity and variability of actual situations [41]. A paradigm that erects barriers, viewing variability as an incidental and always negative element, might make the system more rigid but not necessarily more capable of efficiently managing complexity: in the work situations presented, workers are not required to *belong to* an organization but to actually *contribute to* organizations where flexibility and commitment are crucial to production targets [42, 43].

Consequently, multiple studies emphasize the ecological characteristics of the rules, stating that operators use rules like instruments of reference to perform proper actions. They transform, model (as artefacts), and reformulate the rules according to the situations.

In this light, risk management hinges on a number of successful compromises elaborated over time, seeking a balance between productive, safety-related, individual and collective goals, but also compensating for the contingencies of the work process.

In constantly changing environments like that of party-structures installers or of traffic agents, experienced workers are described as constantly assessing upcoming situations and anticipating possible future actions. They rely on the diagnosis of evolving situations and on the construction of shared operating modes. This collective competence is not an organizational competence in its traditional meaning [31]. It is often about short-lived, high-variable systems in action, where a previous individual experience quickly ends up connecting with that of the group. Other methods that do not share EAWA's receptiveness for ecological observation have difficulties appreciating the value of this process of reformulation of rules aimed at facing the numerous factors interfering with the job carried out in a dynamic environment. In this light, this systemic view of risks, which leads to an ecological adjustment of the rules, is viewed like a mere deviation from the expected procedure.

Above all in the case of organizations where flexibility and commitment are crucial issues, the ability of the organization to manage such processes of experience production becomes critical in achieving the goals of the organization. Therefore, even in the most strictly regulated sectors like that of nuclear reactors, one of the findings of internal self-assessment is the lack of an operational feedback system within the organization.

EAWA proves very useful to organizations in helping them to improve operational feedback and to regulate themselves according to their awareness of actual operational activities. If they are unable to accomplish this, such organizations have no adjustment abilities and, as Perrow [44] suggests, no memory.

EAWA is also useful to organizations in that it enables them to move from a training based on a cognitive model regarding the scientific and technical issues, i.e. the way the system operates, to a training encompassing an operative model dictated by practical concerns about the way the activity is actually performed and focusing on 'how is it managed'.

In its more recent expressions, by documenting the adoption of dynamic and adapting behaviours compared to the sequences of actions established for the execution of a task or procedure, EAWA has remodelled the notion of what it originally used to define as informal, individual practices at work, moving toward the acknowledgment and description of the economic and safety outcome of these collective processes of experience elaboration.

These evolutions clearly justify a collective reflection about the methodological perspectives assumed in the projects developed.

Methodological perspectives

The task of analysing the theoretical and methodological options of a set of research works corresponds to the intention to outline the state of development of a particular paradigm. From this perspective, we can analyse what distinguishes or what brings together the various postures adopted; we can be attentive to certain emerging characteristics in a significant number of studies; we may also want to reveal some of the shared apparent evidences.

- The paradigm that has prevailed in former symposia is still noticeable in many of the papers presented in Maastricht. As an example, we can name Garrigou et al. [45], who created an interactive training module for agriculture workers exposed to pesticides, strengthened by practical knowledge of their activity, and (thanks to EAWA) in which the intention was to teach the workers how to avoid contact with chemicals. With a similar perspective, Brito et al. [46] trained workers in the educational system in Brazil to increase their awareness of the close links between work and health. Results are expressed in terms of hope that the trainees will be able to bring about preventive changes in their work situations.

Even if [47] analysed a posteriori a training experience, including transfer of knowledge on EAWA to trainees, the main idea is again, to project research avenues toward better compliance to the training content in order for the trainees to identify risks in their work.

Nevertheless, a new diversity was noticed in this symposium on the types of association between training and EAWA, due, maybe, to the above mentioned evolutions and of the search for new paths for the resolution of problems up until now uncommon.

We can start by referring, as an example, to the paper of Re et al. [48] who talk about the gradual constitution of an information system for physicians in private practice to enhance their diagnostic skills regarding professional illness. The authors turn to EAWA, not only to preserve the specificity of the work activities, but also to anchor the system in the reality of all the actors taking part in the process of declaration and acknowledgement of professional illnesses. We can consider that the authors use EAWA as a tool to ameliorate the friendliness of the system for the users and, by this means, they facilitate users' training, necessary for their involvement in the collective process.

Another example is the paper of Pelegrin [49] who uses EAWA as a step to achieve better services in a rehabilitation centre. By collecting data about the work done by professionals in contact with handicapped workers, he brings to the fore the evolution of this collaboration. New criteria can then be identified to improve the services and train the professionals to face the new characteristics of the beneficiaries.

Finally, it is also noticeable in the studies presented the increasing presence of authors from neighbour scientific disciplines, particularly sociology [50].

- In successive symposia, the increasing use of co-analysis has been noticed, on a perspective that involves partnership with workers, in conformity with the above-defined tradition. We must, nevertheless, note that the significant increase in videotape recording use –

undeniable methodological support for many projects – has probably been propitious to these developments.

Some interventions reported in former symposia – namely in the one held in Toronto (in 1994) – had already revealed an increasing use of principles of self-confrontation or crossed-self-confrontation with work activity.

This tendency was confirmed. But several papers in this symposium also attest how much the theoretical contribution of Clot [6] reinforced the attention paid to those practices, and the degree to which it aroused interest for the possibility of an 'activity clinic' aimed at the potential development of the subject of action. It is interesting to notice that in this way, we came back to the dilemma faced by the pioneers of ergonomics: to take action over the production conditions or on the individuals. As a matter of fact, experience has shown that it is not possible to achieve one without the other.

- Finally, it was with curiosity and interest that we realized that many authors did not consider it necessary to make explicit and to justify their methodological options, as if we were talking about evidences. If we put this fact in parallel with the changes emerging in the tradition concerned, we can put forth the hypothesis that either by silence, or by making the differences explicit, the initial paradigm is now going through a phase of renewal or of greater internal differentiation.

That is probably the reason why some authors considered obvious the need of more explicit underlying theories of the problem, of the training programme and of the training evaluation process, as much as of their interrelation. As proposed by Berthelette and her team, considering the underlying theory of the training programme will lead to a better follow-up and assessment of the implementation process. Doing so, researchers will be in a better position to evaluate the impact or short-term, mid-term or long-term effects [51, 52].

Could the forthcoming challenge concerning knowledge development on the association between training and EAWA be to be more rigorous in clarifying underlying theory of the problem, the programme and its evaluation?

Questions of evaluation

As previously stated, when we talk about evaluation referring to the relationship between EAWA and training, two main types of works are common: those works in which EAWA is used to evaluate the impact of an intervention comprising training; and those which try to evaluate the internal consistency and the impact of ergonomics-related training programmes. Nevertheless, other questions have been increasingly raised during recent debates around the evaluation theme, concerning the status of evaluation as ergonomic intervention in itself, as an ergonomic catalyser, as a key element to effective transformation in terms of contexts, actors, and practices. Three evaluation-related questions appeared as dominant during the symposium:

(1) The quality of the transposition of the prescribed objectives to the actions previewed and implemented in the programme;
(2) The importance of the use of conceptual frameworks in the analysis of the training programmes and of their (in)success potential;
(3) The importance of considering intervention targets' characteristics and the specificity of work contexts.

(1) Concerning the first question, as many of the papers presented have quite well demonstrated [53, 54] we frequently find either incongruities between the objectives aimed at and the planned actions, or incongruities between these actions and what eventually is implemented. The authors call our attention to these biases, which often lead to conclusions of programme inefficiency without a proper verification of the programmes' implementation degree. Nevertheless, despite the awareness of this bias being common to the different evaluative researches, its interpretation is not. This is because it is strictly connected to different points of view assumed over reality, to the objectives of the evaluation, and to the different underlying conceptual frameworks. It is common for the intention, so valued by Activity Ergonomics, to go beyond the apparent insuccess of a training programme or even beyond its apparent success, as in the cases reported by Cau-Bareille et al. [54] and by Santos and Lacomblez [55], in which an apparent success of the training process hid workers' difficulties, skilfully overcome by the use of individual and collective strategies, of which *discovery* was only possible thanks to the virtues of EAWA.

(2) Consequently, to go beyond the awareness of the biases between what was planned and what was implemented, and to understand the significance of these biases and the implications held within, the theoretical frameworks underlying the analyses assume particular importance. Some examples of this can be found in Leduc et al. [56] through a general evaluative approach, supported by the use of 4 complementary conceptual frameworks, chosen as a function of the nature of the programme under evaluation and of its general theme [57–60]. It can also be seen in Santos & Lacomblez [55], through a more singular approach, where both the real context of programme implementation and its concrete goals contribute to the definition of a theoretical framework suitable for supporting programme evaluation [61]. Either way, it seems clear that the use of theoretical frameworks helps us to better understand why certain training actions do not allow the learning outcomes they were designed for. And it seems also true that the more these frameworks are congruent with the characteristics of the object of analysis, the more efficient they are potentially. EAWA has, in this matter, an important role to play, as underlined by Cau-Bareille et al. [54] and by Santos and Lacomblez [55].

(3) This brings us to the third aspect whose importance is determinant in what concerns interventions evaluation: the consideration for trainees' particular characteristics as well as for the working setting in which they are meant to use the skills developed. In fact, Cau-Bareille et al. [54], demonstrate through a well-designed EAWA-supported evaluation programme, what has been thoroughly emphasized throughout these 15 years of *Ergonomic Analysis of Work Activity & Training* IEA symposia: the learning potential of the elder workers; the importance of their participation in training programmes' planning, implementation and evaluation; the benefits of a close articulation with concrete real work situations, namely in terms of social and symbolic mediators of the learning processes; the respect for workers' former experience and for their individual and collective traditional strategies; the importance of the follow-up of programme impact in context.

PROSPECTS AND CHALLENGES

Despite the interest of the contributions and the reflections we have been explaining throughout this paper, a few stubborn questions keep coming to our minds and thus to our debates when we discuss evaluation matters. For whom are we evaluating? Who are the beneficiaries of these reflections? What did happen to those trainees, to those trainers, to those contexts, to those training promoters and conceivers, as a result of those evaluation processes and of those analyses of those intervention and evaluation processes? If we evaluate and write ourselves a report or a paper, even if we show it, afterwards, to the people involved in the training programmes concerned (promoters, trainers, trainees), what do we expect to transform? Are these expectations congruent with our theoretical frameworks? All these are quite crucial questions if we consider that ergonomics' main goal is to understand in order to transform and to empower. Perhaps a starting point as good as any could be to start seeing our job as evaluators, not only as *evaluative research*, but as *evaluative intervention*, and start planning congruently.

Of course, this would necessarily mean that the 'others', the 'evaluation objects' would have to become co-subjects, co-constructers, co-responsible for the change. Which is nothing that an ergonomist couldn't handle – much to the contrary. But this would also mean that we – the scientists, the evaluative researchers – would become co-objects of the evaluation process. This also means that our valued evaluation criteria should go beyond questions of scientific merit (of internal consistency, of theoretical congruency) and consider also, with similar scientific weight, the matters of worth, of contextual validity, of effective transformation. It is always a matter of the demand, the timings involved, the resources available... But we must also bear in mind that our responsibility, our theoretical and social congruency starts at that very first moment of negotiation, of co-construction of new 'commonplaces'.

REFERENCES

[1] Teiger, C. (2002). Origines et évolutions de la formation à la prévention des risques "gestes et postures" en France. *Ind. Relat. (Relations Industrielles)*, 57 (3), 431–62.

[2] Valverde, C. and Lacomblez, M. (2006). Contributions of the "safety coordinators" training to promote building sector prevention contextualization. In: Pikaar, R.N., Koningsveld, E.A.P., Settels, P.J.M. (eds), *Meeting Diversity in Ergonomics*. Proceedings IEA 2006 Congress. ISSN 0003-6870, Elsevier.

[3] Duwelz, M., Thibault, J.F., and Josse, P. (2006). Long term impact assessment of ergonomic training courses on professional practices. In: Pikaar, R.N., Koningsveld, E.A.P., Settels, P.J.M. (eds), *Meeting Diversity in Ergonomics*. Proceedings IEA 2006 Congress. ISSN 0003-6870, Elsevier.

[4] Santorum, K. and Brito, J. (2006). The surveillance of the working environments and processes as activity of work. In: Pikaar, R.N., Koningsveld, E.A.P., Settels, P.J.M. (eds), *Meeting Diversity in Ergonomics*. Proceedings IEA 2006 Congress. ISSN 0003-6870, Elsevier.

[5] Schwartz, Y. (2000). *Le paradigme ergologique ou un métier de philosophe*. Toulouse: Octarès.

[6] Clot, Y. (2000). La formation par l'analyse du travail: pour une troisième voie. Éducation et Formation. Biennales de l'Éducation et de la formation. In *Manières de pensée, manière d'agir en éducation et formation* (Maggi, B. dir.) Paris: PUF.

[7] Prévot-Carpentier, M. and Wild, M. (2006). Participatory ergonomics and training in the French National Employment Agency. In: Pikaar, R.N., Koningsveld, E.A.P., Settels, P.J.M. (eds) *Meeting Diversity in Ergonomics*. Proceedings IEA 2006 Congress. ISSN 0003-6870, Elsevier.

[8] Frontini, J.M., Modestine, G., Penel, P., and Teiger, C. (1996). Changer de regard sur les gestes et postures de travail pour mieux prévenir les risques: préventeurs et ergonomes, même enjeu? In *Actes du 31ème Congrès de la SELF*, vol. 1. Bruxelles: SISH-ULB, pp. 282–9.

[9] Rabardel, P., Teiger, C., Laville, A., et al. (1991). Ergonomic work analysis and training. Exposé introductif à la session "Ergonomic and training". In: Quéinnec Y. and Daniellou, F. (eds), *Designing for Everyone*. Proceedings of the 11th Congress of the IEA. London: Taylor and Francis, pp. 1738–40.

[10] Proceedings of the 12th Triennal Congress of the International Ergonomics Association, vol. 5. Toronto, 1994.

[11] L'ergonome, le formateur et le travail. *Educ. Permanente* (124), 1995.

[12] Ergonomists, training and occupational health and safety. *Saf. Sci.* Special Issue, 23 (2/3), 1996.

[13] Abrahão, J., Berthelette, D., Desnoyers, L., et al. (1997). General introduction to the symposium: aims, context, concept, methods, practices and problems. *Proceedings of the 13th Triennial Congress of the Internacional Ergonomics Association*, vol. 1. Tempere, 1997.

[14] Contribution à la réflexion sur l'évaluation des actions de formation en milieux de travail. *Perform. Hum. Tech.* Hors Série, Décembre, 1998.

[15] Proceedings of the IEA 2000/HFES 2000 Congress. San Diego, 2000.

[16] Ind. Relat. (Relations industrielles) 56(3), 2001.

[17] Proceedings of the IEA 2000/HFES 2000 Congress. Seoul, 2003.

[18] *PISTES* 6 (2), 2004. http://www.pistes.uqam.ca.

[19] Teiger, C. and Lacomblez, M. (2005). L'ergonomie et la trans-formation du travail et/ou des personnes (1). *Educ. Permanente*, 165, 9–28.

[20] Teiger, C. and Lacomblez, M. (2006). L'ergonomie et la trans-formation du travail et/ou des personnes (2). *Educ. Permanente*, 166, 9–28.

[21] Lacomblez, M. and Teiger, C. (2006). Ergonomia, formações e transformações. In *Ergonomia* (P. Falzon, ed.) São Paulo: Edgard Blücher.

[22] Santos, R., Lopes, P., and Rebelo, F. (2006). Optimization of an industrial production unit: Case study. In: Pikaar, R.N., Koningsveld, E.A.P., Settels, P.J.M. (eds) *Meeting Diversity in Ergonomics*. Proceedings IEA 2006 Congress. ISSN 0003-6870, Elsevier.

[23] Teiger, C. (1993). L'approche ergonomique: du travail humain à l'activité des hommes et des femmes au travail. *Éduc. Permanente*, 71–96.

[24] Delgoulet, C. (2001). La construction des liens entre situations de travail et situations d'apprentissage dans la formation professionnelle. *Rev. Électron. PISTES* 3 (2). http://www.unites.uqam.ca/pistes/.

[25] Montreuil, S., Brisson, C., and Trudel, L. (2006). Integrating into prevention process by and following employee ergonomics training. In: Pikaar, R.N., Koningsveld, E.A.P., Settels, P.J.M. (eds), *Meeting Diversity in Ergonomics*. Proceedings IEA 2006 Congress. ISSN 0003-6870, Elsevier.

[26] Ulin, S.S., Rabourn, R.A., Armstrong, T.J., and Lau, M.H. (2006). Ergonomics support for small companies. In: Pikaar, R.N., Koningsveld, E.A.P., Settels, P.J.M. (eds), *Meeting Diversity in Ergonomics*. Proceedings IEA 2006 Congress. ISSN 0003-6870, Elsevier.

[27] St-Arnaud, Y. (1992). *Connaître par l'action*. Montréal: Les Presses de l'Université de Montréal.

[28] Schön, D.A. (1994). *Le praticien réflexif: à la recherche du savoir caché dans l'agir professionnel*. Montréal: Les Édictions Logiques.

[29] de Terssac, G. and Chabaud, C. (1990). Référentiel opératif commun et fiabilité. In *Les facteurs humains de la fiabilité dans les systèmes complexes* (J. Leplat and G. de Terssac, eds). Toulouse: Octarès.

[30] Leplat, J. (1991). Activité collective et nouvelles technologies. *Rev. Int. Psychol. Soc.*, 4 (3/4), 335–56.

[31] Grosjean, M. (2000). Les communications collectives : un mode d'approche des compétences du collectif. Exemple du collectif hospitalier. *Psychol. Trav. Organ.*, 6 (3–4), 103–30.

[32] Leplat, J. (2000). Compétences individuelles, compétences collectives. *Psychol. Trav. Organ.*, 6 (3–4), 47–73.

[33] Le Boterf, G. (2006). *Construire les compétences individuelles et collectives*. Les Éditions d'organisation.

[34] Chatigny, C. (2001). Les ressources de l'environnement: au cœur de la construction des savoirs professionnels en situation de travail et de la protection de la santé. *PISTES* 3 (2). www.unites.uqam.ca/pistes.

[35] Chatigny, C., Grenier, J., and Boisclair, A. (2006). Becoming and remaining a worker in a woman's shelter. In: (Pikaar, R.N., Koningsveld, E.A.P., Settels, P.J.M. (eds), *Meeting Diversity in Ergonomics*. Proceedings IEA 2006 Congress. ISSN 0003-6870, Elsevier.

[36] Fournier, P.S. and Montreuil, S. (2006). The activity analysis to improve training on truck driver's fatigue prevention. In: Pikaar, R.N., Koningsveld, E.A.P., Settels, P.J.M. (eds), *Meeting Diversity in Ergonomics*. Proceedings IEA 2006 Congress. ISSN 0003-6870, Elsevier.

[37] Jackson, J.M. and Dischinger, M. (2006). Ergonomic work analysis as a tool for reflective practice. In: Pikaar, R.N., Koningsveld, E.A.P., Settels, P.J.M. *Meeting Diversity in Ergonomics*. Proceedings IEA 2006 Congress. ISSN 0003-6870, Elsevier.

[38] Barros Duarte, C. and Lacomblez, M. (2006). Intervention in health at work: Contributions towards a methodology centred on collective reflection. In: Pikaar, R.N., Koningsveld, E.A.P., Settels, P.J.M. (eds), *Meeting Diversity in Ergonomics*. Proceedings IEA 2006 Congress. ISSN 0003-6870, Elsevier.

[39] Zara-Meylan, V. and Cau-Bareille, D. (2006). Can risk management be transmitted to novices? In: Pikaar, R.N., Koningsveld, E.A.P., Settels, P.J.M. (eds), *Meeting Diversity in Ergonomics*. Proceedings IEA 2006 Congress. ISSN 0003-6870, Elsevier.

[40] Sznelwar, L., Lacman, S., and Uchida, S. (2006). Working at streets as public agents – how to learn and create a profession. In: Pikaar, R.N., Koningsveld, E.A.P., Settels, P.J.M. (eds), *Meeting Diversity in Ergonomics*. Proceedings IEA 2006 Congress. ISSN 0003-6870, Elsevier.

[41] Alvarez, D. Figueiredo, M., and Athayde, M. (2006). The activity viewpoint and the limits of the health, safety and environment (HSE) program in the Brazilian offshore oil industry. In: Pikaar, R.N., Koningsveld, E.A.P., Settels, P.J.M. (eds), *Meeting Diversity in Ergonomics*. Proceedings IEA 2006 Congress. ISSN 0003-6870, Elsevier.

[42] Castillo, J. (2006). The role of ergonomics in determining prevention for work systems in Colombia. In: Pikaar, R.N., Koningsveld, E.A.P., Settels, P.J.M. (eds), *Meeting Diversity in Ergonomics*. Proceedings IEA 2006 Congress. ISSN 0003-6870, Elsevier.

[43] Mengolini, A. and Vautier, J.-F. (2006). How to improve safety using a micro–macro ergonomic approach. In: Pikaar, R.N., Koningsveld, E.A.P., Settels, P.J.M. (eds), *Meeting Diversity in Ergonomics*. Proceedings IEA 2006 Congress. ISSN 0003-6870, Elsevier.

[44] Perrow, C. (1999). *Normal Accidents. Living with High-Risk Technologies*. Princeton, NJ: Princeton University Press.

[45] Garrigou, A., Baldi, I., and Rougetet, L. (2006). Developing a training module against the risks associated with the use of phytosanitary substances in the wine-growing industry. In: Pikaar, R.N., Koningsveld, E.A.P., Settels, P.J.M. (eds), *Meeting Diversity in Ergonomics*. Proceedings IEA 2006 Congress. ISSN 0003-6870, Elsevier.

[46] Brito, J., Athayde, M., Neves, M.Y., and Muniz, H.P. (2006). Training, research and intervention relating to work and education workers' health. In: Pikaar, R.N., Koningsveld, E.A.P., Settels, P.J.M. (eds), *Meeting Diversity in Ergonomics*. Proceedings IEA 2006 Congress. ISSN 0003-6870, Elsevier.

[47] Chatigny and Balleux (2006). Instructors in the poultry industry: Interface players in a critical position. In: Pikaar, R.N., Koningsveld, E.A.P., Settels, P.J.M. (eds), *Meeting Diversity in Ergonomics*. Proceedings IEA 2006 Congress. ISSN 0003-6870, Elsevier.

[48] Re, A., Oddone, I., Andéol, M., and Igonet, G. (2006). A general physician-centred system for preventing environmental diseases. In: Pikaar, R.N., Koningsveld, E.A.P., Settels, P.J.M. (eds), *Meeting Diversity in Ergonomics*. Proceedings IEA 2006 Congress. ISSN 0003-6870, Elsevier.

[49] Pelegrin, B. (2006). Who are the beneficiaries of the Pre-Vocational Counselling Centres (CPO) and the Vocational Rehabilitation Centres (CRP). In: Pikaar, R.N., Koningsveld, E.A.P., Settels, P.J.M. (eds), *Meeting Diversity in Ergonomics*. Proceedings IEA 2006 Congress. ISSN 0003-6870, Elsevier.

[50] Gomide Vezzá, F.M. and Zilbovicius, M. (2006). Consultancy practice using ergonomic work analysis. In: Pikaar, R.N., Koningsveld, E.A.P., Settels, P.J.M. (eds), *Meeting Diversity in Ergonomics*. Proceedings IEA 2006 Congress. ISSN 0003-6870, Elsevier.

[51] Bédard, A. and Berthelette, D. (2002). L'évaluation d'intervention: un défi à relever. *Effectif*, 5.

[52] Berthelette, D., Desnoyers, L., Gourde, D., and Authier, M. (1998). Évaluation de l'implantation d'un programme de formation en santé et en sécurité du travail. *Perform. Hum. Tech.*, 96, 21–26.

[53] Berthelette, D., Leduc, N., Bilodeau, H., et al. (2006). From underlying theory to program delivery: The safe patient transfer training program. In: Pikaar, R.N., Koningsveld, E.A.P., Settels, P.J.M. (eds), *Meeting Diversity in Ergonomics*. Proceedings IEA 2006 Congress. ISSN 0003-6870, Elsevier.

[54] Cau-Bareille, D., Delgoulet, C., and Gaudart, C. (2006). When learning difficulties and specificities of older workers stand to show training deficiencies. In: Pikaar, R.N., Koningsveld, E.A.P., Settels, P.J.M. (eds), *Meeting Diversity in Ergonomics*. Proceedings IEA 2006 Congress. ISSN 0003-6870, Elsevier.

[55] Santos, M. and Lacomblez, M. (2006). Assessing a training course: Between legitimacy and its basis. In: Pikaar, R.N., Koningsveld, E.A.P., Settels, P.J.M. (eds), *Meeting Diversity in Ergonomics*. Proceedings IEA 2006 Congress. ISSN 0003-6870, Elsevier.

[56] Leduc, N., Berthelette, D., Bilodeau, H., et al. (2006). Analysis of an ergonomic training program theory through four conceptual frameworks. In: Pikaar, R.N., Koningsveld, E.A.P., Settels, P.J.M. (eds), *Meeting Diversity in Ergonomics*. Proceedings IEA 2006 Congress. ISSN 0003-6870, Elsevier.

[57] Bloom, B.J., Engelhart, D.R., Furst, E.J., et al. (1969). *Taxonomie des objectifs pédagogiques*, t.1. : Domaine cognitif. Montréal: Les Presses de l'Université de Québec.

[58] Weston, C. and Cranton, P.A. (1986). Selecting instructional strategies. *J. Higher Educ.*, 57 (30), 259–88.

[59] Organisation mondiale de la santé, Santé et bien-être social Canada, Association canadienne de santé publique (1986). Charte d'Ottawa pour la promotion de la santé. *Can. J. Public Health*, 77, 425–30.

[60] Bandura, A. (1986). *Social Foundations of Thought and Action: A Social Cognitive Theory*. New Jersey: Prentice-Hall.

[61] Rabardel, P. (1995). *Les hommes et les technologies: approche cognitive des instruments contemporains*. Paris: Armand Colin.

Part III
Comfortable Products

Meeting Diversity in Ergonomics (2007)
Pikaar, Koningsveld, and Settels (Eds)

9

Ergonomics in Design: Accomplishments, Directions and Future Challenges

Pierre-Henri Dejean

Université de Technologie de Compiègne, Compiègne, France; IEA Ergonomics in Design Technical Committee

Lina Bonapace

Ergosolutions, Consultants in Ergonomics for Design, Milan, Italy; IEA EQUID Ergonomics Quality in Design Standing Committee

Abstract. This text is based on a summary presentation given by Lina Bonapace at the closing session of the triennial IEA World Congress 2006. It deals with product design and its potential links to ergonomics, although these links are not as yet very well developed. The paper provides the fundamentals of product design and product ergonomics to help form a basis for co-operation. It then considers the possible contributions of a renewed form of ergonomics – including cultural factors, diversity, enjoyment – as well as the challenges to be met – globalization, safety and pluridisciplinarity; all culminating to form a complex and interesting field.

Keywords: Product ergonomics

ERGONOMICS, DESIGN AND PRODUCT: A FEW BACKGROUND CONCEPTS

Product ergonomics: a role to be defined

We live in a world full of products that we use in most of our everyday activities. From its very beginnings, ergonomics has always been closely associated with the creation of all forms of artefact, from the design of airplane cockpits, to tools and furniture, and finally the design of such interactive products as portable telephones. At the same time, ergonomics has been involved in many other areas in which design was not a priority. It was thus very rarely involved in the original planning of a product.

A new outlook on both ergonomics and product evolution appears to be necessary for ergonomists to move into new areas of activity. The world of products is both vast and changing quickly, with new technologies and commercial strategies as well as economic and social change. For ergonomics to play a role in this complex world, it must keep up with it. We will provide an overview of ergonomics' place and role in this world. To do so, we will first review the fundamentals of 'product' and then the new elements ergonomics must deal with if it is to remain relevant in the area of product design.

What is a product?

'Product' has become a generic term covering a wide range of quite disparate goods, from cleaning and food products to more technical wares. We also speak of banking products and insurance products, while a number of services are built around various products. We must therefore give a clear definition of the kind of products ergonomics could help improve. The ergonomist's vision is a first filter, linking the 'product/artefact' (i.e. any object made by man) to the activity for which it was made (Random dictionary). The second filter is the industrialist's vision linking it to design. Deforge [1] divided the world of objects into two: unique objects (such as works of art) and products. A product often involves the multiplication of identical samples of the same object, resulting from industrial mass production. As we go from unique objects to products, we go from artist and craftsman to wage owner, and from autonomous creativity to imposed means of production and the separation of roles and responsibilities. The main feature of an industrial product is that it is designed by one group – engineers and designers – but actually manufactured by another – engineers and workers. A third group, made up of advertisers and retailers, handle its sale, and finally it is bought and used by a fourth group – the consumers. The design of a product must therefore solve all the problems that could arise during and after its manufacturing before that manufacturing has actually begun. The risks are important enough to justify preliminary studies and thus anticipate the many problems that could arise. We will give here a brief overview of the problems involved. For manufacturers and retailers, cancelled orders and an avalanche of after-sales service calls can become catastrophic in terms of their public reputation and the impact on costs and the organization itself. Defective products also have negative consequences for consumers, both in terms of money and well-being; there is waste, loss of time and temper, and inefficiency as well as a loss of motivation for the task involved. These consequences are even worse if we consider the number of domestic accidents and their human and economic costs for society. It is interesting to note that in affluent societies there are more domestic accidents or car wrecks than accidents at work, where ergonomics is involved.

Economic and social issues

This first overview brings out two fundamental points. We must react to this situation by giving ergonomics its rightful place in product design. It is a social and economic need. This is all the more important because there has been a general trend towards making citizens, manufacturers and business persons more aware of the consequences of their actions. Fair business practices, along with social and economic justice, are signs of an assessment of the responsibility of manufacturers and business persons in terms of sustainable development and company policy. This reaction is all the more interesting because it has happened independently of the state. Ergonomics has to find its place in this movement. That is how we view the efforts of the International Ergonomics Association (IEA; www.iea.cc) with the EQUID initiative.[1] If the usefulness and legitimacy of ergonomics in product design is thus established, what criteria and which people should we take into account, and how can all these factors be integrated in a process involving tools that are compatible with the constraints of industrial design: quality, cost and time frames?

[1] This IEA EQUID (Ergonomics Quality in Design) initiative is to help the public make more informed decisions about the ergonomic quality of products and to promote the integration of ergonomics into the design process. Through collaboration with stakeholders involved in the design development process, and considering the role of the ergonomics/human factors profession, this project hopes to promote awareness, guidance and recognition of ergonomics in design.

The purpose of product ergonomics

Ergonomics defines criteria that will enable assessment of the acceptability, or otherwise, of a product.

Changes in the world indicate which fundamental criteria remain operational and which new ones should be introduced. Following this discussion, we will return to the changes these criteria bring to the overall knowledge of our subject.

Safety

This is the only criterion with priority over the others. All product components are involved, from physical aspects such as shape, weight and dimensions, to the interactional effects of the product on the consumer, and the idea he or she has of it, including cognitive and cultural implications. Two levels of safety are implied, just as in industrial ergonomics (workplace and work environments, etc.): the short term, with accidents and therefore risk prevention, and the long term with illnesses that could occur when the product is used. Here again, exposure and length or repetition of use are crucial, but estimates are imprecise, and difficult to verify. We must therefore consider the cumulative effect of repercussions from one product to another, and also consider age groups that are much broader than among working people. Safety should be considered as preventive. It can then become beneficial beyond the mere suppressing of risks. Scandinavians have led the way in this regard by producing seats that not only avoid backache but also take pressure off the spine, and in sustainable development in the housing area, with residences designed to produce more energy than they consume. Proactive design is therefore a trend that is developing well beyond our subject of interest.

Efficiency

Efficiency and its corollary, product usefulness, bring us back to basic notions:

1. Service quality: does it work as it was expected to, or even better? Does it go beyond consumer expectations such as in the advertisement 'I dreamt it, Sony made it'?
2. The quality of implementation required to attain the purpose – usability.

In both cases, ergonomics can help through knowledge of the precise needs of users and of their skills: this requires close co-operation with marketing.

The variability factor: dealing with individuals and use contexts

A classic anthropometric factor, variability means a broadening of the field due to an aging population and greater attention given to the disabled. This growing care for populations' physical diversity is shown in the development of universal design [2], a notion shared by both design and ergonomics. Variability deals with cognitive styles and cultural differences that lead different people to adopt different attitudes and behaviours when faced with the same product. This leads to two ways of improving design: a better understanding of differences and the integration of differences and their consequences through a greater tolerance of errors. Accounting for the most probable and most serious errors is crucial for efficient use.

Comfort

This is not an easy notion to define in an absolute sense and it is usually measured by comparing it to discomfort. A comparative study with the other criteria reveals new insights and leads to anticipating possible conflicts. An efficient and safe situation may not be the most comfortable, and what is comfortable in the short term may prove to be quite uncomfortable over a longer period. In car manufacturing, for instance, we find two different approaches to comfort: hard comfort that lets the driver feel every reaction of the car to the state and nature of the road, and soft comfort that softens or cancels these reactions. The first approach supposes a sporting way of driving, keeping the driver in permanent awareness of the road and in a state of greater vigilance, while the second supposes a calmer, slower driving style. To give another example: a deep enveloping armchair may at first give a great sensation of comfort, while prolonged use may lead to problems of blood circulation, articular pains and difficulty when changing positions. All this eliminates the feeling of well-being without the user suspecting that his problems were caused by what appeared to be a comfortable chair.

Enjoyment

Jordan (1999) introduced pleasure as a criterion to assess. According to Jordan, enjoyment touches on several aspects: physiological, sociological, psychological or ideological, and interacts with many other criteria. But it must also be linked to more basic notions such as the user's experience, or to more recent ones such as emotional design and well-being. Branaghan et al. [3] summarizes the effects of a pleasurable experience on the user as a lasting and wholly satisfactory sensation in thought, feeling and emotions that one keeps in mind and talks about with pleasure.

PROVIDING THE 'PEOPLE PUSH' IN DESIGN

Speaking up for users

The above criteria tend to justify the intervention of ergonomics in product design and define its role among the other intervening disciplines. This role goes beyond simply keeping in touch with users. That remains important, but it is not a role that belongs exclusively to ergonomics. A lawyer, for instance, must analyse, understand, advise and even go beyond what a client has to say in order to undertake his defence. He ends up teaching his client new concepts. The ergonomist should adopt a similar technique for product design. He must teach users new attitudes, new ways of living, new ideas. We must speak up for users.

Promote new consumer habits

Sustainable development teaches us the necessity not only of working on the composition of a product but also of showing consumers how to adapt their behaviour to a new situation. This adaptation will come much more quickly if the product itself and its environment (packaging, advertising) help. Ergonomics thus acquires another role, that of briefing the designer to ensure that the objects designed will lead users to adopt an ecologic attitude 'as naturally or intuitively as possible'. This can be achieved by capturing users' needs and requirements as a basis of successful products.

DESIGNING FOR DIVERSITY: ACQUIRING NEW KNOWLEDGE

Broadening the definition of user

We shall start with immediate users, those who are in personal contact with the product at some point in its existence (lifecycle). Besides customers and consumers, there are employees of the manufacturer, retail business, and maintenance or reprocessing companies. Obviously their interests are slightly different from those of users, although they all share the same interest in the product's success. For the first group, this means that their needs are met; for the second group, that they will keep their jobs. But we can also talk of indirect users. In this regard, a Chinese proverb aimed at architects states: 'A house is not built just for those who live in it but also for those who look at it'. Indirect users have no voluntary links to the product, but they either enjoy or endure its existence. To determine who these indirect users are, all we have to do is follow the lifecycle of the product. An indirect user may live in a house close to the road and suffer the irritation caused by traffic, noise and pollution. The innocent bystander can very well admire the product and feel a positive emotion which helps his well-being and the popularity of the product, the manufacturer and the country in which it was made. Watching beautiful cars go by, appreciating the elegance of a dress or an accessory, admiring the ingenuity, ease of use and finish of a mechanical component are all part of the small pleasures of life. But the indirect user may also have a negative feeling about the same product: enduring the noise and smell of the traffic, for instance. The product can then bring on bad moods or even depression, not to mention the less frequent but more serious negative effects, such as accidents. Hence the necessity for car designers to address not only the driver's and passengers' safety (direct users) but also that of the pedestrian who might become a quite unwilling indirect user!

Consider cultural differences

The problem caused by cultural factors stems for a large part from globalization. How can a company manufacture products that are attractive to such a wide range of people with such a great variety of cultural backgrounds? Culture, after all, can exert considerable influence on product design and acceptance. Conceptors (designers, engineers) were at a loss in this regard. The business world for its part took a variety of empirical measures. It tried to have products designed by teams from the targeted market – the automobile industry, multinational makers of electrical appliances – and created design centres around the world. Ergonomists became aware of this phenomenon and brought important contributions through the increasing number of special sessions held during major events (Seoul in 2003, Las Vegas in 2005, Maastricht in 2006). Much coordination and dissemination work remains to be done and this will become part of the objectives of the IEA Ergonomics in Design Technical Committee.[2]

Emotion and reason as consumer motivators

Until recently, ergonomics considered only negative emotional situations – urgency, extreme stress – in which emotion overtook reason. The unforeseeable and uncontrollable nature of these situations were a nightmare for the manufacturer who was used to rationality, regulation,

[2] The main goals of the IEA TC on Ergonomics in Design are: to improve and increase ergonomics knowledge for application in design; to improve communication and transfer of knowledge between the fields of ergonomics for workers and ergonomics for consumers (from avoiding pain to gaining pleasure); to improve knowledge transfer from ergonomics in design research; to practise and to increase feedback from the applications to the academic community; to improve the relationship between ergonomists and other actors involved in design: product design, interaction design, engineering, marketing, management, sustainability, etc.; to promote ergonomics in the design process; to promote ergonomics in design to the public at large.

control of the uncertain and the reliability of forecasts. It was largely to avoid, or at least try to control, these alarming situations that industries turned to ergonomics without concerning themselves too much with the means to be used (simulators, the setting up of experiments). The world of products and services is more focused on positive emotions. Can ergonomists go along with this and do they have a part to play in these situations? When designers speak of emotion they mean the feeling that links the individual to the product: will he like it? Will he like it just a little, a lot, passionately, not at all? For those involved in marketing and sales, it is a positive attraction that will bring the individual to buy the product [4]. So where do ergonomists fit it? The ENGAGE project, associating ergonomists and the other professionals involved in the product (Porter, [5]; ENGAGE web site: www.designandemotion.org) has allowed this approach to move forward. As far as positive emotion is concerned, ergonomists must learn and contribute. There is no doubt that it is important for a product to be emotionally attractive, particularly if it provides a quality service and represents a break with what already exists. All those involved in safety know that the use of many forms of security equipment, such as harnesses, safety hats and boots, is harder to impose if the user feels the least bit ridiculous or embarrassed by wearing them. But ergonomists also know that using the product has an impact on its emotional value. The service it provides and how it is provided can have a positive or negative impact. The fact that a product is used can in itself tend to reconcile the emotional and rational approaches, or at least shed light on why a product is liked or not.

THE CHALLENGES TO BE MET

Evolving context for ergonomics in design

Over the last ten years, the arrival of a number of new significant elements has modified the context in which product design has traditionally evolved and to which it is still not totally adapted. Ergonomics must become involved in this movement. One of the aims of IEA Technical Committee is to point out these issues and start finding answers by making existing knowledge accessible to a wider public. We are referring here to globalization, high technologies, information technologies, trade and consumer regulations and rights.

These are emerging issues.

Product safety sheds light on the social stake in product design and how industry trails labour in this area. Accidents are more frequent outside the work place, during domestic activities and travelling, than they are at work. The resulting human dramas also have considerable economic consequences on the finance of individuals, insurance companies and the community. The fact that these costs are shared among these various entities results in their being widely underestimated, and these same entities try to avoid them rather than to eliminate their cause. If certification and consumer information are making headway, the suggestion and establishment of design tools is still in its infancy and lags behind what is done in professional risk prevention. A glance at other areas of ergonomics can indicate what can be transferred from the world of work to the general public, as well as the limits in this regard. This situation will evolve rapidly and the courts will soon become involved, and the arrival of notions such as the precautionary principle will push things forward.

The integration of cultural diversity in product design, although at first less evident, is now progressing rapidly and is becoming an important issue for ergonomics. We must now move from research that gathered appropriate information to the practical application of that information, and particularly to the development of tools and methods. Product ergonomics appears ahead of general ergonomics in this regard, and as was the case for enjoyment, some transfers seem relevant – designing work situations that integrate enjoyment and cultural diversity in the same way as we design products.

Adopting innovative language and attitudes

These new contexts form part of the challenges that ripe markets impose on marketing, in which the needs expressed are generally met and where growth is assured by more rapid product replacement (repurchasing). This is particularly true for markets which have reached the point of saturation, such as the automobile market, and for those that are close to reaching it, such as the portable telephone market. To counter this, companies have come up with two strategies: the more immediate one consists of programmed obsolescence of the product through a deficiency in one of its components – something that cannot be repaired. This is denounced by consumer associations and goes against the fight for quality and the undercurrent of sustainable development. The second attitude consists of offering new services, significantly better performances, or such enjoyment that it alone will justify replacement, to the point where the word replacement itself is no longer appropriate. It is in this context that designing products using only as reference those produced by the competition becomes obsolete. More precisely, in the trend towards improvement in service provision, it is obvious that technology is central to the hopes of product promoters and to innovation, although it must be used wisely. Improved performance and the dramatic drop in costs in the field of electronics have revolutionized the product world by the massive invasion of almost all products by electronic components. The success of what is basically a technological fad has somewhat diminished, leading project managers to act a little more carefully and the retailers of electronic components to be a little more modest. Meanwhile, among other instances of costly failures, the first wave of domotics lasted not much longer. Features that are not strictly necessary although not superfluous, or the difficulty of access to these features, or both, bring us back to the two fundamentals of product and quality of service, to which should be added the quality of the manner of service provision. In terms of methods, tools and knowledge, we must return to value analysis and to product and interface ergonomics. Value analysis[3] and ergonomics, in different but complementary ways, help identify and prioritize the features a product must offer for the service to be acceptable. The service the potential user is expecting is qualified by features that will be assessed in terms of satisfaction. Interface ergonomics helps adjust the way the service is provided and therefore the appropriate way of assuring each feature. These contributions are meant to lessen the effect of fads or technological emotion and thus produce real innovations, those accepted by the consumer and thus recognized in the marketplace. Only through such interaction between technology and social science can logical choices be made amid the abundance of technological supply – a relatively new phenomenon but certainly one that will last, having led the designer of a research mechanism from a world where all was lacking, or at least where choice was limited, to one where supply seems endless. This leads to in-depth work and technological monitoring, along with monitoring of the uses that ergonomics can introduce.

[3] Value analysis is a industrial method currently used in engineering and design. Its main goal is to identify, define and assess the product functions to be offered to the user. The main characteristic of this method is that the focus is on the functions and not the existing solutions. Ergonomics can help to assess the usability of these functions. (Techniques of value analysis and engineering. Lawrence D. Miles, www.wisc.edu/wendt/miles/milesbook.html).

Reinforcing practices and knowledge

Although computer-assisted design has grown in importance, there are still numerous fields to investigate, and the tool will not replace the knowledge it assists. Virtual reality and rapid prototyping provide new ways of validating projects as they proceed. The cost of these tools and the difficulties they present in terms of use and development are such that much still has to be done for average to small projects. Measuring products always requires more anthropometric and biomechanical knowledge. This basic product design task has taken on greater importance with an aging population and the development of musculoskeletal disorders. Materials are chosen according to their mechanical possibilities, and typical engineering work must now involve ecology, and quite often sensorial analysis. We will close this part with a mention of consumer tests, something that has almost become a constant and that must be taken into account by designers. Consumerism uses ergonomics but actually goes beyond it. Here again, the role of ergonomics is not exclusive. It must prove its skills and deal with other disciplines such as technology, law and social science.

Application of activities throughout the design process

Basic ergonomics knowledge has been available for quite some time to improve product design. We have seen how it might be achieved, but we must also consider the problem of its dissemination, and even more importantly of its availability to designers. Indeed, the designers' work has evolved considerably in terms of the tools used, tools that have an impact on the operating mode and even on the cognitive mechanisms that are part of the design process. We are now faced with an irreversible move towards instantaneous demands for available information in the workplace. In this regard, the Internet, through search engines, has become a focal point. We must now learn to manage this modus operandi to ensure that it provides greater coherence in research and data management and avoids dispersions or dead-ends, either of which can jeopardize the quality of the designed product.

To these general principles we must add implementation and follow-up techniques, the latter going beyond the public launch of the product, with the introduction of experience feedback and design traceability which help link the consequences noted in the use of the product to the cause in the design. Quality contributes to bringing coherence and also acts as an integration vector for new disciplines in industry, including design and ergonomics. Therefore quality should be perceived as assistance rather than as a constraint.

Ergonomics directed towards pluridisciplinarity

Few works about design [6] deal with the various trades with which the ergonomist must be familiar in order to assist in product design. Knowledge of these different skills does not mean taking the place of the stakeholders, but rather understanding what each is required to produce. This knowledge of trades shows that the role and the tasks of each person go well beyond what is generally understood. Ergonomics is not limited to measuring objects but must also act on the functions that the product must ensure. The role of design is not limited by the stroke of a pencil. Design must intervene in the structure of the product, in how it will be made. First, all trades do not understand one another, as in the case of manufacturing or of activities which always deal with the same objects, such as construction or the manufacturing industry. The result of this is that the identification of the roles of each as seen by the others can differ, and this can lead to tension. Arbitration will therefore often be necessary at least on two points: firstly, competence as to who does what, and secondly as to means and time.

Regarding the information we can provide on our discipline and the promotion that still has to be done, we hope that we have provided the elements allowing arbitration to be carried out in a way that is neither arbitrary nor draconian.

We have already seen that arbitration is necessary regarding tasks to be performed. Product design has grown considerably in size and complexity. It has therefore become increasingly necessary to set limits to a design project otherwise it will become unrealistic. This kind of work involves project management techniques which will help to determine the feasibility of a project of sufficient quality, within reasonable time frames and costs, and within the means of the company and the market context. It also requires the strategic involvement of the company, and each of the stakeholders must strive to inform that company in this regard. If these requirements are not met, ergonomics will only play a minor role in product design. So what should be done?

Moving from knowledge to tools

To respond within acceptable time frames and costs, particularly in the case of projects of medium importance, the professional ergonomist must have performing tools and data. Prompt responses, availability, and data reliability are what designers appreciate the most. That is also one of the reasons they call for standards, the limits of which we are very much aware. One of the challenges of ergonomics is to meet that demand. Much of the knowledge is there, but its transformation into sufficiently autonomous tools not requiring the systematic assistance of a specialist is still quite insufficient. It is therefore in that area that efforts must be made. The success of a program like EQUID will only be assured if, besides the overall method leading to certification, there is also a choice of performing tools to implement it. This opens a vast area of research development prior to study bureaus but directed towards them. It is hoped that the tools developed will be so well-performing and so easy to use that they will become products.

CONCLUSION

As explained above, if product ergonomics is to find its proper place and play its role in sustainable development, it will have to increase its awareness of new problems as well as its knowledge of methodology and application tools.

In the introduction, we mentioned the complexity of product design. It is a fact of life with which the ergonomist must deal. On the other hand, we would find it regrettable if very real complexity was considered as nothing more than a negative element, when in fact it is rich in the extraordinary variety of knowledge it reveals about the product and the great amount of new knowledge that it helps ergonomics to find and develop. All those who have tried to deal with the issue or were simply interested in it quickly understood that this complexity is anything but dull . . .

REFERENCES

[1] Deforge, Y. (1990). *L'oeuvre et le Produit*. Paris: Maloine.

[2] Dong, H., Bobjer, O., McBride, P., and Clarkson, P.J. (2006). Inclusive product design: Industrial case studies from the UK and Sweden. *Contemporary Ergonomics*. New York: Taylor & Francis.

[3] Branaghan, R.J., et al. (eds) (2001). Design by people for people. *Essays on Usability*. Chicago, IL: Usability Professionals Association.

[4] Khalid, H. (2006). Embracing diversity in user needs for affective design. *Appl. Ergon., Special Issue: Meeting Diversity in Ergonomics*, 37 (4).
[5] Porter, C. (2006). ENGAGE: Designing for emotion. In: Pikaar, R.N., Koningsveld, E.A.P., Settels, P.J.M. (eds), *Meeting Diversity in Ergonomics*. Proceedings IEA 2006 Congress. ISSN 0003-6870, Elsevier.
[6] Quarante, D. (2001). Eléments de Design Industriel Politechnica 3 eme edition, Paris.
[7] Green, W.S. and Jordan, P.W. (2002). *Pleasure with Products, Beyond Usability*. New York: Taylor & Francis.

Meeting Diversity in Ergonomics (2007)
Pikaar, Koningsveld, and Settels (Eds)

10

Applying Axiomatic Method to Icon Design for Process Control Displays

Sheau-Farn Max Liang

Department of Industrial Engineering & Management, National Taipei University of Technology, Taipei 106, Taiwan, ROC

Abstract. Graphical User Interfaces (GUIs) have been extensively applied to the user interfaces of almost every computer system. An important feature of GUIs is the icon. While it is not difficult to find relevant information for the design of icons from handbooks, industry standards or guidelines, it is surprising to find many inconsistencies in the information. Hence, it is necessary to find a new and improved method for the design of icons. An axiomatic method was applied to icon design. This method was based on the two axioms of the Axiomatic Design (AD) principles – The Independence Axiom and the Information Axiom. From the viewpoints of semiotics and human information processing, visual distinctiveness and the appropriateness of representation are two key factors in icon design. It was proposed that the discriminability of icons could be analyzed through the Independence Axiom, and the meaningfulness of icons could be evaluated through the Information Axiom. From a previous study, a set of icons used in a Distributed Control System (DCS) product for ASEAN market was reviewed. The review was used as an example to show how the axiomatic method could be applied as a framework for the design of icons in process control displays.

Keywords: Axiomatic Design, Icon Design, Process Control Displays

INTRODUCTION

Icons have become an important feature of current widespread GUIs for almost every interactive computer system due to their aesthetic attractiveness, possibility of rapid recognition, and international potential. At the same time, evidence of confusion and misinterpretation about icons suggests that their design is not as simple as replacing text with graphics. In the following sections, issues about icon design are described from the viewpoints of semiotics and human information processing. Research on cognitive characteristics of icons, and relevant design guidelines and standards are also reviewed.

Semiotics of icons

Semiotics is a field of study concerned with the making and representation of meanings in everything that can be taken as a sign [1]. In a semiotic sense, an icon is a form to represent its referent and to let the interpreter make sense of this representation. In Peirce's terms, the *Representamen* is the form that the sign takes, the *Interpretant* is the sense made of the sign, and the *Object* is the referent of the sign [1]. This Peircean model is illustrated in Fig. 1.

There has been an effort to studying typologically the relationship between a *Representamen* and its *Object* – i.e., the representation. Peirce's famous typology of signs [1] defined a sign with three different modes as the following

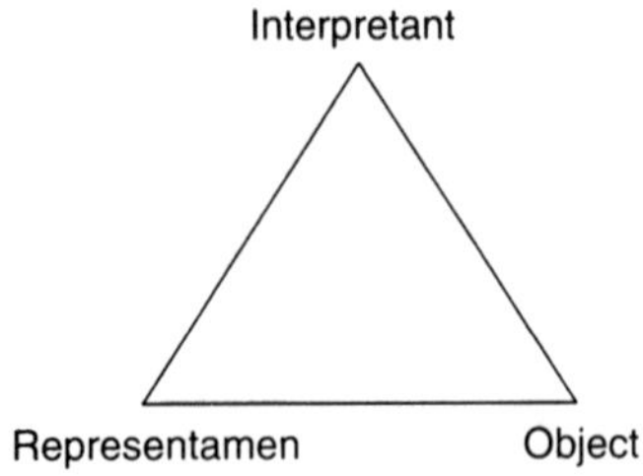

FIGURE 1. The Peircean Model

Symbolic mode

The representation is primarily based on the arbitrary or conventional relationship between a *Representamen* and its *Object* so that the relationship must be learned, such as texts to represent their meanings in any language or the color code in a traffic light.

Iconic mode

The representation is predominantly based on the resemblance between a *Representamen* and its *Object*, such as portraits or metaphors.

Indexical mode

The representation is mainly based on the direct connection between a *Representamen* and its *Object*, such as indexical words – e.g., that, this, here, there, etc., – or medical symptoms to represent their causes.

Note that these three modes are not mutually exclusive and they could coexist within a sign. Usually there is a dominant mode over the other two, and this dominance depends on the context of use [1]. For example, the human figures on a toilet sign illustrated in Fig. 2 are in the iconic mode. Conventionally, the color code of blue for the male and red for the female is in the symbolic mode. Eventually, this sign represents neither a male nor a female, but dominantly indicates the location of a toilet as in the indexical mode.

The modes could also evolve with time, from the iconic mode to the symbolic mode, such as Chinese ideographic characters and ancient Egyptian hieroglyphs. Some examples of Chinese ideographic characters are illustrated in Fig. 3. Characters look more similar to their associated natural objects in their ancient forms than in current symbolic ones.

Under Peirce's semiotic triangle structure, an icon is defined as a *Representamen*; its referent is defined as an *Object*. For a user interface of a computer system, an icon could be a pictograph, symbol, abstract shape, or alphanumerical character, whereas its referent could be

FIGURE 2. A Toilet Sign with the Left Figure in Red and the Right in Blue

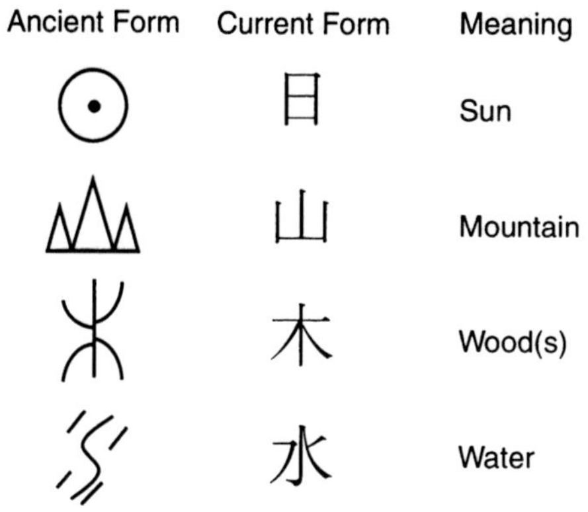

FIGURE 3. Examples of Chinese Ideographic Characters

an object, concept, status, or function. According to the Peircean model, the appropriateness of representation depends on the sense made of the icon, its referent, and their relationship.

Human information processing of icons

The recognition of icons is influenced by human visual stimulus processing – bottom-up processing – as well as by experience and knowledge of icons – top-down processing [2]. From a human information processing perspective, well-designed icons should be visually distinguishable from each other and activate the appropriate meanings in the mind of the interpreter [2, 3]. A survey confirmed that the discriminability and meaningfulness of icons had been identified by the designers as the most important factors for the design of icons [4]. The discriminability of icons is mostly affected by bottom-up processing and the meaningfulness of icons is primarily influenced by top-down processing.

Cognitive characteristics

Some cognitive characteristics about icons have been studied, such as abstractness, concreteness, visual complexity-simplicity, visual distinctiveness, meaningfulness, familiarity, and semantic or articulatory distance.

Contrary to abstract icons, it has been found that concrete icons tend to be more visually obvious since they represent more actual or physical reality, such as real objects, materials, or people [5, 6]. Simplicity has been endorsed as one of the important principles for icon design [e.g., 7, 8]. It has been reported that the amount of detail or intricacy within an icon affects search efficacy. The simpler the design is, the shorter the search time [9]. Visual distinctiveness of icons is related to physical conditions, such as visual acuity and contrast sensitivity, as well as perceptual conditions, such as discriminability between icons, and family resemblance among related icons [10]. The meaningfulness of an icon can be measured by its uniqueness, completeness, and clarity from subject's ratings [6, 10]. Familiarity could be a binary measure [10] or the frequency with which icons are encountered [6]. Semantic, or articulatory, distance is a measure of the closeness of the relationship between an icon and its referent [6]. The percentage of correct responses of the relationship could be used as the measure under the assumption of a linear relationship between the icon and its referent [11].

Design guidelines and standards

Current icon design practice is usually guided by design guidelines and standards. While it is not difficult to find relevant information for the design of icons for process control displays from handbooks, industry standards or guidelines [12, 13], it is surprising to find many inconsistencies in the information with respect to the use of colors, shapes and layout orientations. Hence, there are no universal standards or guidelines for the design of icons for process control displays.

The design guidelines from Marcus [14] suggest that icons can be distinguished by using large objects, bold lines, and simple areas, and designs can be evaluated by showing them to potential viewers. The design guidelines from the U.S. Nuclear Regulatory Commission [15] emphasized the importance of icon distinguishability and the appropriate use of icons. The design guidelines from Shneiderman [16] recommend making each icon distinctive from every other icon, and to represent the object or action in a familiar and recognizable manner. The ISO standard 11581–1 [17] recommended that all available icons should be comprehensible, learnable, and discriminable. According to the SEMI standard E95–1101 [18] the purpose of an icon is to show different types of objects, improve operability, and help the user better understand the functionality.

However, these guidelines and standards only point out 'what' is important to be achieved but lack a clear description about 'how' to achieve it. There is no systematic means for designers to analyze and evaluate the visual distinctiveness of icons and the appropriateness of representation. As a result, some critical design problems are only found and remedied after user-testing at the end of the design process. Design therefore becomes a trial-and-error process and tends to be time-consuming and costly. Hence, an axiomatic method is proposed to provide a systematic framework for the design of icons.

AXIOMATIC DESIGN

The AD approach [19, 20] is a tool for designers to construct and understand design problems, as well as to find possible solutions. AD has been widely applied in the designs of software applications, consumer products, manufacturing systems, and decision support systems [21]. There are four central concepts for the AD: Domains, Hierarchies, Zigzagging, and Design Axioms.

AD views the design process as a series of mappings between four domains: the customer domain, functional domain, physical domain, and process domain. The first mapping is from the Customer Attributes (CAs) in the customer domain to the Functional Requirements (FRs) in the functional domain. The second mapping is from the FRs in the functional domain to the Design Parameters (DPs) in the physical domain. The third and last mapping is from the DPs in the physical domain to the Process Variables (PVs) in the process domain [22]. The relationship is presented in Fig. 4.

The CAs, FRs, DPs, and PVs could be decomposed into a hierarchy within each correspondent domain. It is a part-whole relationship between the lower level and the higher level in the hierarchy. The mappings between two domains are through a zigzagging way from the top of the hierarchy in the preceding domain to the bottom of the hierarchy in the following domain. In Fig. 5, for example, zigzagging mappings between the functional domain and the physical domain start from the FR on the top of the FR hierarchy in the functional domain to the DP on the top of the DP hierarchy in the physical domain. The next mapping is from the DP to the second level of the FR hierarchy i.e., FR_1 and FR_2, and the subsequent mapping is

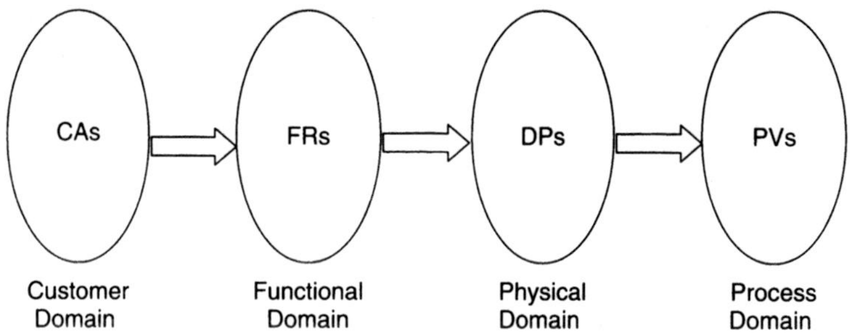

FIGURE 4. Four Domains of AD

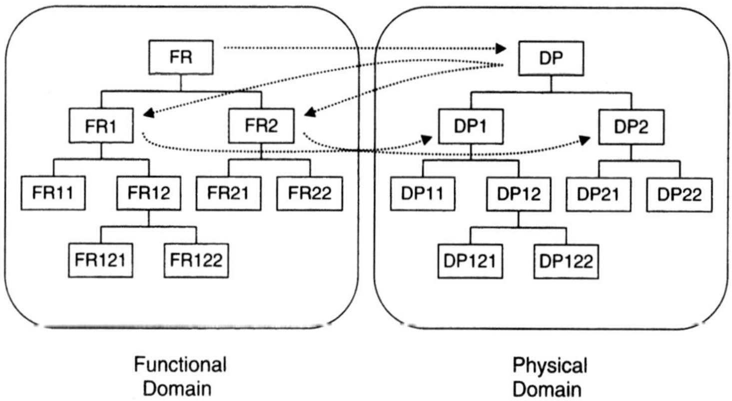

FIGURE 5. Zigzagging Mappings between Functional Domain and Physical Domain

from the second level of the FR hierarchy to the same level of the DP hierarchy i.e., DP_1 and DP_2. This procedure continues until the mapping reaches the bottom of the DP hierarchy [22].

The objective of the AD is to establish a scientific foundation for design activities by two axioms [22]:

Axiom 1

The Independence Axiom – Maintain the independence of FRs.

Axiom 2

The Information Axiom – Minimize the information content of the design.

Independence axiom

The mapping between the FRs in the functional domain and the DPs in the physical domain has been frequently used for the design and usually shown in a matrix format as below:

$$\{FR_n\} = [A]_{nm} \bullet \{DP_m\} \tag{1}$$

$\{FR_n\}$, the n-vector of FRs, represents a set of n design goals i.e., 'what' to be achieved. $\{DP_m\}$, the m-vector of DPs, represents a set of m design features to achieve the $\{FR_n\}$ i.e., 'how' to achieve them. $[A]_{nm}$, the design matrix, represents the mapping relationship between the $\{FR_n\}$ and $\{DP_m\}$. If we use a_{ij} ($i = 1, 2 \ldots n$, and $j = 1, 2 \ldots m$) to represent the elements in the design matrix $[A]_{nm}$, then

$$a_{ij} = \frac{\partial FR_i}{\partial DP_j} \tag{2}$$

Eqn. 2 shows that when the change of DP_j derives the change of FR_i, a_{ij} has a nonzero value. When the change of DP_j has no influence on the FR_i i.e., $\partial FR_i = 0$, a_{ij} equals to zero. This mapping relationship can then be simply represented by the binary values of elements in the design matrix: While the value of zero denotes no relationship between associated FR and DP, a cross ($\times$) stands for a nonzero value representing a non-negligible relationship between the FR and DP.

According to the mapping relationship, the design is categorized into three types: the uncoupled design, decoupled design, and coupled design. A 3×3 design matrix shown as Eqn. 3 is used to illustrate these three design types. Note that a_{ij} ($\mathrm{i, j} = 1, 2$, or 3) represents the elements in the design matrix.

$$\begin{Bmatrix} FR_1 \\ FR_2 \\ FR_3 \end{Bmatrix} = \begin{bmatrix} a_{11} & a_{12} & a_{13} \\ a_{21} & a_{22} & a_{23} \\ a_{31} & a_{32} & a_{33} \end{bmatrix} \begin{Bmatrix} DP_1 \\ DP_2 \\ DP_3 \end{Bmatrix} \tag{3}$$

When $a_{\mathrm{ij}} \neq 0$ for all $\mathrm{i} = \mathrm{j}$, and $a_{\mathrm{ij}} = 0$ otherwise, the design is an uncoupled design illustrated as Eqn. 4.

$$\begin{Bmatrix} FR_1 \\ FR_2 \\ FR_3 \end{Bmatrix} = \begin{bmatrix} a_{11} & 0 & 0 \\ 0 & a_{22} & 0 \\ 0 & 0 & a_{33} \end{bmatrix} \begin{Bmatrix} DP_1 \\ DP_2 \\ DP_3 \end{Bmatrix} \tag{4}$$

In Eqn. 4, $FR_1 = a_{11}DP_1$, $FR_2 = a_{22}DP_2$, and $FR_3 = a_{33}DP_3$. When $a_{\mathrm{ij}} \neq 0$ for all $i \geq j$, and $a_{\mathrm{ij}} = 0$ otherwise, the design is a decoupled design illustrated as Eqn. 5.

$$\begin{Bmatrix} FR_1 \\ FR_2 \\ FR_3 \end{Bmatrix} = \begin{bmatrix} a_{11} & 0 & 0 \\ a_{21} & a_{22} & 0 \\ a_{31} & a_{32} & a_{33} \end{bmatrix} \begin{Bmatrix} DP_1 \\ DP_2 \\ DP_3 \end{Bmatrix} \tag{5}$$

In Eqn. 5, $FR_1 = a_{11}DP_1$, $FR_2 = a_{21}DP_1 + a_{22}DP_2$, and $FR_3 = a_{31}DP_1 + a_{32}DP_2 + a_{33}DP_3$. If a design is neither an uncoupled design nor a decoupled design, then it is a coupled design. An example of a coupled design is illustrated as Eqn. 6.

$$\begin{Bmatrix} FR_1 \\ FR_2 \\ FR_3 \end{Bmatrix} = \begin{bmatrix} a_{11} & 0 & a_{13} \\ a_{21} & a_{22} & a_{23} \\ a_{31} & a_{32} & 0 \end{bmatrix} \begin{Bmatrix} DP_1 \\ DP_2 \\ DP_3 \end{Bmatrix} \tag{6}$$

In Eqn. 6, $FR_1 = a_{11}DP_1 + a_{13}DP_3$, $FR_2 = a_{21}DP_1 + a_{22}DP_2 + a_{23}DP_3$, and $FR_3 = a_{31}DP_1 + a_{32}DP_2$. Note that only the uncoupled design satisfies the Independence Axiom. That is, each FR can be achieved independently through the associated DP without affecting other FRs in the same set. For a decoupled design, there is a unique sequence to achieve the entire FRs one by one. For example, in Eqn. 5, the FR_1 has to be achieved first by adjusting the DP_1. Based on the adjusted DP_1, the FR_2 can only be achieved next by adjusting the DP_2. Finally, with the adjusted DP_1 and DP_2, the FR_3 can be achieved only by adjusting the DP_3. Other sequences would have to adjust more than one DP simultaneously for achieving one FR, which violates the Independence Axiom. For a coupled design, it is not possible to achieve any FR without adjusting more than one DP. Therefore, it also contravenes the Independence Axiom. Since a coupled design is quite common practice, especially for a large and complex design, the binary values of elements (a_{ij}) in the design matrix have been extended to values between 0 and 1 for measuring the strengths of coupling in order to find a possible quasi-uncoupled design [23].

The uncoupled design is claimed by the Independence Axiom as a better design than the other two types in terms of its simplicity, robustness, and ease of change [22].

Information axiom

Relevant to information theory [24], the Information Axiom indicates that the best design is the design with minimum information content, i.e. in statistical terms, the highest probability of success to achieve the design goals i.e., the FRs. For example, to minimize the information content for the design of a schematic display in process control rooms, the designer may use standardized symbols and lines in a consistent manner for representing the equipment and flow lines.

The information content is measured by its information amount. The information amount is defined as the probability of achieving a certain FR. For example, if the probability of achieving the FR_i is P_i, then its information content, I_i, is defined as Eqn. 7:

$$I_i = \log_2 \frac{1}{P_i} = -\log_2 P_i \tag{7}$$

If the probability of achieving the FR_i is one ($P_i = 1$), then the information content is zero ($I_i = 0$). It requires no information to achieve the FR_i. In contrast, if the value of P_i approaches to 0, then the value of I_i approaches to infinity. It needs more information to achieve the FR_i if the probability is small. For an uncoupled design, the overall information content, shown as Eqn. 8, is the summation of all information contents in the design.

$$I_{Overall} = \sum_{i=1}^{n} I_i = -\sum_{i=1}^{n} \log_2 P_i \tag{8}$$

The probability of achieving FR could be derived from the concepts of system range, design range, and common range illustrated in Fig. 6 [25].

In the context of manufacturing, the system range is defined as the capability of a system, and is presented by the Probability Density Function (PDF) of the system. The design range is the specification requirement of the design, and is presented as an interval of an FR. The common range is the overlap between the system range and the design range, and is presented as the overlapping area (A_{cr}). Since the area is the probability of achieving FR, the Eqn. 7 can then be revised as below:

$$I = \log_2 \frac{1}{A_{cr}} = \log_2 \left(\frac{System\ Range}{Common\ Range} \right) \tag{9}$$

In a study which applied the axiomatic method to an anthropometric design of a microscope workstation in semiconductor manufacturing [26], the information content was derived from

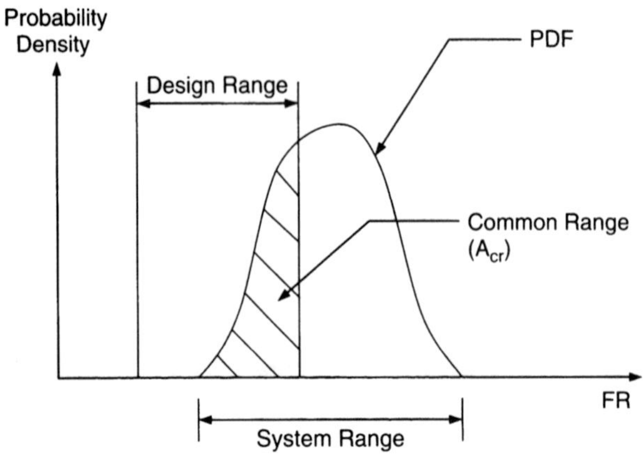

FIGURE 6. Relationship among System Range, Design Range, and Common Range [25]

the concepts of desired range, supplied range, and common range. For an ergonomic design, the desired range is usually set by the 5th–95th percentile anthropometric measures, and the supplied range is the range supplied by the manufacturer. The common range is the overlap between the desired range and the supplied range [26]. The information content is defined as Eqn. 10:

$$I = \log_2 \left(\frac{Desired\ Range}{Common\ Range} \right) \tag{10}$$

These two design axioms can be applied to the new design of products, manufacturing processes, or systems, as well as to the evaluation and improvement of existing designs. The procedure is first to eliminate any decoupled or coupled design by applying the Independence Axiom. If more than two alternatives still remain, the second step is to select the design with the minimum information content by applying the Information Axiom.

AXIOMATIC METHOD FOR ICON DESIGN

Discriminability of icons

Usually, an icon is not designed in isolation but along with other icons. For those icons related to each other, a certain level of family resemblance among them has to be assured in the design. On the other hand, an icon has to be distinguishable from other icons when they are displayed together. Confusion, the failure to discriminate, occurs when similar stimuli represent different concepts [2]. The perceived similarity between two icons is positively affected by the number of common features and negatively affected by the number of distinctive features [27]. The features of icons include pictograph, symbol, shape, color, or other visual elements. When an icon has more than one feature, the dimensional relations – integral or separable [28] have to be considered.

To analyze the discriminability of icons under the Independence Axiom, we can use $\{FR_n\}$ as a set of independent concepts or functions that icons represent, and $\{DP_m\}$ as a set of visual features. By examining the design matrix, $[A]_{nm}$, the discriminability of a set of icons

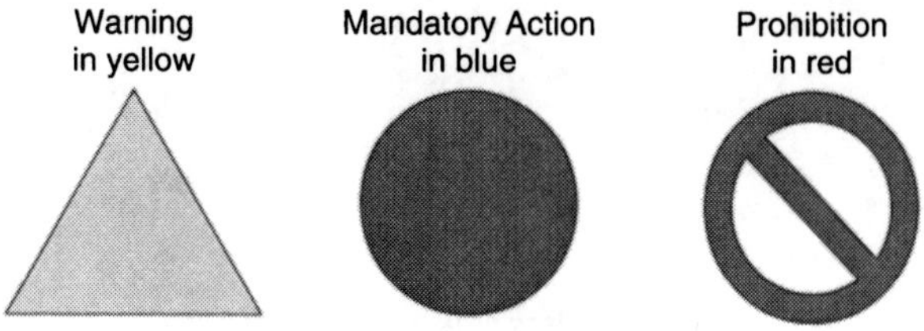

FIGURE 7. ISO-3864

can be realized. A good design is the design satisfying the Independence Axiom, that is, an uncoupled design.

As an example, shown in Fig. 7, in the ISO–3864 standard [29] a yellow triangle is used to represent the warning sign, a blue round shape is used to represent the sign of mandatory actions, and a round shape with a red circle and backslash is used to represent the prohibition sign.

If we consider the shapes and colors as the two separable visual features (DPs) and the three types of signs as independent referents (FRs), then the mapping relationship can be expressed as Eqn. 11

$$\begin{Bmatrix} \text{Warning} \\ \text{Mandatory} \\ \text{Prohibition} \end{Bmatrix} = \left[\begin{array}{cc|ccc} \times & 0 & \times & 0 & 0 \\ 0 & \times & 0 & \times & 0 \\ 0 & \times & 0 & 0 & \times \end{array}\right] \begin{Bmatrix} \text{Triangle} \\ \text{Round} \\ \hline \text{Yellow} \\ \text{Blue} \\ \text{Red} \end{Bmatrix} \tag{11}$$

Equation 11 shows that the three types of signs can not be discriminated from only the two different shapes, but they can be exactly discriminated from the three different colors. It points out that viewers might fail to discriminate a mandatory action sign from a prohibition one, for example in a dim environment.

Meaningfulness of Icons

Modified from Peirce's semiotic triangle [1], a dual triadic model shown in Fig. 8 was proposed to address issues regarding the meaningfulness of icons.

In Fig. 8, 'Icons' represents a set of icons, whereas 'Referents' denotes a set of referents. The interpreter is either the designer or the user. Designers know what referents should be represented in the design, and are familiar with the visual features to be used to design the icons – as the two solid lines in Fig. 8 indicate. However, designers' concerns about the meaningfulness of icons are affected by the appropriateness of representation, users' familiarity

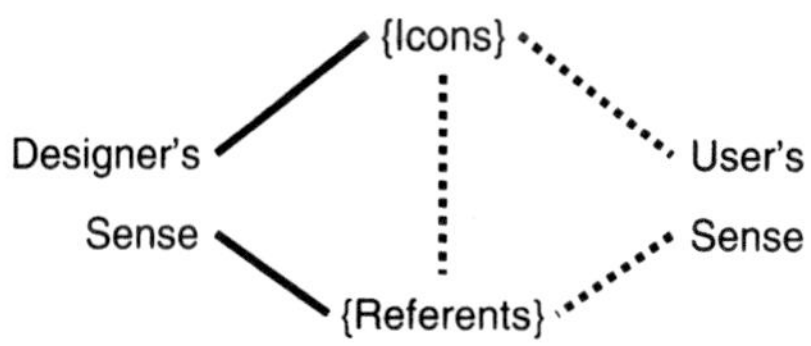

FIGURE 8. Dual Triadic Model for Icon Design

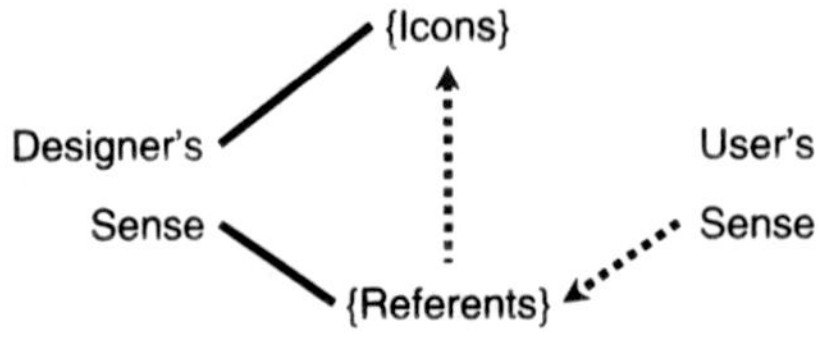

FIGURE 9. Comprehension of Referents

with the icons, and users' comprehensibility of the referents – as indicated by the three broken lines in Fig. 8.

For the design of interactive computer systems, the referents may be the systems' functions or status. These are domain-specific and task-dependent. Users have to first learn or comprehend these referents before they are able to make sense of associated icons. For example, to understand the icons displayed on an alarm summary screen in a process control console, the operator has to comprehend the alarm function in the system.

This process of comprehension is assumed to be from the referents to their representing icons as illustrated in Fig. 9.

On the other hand, familiarity with icons depends not only on the user's domain expertise but also on his or her cultural background, for example a color may symbolize different meanings in different cultures. This process of familiarity is assumed to be from the icons to their associated referents as presented in Fig. 10.

A cross-cultural issue about the design of icons is the design for internationalization and localization. Although there are several sources of reference on the design for internationalization [e.g., 30–33], the suggestions are typically more artistic from the perspectives of design and marketing, rather than systematic and data-driven. Some efforts have been made in Hong Kong [34], Singapore [35] and China [36] to investigate user interface designs related to the use of icons in process control displays and to suggest the use of relevant icons for internationalization, thereby improving the usability of process control displays.

As a user's interpretation of icons may be interfered with or facilitated by their cultural background, and because it is difficult to predict cultural influence on the user's interpretation of the icon, user testing is a common tool employed by designers to assess the appropriateness of the representation. It is suggested here that the information content, I_i, can be measured before user testing through the probability of successful associations between to-be-used visual features and their referent concepts. The best design can then be selected according to the Information Axiom.

For example, a shape-concept association survey conducted in Malaysia [36] revealed that from a set of shapes in Fig. 11, the shape Square and Circle were the two popular shapes to represent the Normal condition with 35% and 28%, respectively. For the Caution concept, the Triangle and Star were the top two choices with 45% and 16%, respectively.

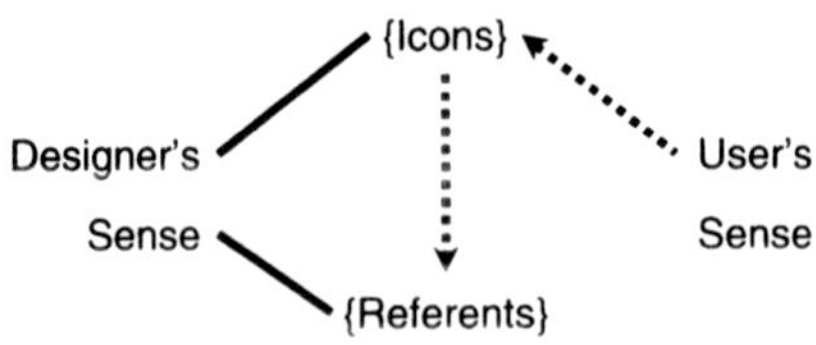

FIGURE 10. Familiarity with Icons

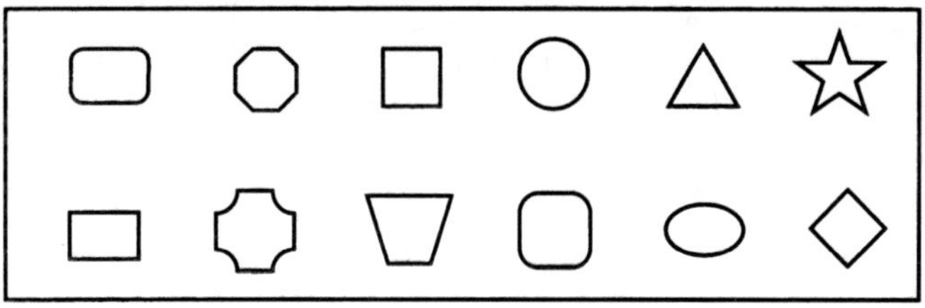

FIGURE 11. The Shapes for the Shape-Concept Association Survey

If we use the Square to represent the Normal condition, then its information content is

$$I_{Normal} = -\log_2 0.35 = 0.46 \tag{12}$$

However, if we use the Circle to represent the Normal condition, then its information content is

$$I_{Normal} = -\log_2 0.28 = 0.55 \tag{13}$$

In this case, the Square is superior to the Circle for representing the Normal condition due to its less information content. It can be further illustrated as in Fig. 12. There are several options – i.e., Shape 1–5 in Fig. 12 – that can be chosen as possible design solutions for representing the referent, but there may be only one option with the largest success percentage of representation as it has the least information content i.e., Shape 3 in Fig. 12.

A larger system usually has more referents to be represented, for example Normal and Caution conditions. One design uses the Square and Star to represent these two conditions, whereas an alternative design uses the Circle and Triangle. Since these two designs are both uncoupled designs, by applying Eqn. 8, the information content of the first design is

$$I_{1st} = -\log_2 0.35 - \log_2 0.16 = 1.25 \tag{14}$$

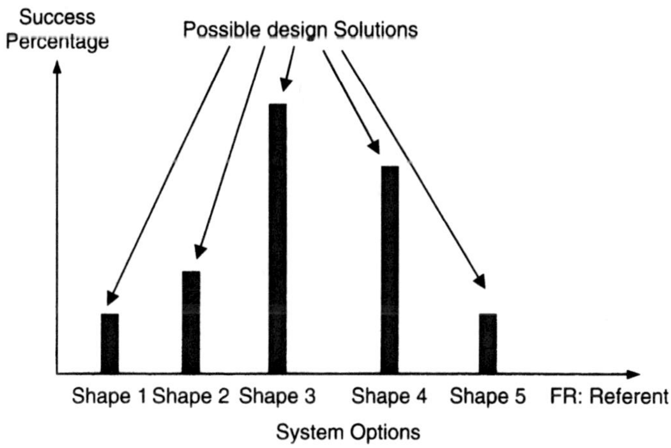

FIGURE 12. Relationship among System Options, Possible Design Solutions, and the Best Solution

The information content of the second design is

$$I_{2nd} = -\log_2 0.28 - \log_2 0.45 = 0.90 \tag{15}$$

Comparison between Eqn. 14 and Eqn. 15 shows that the second design, with less information content, is the better choice.

CASE: ICON DESIGN FOR ALARM SUMMARY DISPLAY

From a previous study [39], a set of icons used in a DCS product for the ASEAN market was reviewed and used as an example to show how the axiomatic method can be applied as a systematic framework for the design of icons in alarm summary operator interfaces.

An alarm is generated whenever an abnormal condition occurs. Alarms are typically associated with points – for example, the value of an analog point may be above or below the acceptable range. Alarms may also be generated when any important event occurs, such as a communication failure. By clicking the 'Alarm Summary' button on the tool bar of the system home page, the 'Alarm Summary' page would be shown on the screen. Typically, 12 alarms could be displayed simultaneously on a single screen. An example of an Alarm Summary page is shown in Fig. 13.

On the 'Alarm Summary' screen, there are four icons in the legend to represent four different conditions of alarm

- Acknowledged and in Alarm
- Unacknowledged and in Alarm
- Unacknowledged and Disabled
- Unacknowledged and Returned to Normal

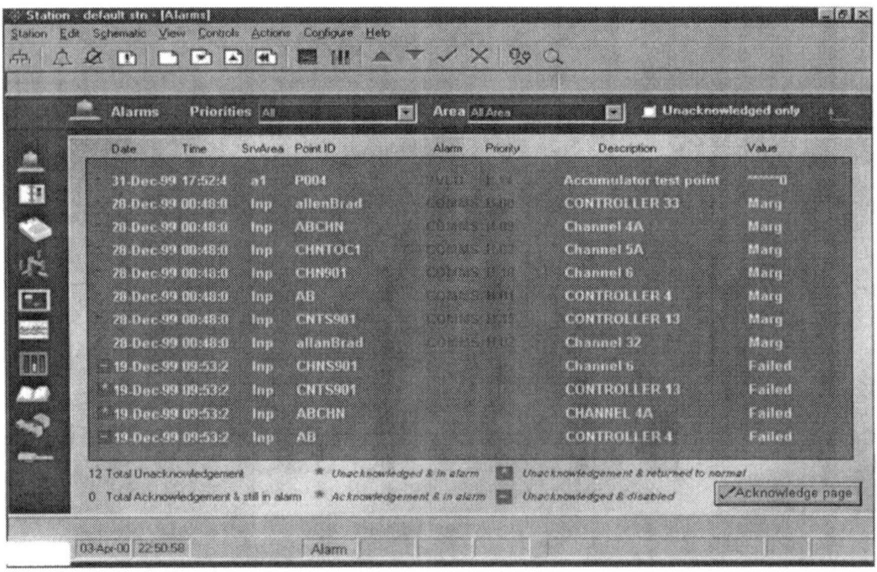

FIGURE 13. An Alarm Summary Page

The information for each alarm was listed horizontally on the 'Alarm Summary' screen with the associated icon shown on the left side of each alarm. An unacknowledged alarm could be acknowledged by clicking on the line of the alarm to highlight it, followed by clicking the 'Acknowledge Alarm' button on the tool bar.

The first step of applying the axiomatic method is to define a set of independent referents, i.e., the FRs. After analysis of the intended meanings of the four alarm conditions, the alarms were classified by two sets of independent FRs. The first set was in terms of Alarm Acknowledgement with two FRs

- FR_1 — Acknowledged
- FR_2 — Unacknowledged

The second set was in terms of Alarm Status with three FRs

- FR_1 — In Alarm
- FR_2 — Disabled
- FR_3 — Returned to Normal

The second step of applying the axiomatic method is to design the icons from a set of visual features, or to review the features used in current design.

The blink was used for the current alarm icon design to distinguish the acknowledged alarms and unacknowledged alarms. However, the distinctiveness among the conditions of 'In Alarm', 'Disabled' and 'Returned to Normal' were not clear. Alarm icons that represented these different conditions shared the same color e.g., red and shapes e.g., square and asterisk, and that might confuse users. Current design is shown in Table 1.

The visual features of current design and their further decompositions are listed as follows

- DP_1: Animation
- DP_{11}: No Blink
- DP_{12}: Blink
- DP_2: Symbol
- DP_{21}: Asterisk

TABLE 1. Current Design of Icons

Intended Meaning	Current Design	Description
Acknowledged & in Alarm		Red asterisk
Unacknowledged & in Alarm	 *blink*	Blinking red asterisk
Unacknowledged & Disabled	 *blink*	Blinking red square with a white dash inside
Unacknowledged & Returned to Normal	 *blink*	Blinking red square with a white asterisk inside

- DP_{22}: Dash
- DP_{23}: Square
- DP_3: Color
- DP_{31}: Red
- DP_{32}: White

Now we could apply the Independence Axiom for the current design:

$$\begin{Bmatrix} \text{Acknowledged} \\ \text{Unacknowledged} \end{Bmatrix} = \begin{bmatrix} \times & 0 \\ 0 & \times \end{bmatrix} \begin{Bmatrix} \text{No Blink} \\ \text{Blink} \end{Bmatrix} \tag{16}$$

$$\begin{Bmatrix} \text{In Alarm} \\ \text{Disabled} \\ \text{Return to Normal} \end{Bmatrix} = \begin{bmatrix} \times & 0 & 0 \\ 0 & \times & \times \\ \times & 0 & \times \end{bmatrix} \begin{Bmatrix} \text{Asterisk} \\ \text{Dash} \\ \text{Square} \end{Bmatrix} \tag{17}$$

$$\begin{Bmatrix} \text{In Alarm} \\ \text{Disabled} \\ \text{Return to Normal} \end{Bmatrix} = \begin{bmatrix} \times & 0 \\ \times & \times \\ \times & \times \end{bmatrix} \begin{Bmatrix} \text{Red} \\ \text{White} \end{Bmatrix} \tag{18}$$

The results showed that only Eqn. 16 was an uncoupled design. Both Eqn. 17 and 18 were coupled design. A new design shown in Table 2 was then proposed to replace the current one.

For the new design of alarm icons, the blinking for the icons was removed since it contravened the design principle regarding conservative use of blink [8]. Users may feel annoyed when there are too many blinks on the screen. All the icons were designed as bells since they were used to represent alarms. Following the two sets of FRs, the icon for acknowledged alarm had a tick on it to indicate that it had already been acknowledged while the question mark on the bell represented the alarm was unacknowledged. The associated visual features are listed as follows

- DP_1: Symbol
- DP_{11}: Tick
- DP_{12}: Question Mark

Compared to the 'return to normal alarm' icon, those icons representing 'in alarm' had shaking lines to indicate that they were in alarm and needed immediate attention. The 'disabled alarm' icon had a crack on the bell to show that it was not functioning normally. The associated visual features are listed as follows

TABLE 2. New Design of Icons

Intended Meaning	New Design	Description
Acknowledged & in Alarm		Red shaking bell with a white tick
Unacknowledged & in Alarm		Red shaking bell with a white question mark
Unacknowledged & Disabled		Orange broken bell with a white question mark
Unacknowledged & Returned to Normal		Green bell with a white question mark

- DP_2 – Pictograph
- DP_{21} – Shaking Bell
- DP_{22} – Broken Bell
- DP_{23} – Bell

In the new design, the red color was reserved for those important alarms – i.e., 'Acknowledged and in alarm', 'Unacknowledged and in alarm'. This could help the users to easily identify higher priority alarms. For the 'returned to normal' alarm, green color was used since it is a common color for representing normal condition [36]. The orange color was used to represent the 'disabled' alarm so as to distinguish it from the colors used for other alarms. Note that the colors were not the primary but a redundant feature to indicate the alarm conditions. The associated visual features are listed as follows

- DP_3 – Color
- DP_{31} – Red
- DP_{32} – Orange
- DP_{33} – Green

Now we could apply the Independence Axiom for the new design

$$\begin{Bmatrix} \text{Acknowledged} \\ \text{Unacknowledged} \end{Bmatrix} = \begin{bmatrix} \times & 0 \\ 0 & \times \end{bmatrix} \begin{Bmatrix} \text{Tick} \\ \text{Question Mark} \end{Bmatrix} \tag{19}$$

$$\begin{Bmatrix} \text{In Alarm} \\ \text{Disabled} \\ \text{Return to Normal} \end{Bmatrix} = \begin{bmatrix} \times & 0 & 0 \\ 0 & \times & 0 \\ 0 & 0 & \times \end{bmatrix} \begin{Bmatrix} \text{Shaking Bell} \\ \text{Broken Bell} \\ \text{Bell} \end{Bmatrix} \tag{20}$$

$$\begin{Bmatrix} \text{In Alarm} \\ \text{Disabled} \\ \text{Return to Normal} \end{Bmatrix} = \begin{bmatrix} \times & 0 & 0 \\ 0 & \times & 0 \\ 0 & 0 & \times \end{bmatrix} \begin{Bmatrix} \text{Red} \\ \text{Orange} \\ \text{Green} \end{Bmatrix} \tag{21}$$

The results showed that Eqn. 19, 20 and 21 were uncoupled design.

Results from the user testing in the previous study [37] confirmed the analysis of the axiomatic method. In the previous study, 30 students from the Diploma in Chemical Engineering of a polytechnic school in Singapore were recruited as the subjects. They were recruited due to the relevance of their background in industrial process control and their potential for being the users of process control products. Participation was voluntary and each subject was given a gift for participating in the study. Among the 30 subjects, 18 were male and 12 were female. Twenty-three were in their second year of school (14 males and 9 females) whereas seven were in their first year (4 males and 3 females). The range of experience with computers was between 1.5 years to 10 years with the median at 5 years. All the subjects were between 17 to 22 years old with the median at 19 years old. Subjects were randomly separated into two test groups. One was with the current icon design and the other was with the new design. The subjects then answered a paper-and-pencil questionnaire requiring them to match each alarm icon with its meaning. Compared to an average of only 35% correct responses for identifying all the current icons, an average of 83% correct answers was achieved for the new icons. The percentages of correct responses for each current and newly-designed icon are listed in Table 3.

Note that the use of the animation (DP_1) in the current design and the use of the pictograph (DP_2) in the new design for representing the alarm acknowledgement were both uncoupled

TABLE 3. Percentages of Correct Responses

Intended Meaning	Current Design	New Design
Acknowledged & in Alarm	7%	80%
Unacknowledged & in Alarm	33%	80%
Unacknowledged & Disabled	53%	80%
Unacknowledged & Returned to Normal	47%	93%
Average	35%	83%

designs. However, their information contents could not be derived from the results of the previous study, therefore, the Information Axiom was not applied in this case study.

DISCUSSION AND CONCLUSION

Results from the case study suggest that the axiomatic method is an effective tool for the analysis and evaluation of the discriminability and meaningfulness of icons. Further research direction is twofold: to refine the method, and to extend its application.

Refinement of the method

Comparison with other relevant methods would be a useful way to refine the current method. The AD process has been compared to methods such as Optimization Design, Reliability Design and Design for Assembly [38]. It has been also compared to the Theory of Inventive Problem Solving (TRIZ) method [39, 40].

It is worthwhile to look at the development of the fundamental-objectives hierarchy in decision-making [41] for the derivation of FRs in the Independence Axiom.

While the discriminability of icons is mainly affected by the distinctiveness of visual features, the meaningfulness of icons is influenced by a user's knowledge and experience of the icons, their referents, and the associations between them. Thus, compared to the discriminability of icons, the meaningfulness of icons is relatively subjective and culturally linked. It would be worthwhile to further explore the cultural effects of the meaningfulness of icons under the framework of the Information Axiom.

Extension of the application

The axiomatic method has also been applied to a usability evaluation of the control buttons on an MP3 player [42]. Further applications may evaluate different types of displays and controls in user interfaces. Another interesting application is to link customer requirements with manufacturing capabilities in a supply/demand chain under the series of mapping between the four domains of the axiomatic method. A research project is being carried out by the author to integrate the picture quality information in a Thin Film Transistor-Liquid Crystal Display (TFT-LCD) supply/demand chain with the application of the axiomatic method.

Since the refinement of the method is relatively long-term and methodology-oriented, it may be more suitable for academia. On the other hand, the extension of the application is relatively short-term and solution-focused and therefore it may be more suitable for industry.

As the collaboration between academia and industry cannot be overemphasized, the success of a collaborative research project depends on a consensus on the objective and time frame of the project.

ACKNOWLEDGEMENTS

The author would like to express his gratitude to the National Science Foundation in Taiwan (NSC 95-2221-E-027-081-MY3) for its support.

REFERENCES

[1] Chandler, D. (2002). Semiotics. *The Basics*. New York: Routledge.

[2] Wickens, C. D., Lee, J. D., Liu, Y., and Gordon Becker, S. E. (2004). *An Introduction of Human Factors Engineering*, 2nd ed. New Jersey: Pearson Prentice Hall.

[3] Raskin, J. (2000). *The Humane Interface*. Massachusetts: Addison-Wesley.

[4] Huang, S., Shieh K., and Chi, C. (2002). Factors affecting the design of computer icons. *International Journal of Industrial Ergonomics*, 29, 211–8.

[5] Garcia, M., Badre, A.N., and Stasko, J.T. (1994). Development and validation of icons varying in their abstractness. *Interacting with Computers*, 6 (2), 191–211.

[6] McDougall, S.J.P., Curry, M.B., and De Bruijn, O. (1999). Measuring symbol and icon characteristics: Norms for concreteness, complexity, meaningfulness, familiarity, and semantic distance for 239 symbols. *Behavior Research Methods, Instruments, & Computers*, 31 (3), 487–519.

[7] Wright S. (1997). Validating a predictive model for computer icon development. *Computers and Industrial Engineering*, 33 (1–2), 189–92.

[8] Horton, W.K. (1994). *The Icon Book: Visual Symbols for Computer Systems and Documentation*. New York: Wiley.

[9] Forsythe, A., Sheehy, N., and Sawey, M. (2003). Measuring icon complexity: An automated analysis. *Behavior Research Methods, Instruments, & Computers*, 35 (2), 334–42.

[10] Kurniawan, S.H. (2000). *A Rule of Thumb of Icons' Visual Distinctiveness*. Proceedings of the Conference on Universal Usability, pp. 159–60.

[11] Rogers, Y. (1989). Icon design for the user interface. *International Review of Ergonomics*, 2, 129–54.

[12] Moray, N. (1997). Human factors in process control. In *Handbook of Human Factors and Ergonomics* (G. Salvendy, ed.), 2nd ed. New York: John Wiley & Sons, pp. 1944–71.

[13] Stanton, N.A. (ed.)(1994). *Human Factors in Alarm Design*. London, UK: Taylor & Francis.

[14] Marcus, A. (1992). *Graphic Design for Electronic Documents and User Interfaces*. New York: ACM Press.

[15] O'Hara, J. M., Brown, W.S., Lewis, P.M., and Persensky, J.J. (2002). *Human-System Interface Design Review Guidelines (NUREG-0700, Rev. 2)*. U.S. Nuclear Regulatory Commission, Washington, DC.

[16] Shneiderman, B. (1998). *Designing the User Interfaces*, 3rd ed. Reading, MA: Addison-Wesley.

[17] International Organization for Standardization (ISO). (2000). *ISO/IEC 11581-1: Information Technology-User System Interfaces and Symbols- Icon Symbols and Functions. Part 1, it Icons-General*. ISO, Geneva.

[18] Semiconductor Equipment and Materials International (SEMI). (2001). *SEMI E95-1101: Specification for Human Interface for Semiconductor Manufacturing Equipment*. San Jose, CA: SEMI.

[19] Suh, N.P. (1984). Development of the science base for the manufacturing field through the axiomatic approach. *Robotics and Computer Integrated Manufacturing*, 1 (3/4), 399–455.

[20] Suh, N.P., Bell, A.C., and Gossard, D.C. (1978). On an axiomatic approach to manufacturing systems. *Journal of Engineering for Industry*, 100 (2), 127–30, Transactions of American Society of Mechanical Engineers.

[21] Suh, N.P. (2001). *Axiomatic Design: Advances and Applications*. New York: Oxford University Press.

[22] Suh, N.P. (1990). *The Principles of Design*. New York: Oxford University Press.

[23] Su, J.C.-Y., Chen, S.-J., and Lin, L. (2003). A structured approach to measuring functional dependency and sequencing of coupled tasks in engineering design. *Computers & Industrial Engineering*, 45, 195–214.

[24] Shannon, C.E. and Weaver, W. (1949). *The Mathematical Theory of Communication*. Urbana, IL: University of Illinois Press.

[25] Suh, N.P. (2005). *Complexity: Theory and Applications*. New York: Oxford University Press.

[26] Helander M.G. and Lin, L. (2002). Axiomatic design in ergonomics and an extension of the information axiom. *Journal of Engineering Design*, 13 (4), 321–39.

[27] Tversky, A. (1977). Features of similarity. *Psychological Review*, 84, 327–52.

[28] Garner, W.R. (1974). *The Processing of Information and Structure*. Erlbaum, NJ: Hillsdale.

[29] International Organization for Standardization (ISO) (2002). ISO 3864–1: *Graphical Symbols- Safety Colors and Safety Signs. Part 1: Design Principles for Safety Signs in Workplaces and Public Areas*. ISO, Geneva.

[30] Dreyfuss, H. (1984). *Symbol Sourcebook: An Authoritative Guide to International Graphic Symbols*. New York: Van Nostrand Reinhold Co.

[31] Fernandes, T. (1995). *Global Interface Design: A Guide to Designing International User Interfaces*. Massachusetts: AP Professional.

[32] Miller, A.R., Brown, J.M., and Cullen, C.D. (2000). *Global Graphics: Symbols- Designing with Symbols for an International Market*. Massachusetts: Rockport Publishers.

[33] Peterson, L.K. and Cullen, C.D. (2000). *Global Graphics: Color: Designing with Color for an International Market*. Massachusetts: Rockport Publishers.

[34] Luximon, A., Lau, W.C., and Goonetilleke, R.S. (1998). *Safety Signal Words and Color Codes: The Perception of Implied Hazard by Chinese People*. Proceedings of the 6th Pan-Pacific Conference on Occupational Ergonomics, pp. 30–3.

[35] Liang, S.-F.M., Plocher, T.A., Lau, W.C. et al. (2000). *Perception of Colors and Graphics in Process Control Workstations*. Proceedings of the 4th Asia-Pacific Conference on Computer-Human Interaction, pp. 120–4.

[36] Liang, S.-F.M., Khalid, H.M., Taha, Z., and Plocher, T. (2004). *In Search of Internationalized Operator Interface Displays in Process Control: A Comparison Among Malaysian, Singaporean and Chinese*. Proceedings of the 7th International Conference on Work With Computing Systems, pp. 253–8.

[37] Liang, S.-F.M., Plocher, T.A., Lau, W.C. et al. (2000). *Usability Evaluation on Honeywell Process Control Workstations*. Proceedings of the 5th Annual International Conference on Industrial Engineering-Theory. Applications and Practice, ID 202 (CD-ROM), pp. 6.

[38] Chen, K.-Z. (1999). Identifying the relationship among design methods: Key to successful applications and developments of design methods. *Journal of Engineering Design*, 10 (2), 125–41.

[39] Mann, D. (2002). *Axiomatic Design and TRIZ: Compatibilities and Contradictions*. Proceedings of the Second International Conference on Axiomatic Design. ICAD 011, p. 7.

[40] Yang, K. and Zhang, H.A. (2000). *Comparison of TRIZ and axiomatic design*. Proceedings of the First International Conference on Axiomatic Design. ICAD56, p. 9.

[41] Keeney, R.L. (1992). *Value-Focused Thinking*. Cambridge, MA: Harvard University Press.

[42] Liang, S.-F.M. and Lin, C.-W. (2006). *Applying Axiomatic Method to Consumer Product Usability Evaluation: A Case Study of a Developing MP3 Player*. Proceedings of the 7th Asia-Pacific Conference on Computer-Human Interaction, (CD-ROM), p. 10.

Meeting Diversity in Ergonomics (2007)
Pikaar, Koningsveld, and Settels (Eds)

11

Human Factors of Virtual Reality – Where are We Now?

Sarah Sharples[1], Alex W. Stedmon, Mirabelle D'Cruz, Harshada Patel, Sue Cobb, Thomas Yates, Rossukorn Saikayasit & John R. Wilson

Virtual Reality Applications Research Team (VIRART),
School of Mechanical, Materials and Manufacturing Engineering,
University of Nottingham, University Park, Nottingham, NG7 2RD, UK

Abstract. Considerable changes have taken place in the focus of research into human factors issues associated with the use of Virtual Reality (VR) technologies. This has coincided with developments in VR display and programming technologies and increased applied use of VR in industrial contexts. This chapter presents a set of case studies that illustrate the human factors issues identified from a diverse set of VR applications over recent years. Case studies are presented from six different applications – a Virtual Factory simulation, Rail simulation, development of visual interaction metaphors for design visualisation, development of collaboration technologies to support industrial decision-making, representation of material properties in Virtual Environments (VE), and development of speech interaction for VR. For each of these case studies, the human factors issues of interest are discussed. The changing research priorities for VR research are then presented in a future research agenda for human factors of VR.

Keywords: virtual reality, simulation, usability, presence, visualisation technologies

INTRODUCTION

Over the past 15 years, changes have taken place in the focus of research into human factors issues associated with the use of VR and the types of VR technology used. This chapter uses work conducted at the University of Nottingham by the Virtual Reality Applications Research Team (VIRART) as a basis for identifying the changing trends in VR research priorities over recent years.

In 1995, a programme of work was conducted to identify the potential health and safety implications of VR in industrial applications. This work was motivated by concerns about the potential implications of VR being introduced into UK workplaces, and a programme of work was conducted to investigate the types of effects that were experienced by VR participants.

At this time, it was thought that the primary method of delivering VR displays in the future would be via head-mounted displays. These displays at the time were based on wired technology, so the participant was physically tethered to a central processing unit. CAVE-type displays and projection systems were also in use, but these were hindered by the then very high cost of projection technologies. Typical VR applications at the time ran at relatively slow frame rates of as few as 4–10 frames per second and the quality of display resolution was also a concern; however, there was general excitement about the direct interaction and manipulation activities that were possible in VEs. Cobb et al. [1] proposed the concept of

[1] Neé Nichols

Visual	Musculoskeletal	Psychological		Physiological
		Behavioural	Cognitive	
Changes in heterophoria Blurred vision Eyestrain	Physical discomfort Equipment fit and safety Posture demands	Stress and mood change Addiction Isolation Hallucinations and visual flashbacks	Perceptual shifts and disorientation Changes in perceptual judgement Change in psychomotor performance	Simulator sickness Postural instability Cardiovascular change Respiratory change Gastrointestinal change Biochemical change

FIGURE 1. Suggested effects from VR use (Cobb et al., 1995; Nichols et al., 1997).

Virtual Reality Induced Symptoms and Effects (VRISE) as a framework for the examination of all types of effects, positive and negative, that might occur after a period of exposure to VR. Cobb et al. [2] and Nichols et al. [3] identified that the influencing factors on VRISE can be broken down into design of the VR technology, VE design, task circumstances and individual user characteristics. In addition, the types of effects experienced were categorised as visual, musculoskeletal, psychological and physiological effects, as shown in Fig. 1.

There are some notable absences from this list. The first is that presence is not included as a suggested effect. At this time, there was a large amount of discussion in the literature regarding the concept of presence, and some subjective assessment tools did exist to measure presence (e.g., Witmer & Singer [4]). However, the importance of presence for the overall experience of the user had not yet been established by empirical research.

The second notable absentee is usability. At this point, there were a number of fundamental considerations regarding the comfort of the participant, and the performance of the system. This can be likened to the known phenomenon in product usability, where usability is only of importance to the participant once the basic functionality and performance requirements have been met.

Finally, at the stage of developing the initial set of effects of VR use, the main circumstance of use being considered was single users using an individual VE in which they were the only participant. Whilst work was developing on collaborative VE within the computer-supported co-operative work community (e.g., Benford et al. [5]) this was very much at the developmental stage, and technology relating to speed of connections was still not sufficient to allow complex collaborative environments.

This paper presents a series of case studies that have been conducted by VIRART in the last 15 years. These cases have been selected to represent the types of industrial applications and technologies that have been developed and evaluated during this period. In this chapter, they are used as a basis for identifying the key human factors challenges facing current and future applications of VR.

The case studies cover the following industrial applications:

1. Industrial factory design
2. Representation of material properties for product design

3. Simulation for human factors evaluation
4. Interaction metaphors – input devices
5. Interaction metaphors – speech
6. Collaboration technologies for industrial design

CASE 1: INDUSTRIAL FACTORY DESIGN

Aim

This application had the primary aim of demonstrating how VR could be used to visualise the impact of different factory designs [6]. These different designs could then be evaluated from the perspective of the impact of workstation design on physical ergonomics, or the consequences of different layouts of material flow, waste and production efficiency.

VR technology, VE and circumstances of use

The VR factory demonstrator could be viewed either on a projection screen with shutter glasses, or via a V8 Head-Mounted Display. The interaction devices available included joy-sticks, mice and a motion-tracked data glove [7]. Fig. 2 shows a view of the virtual factory environment.

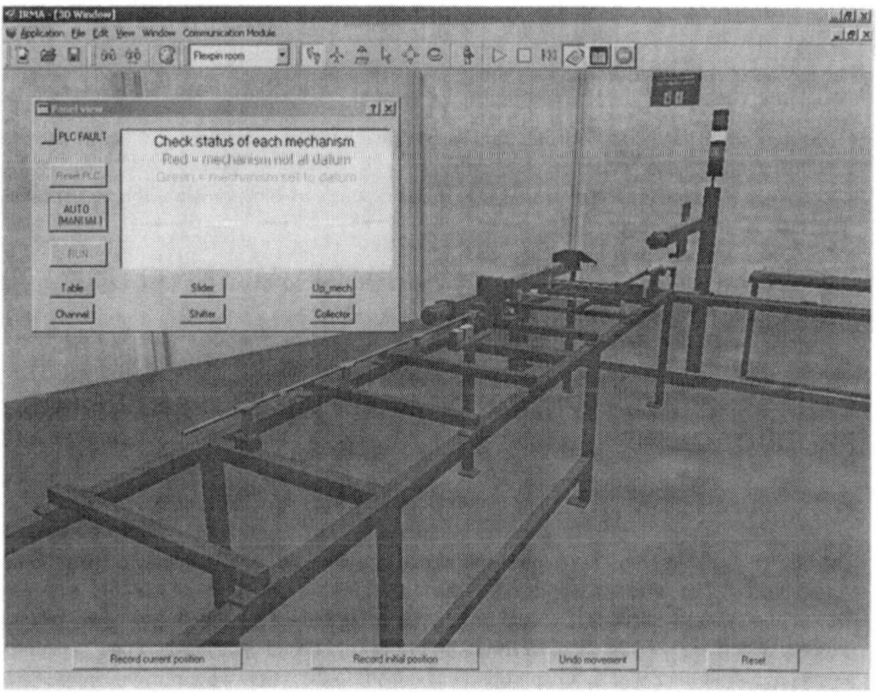

FIGURE 2. View of virtual factory environment.

Summary of human factors evaluation

For this study, a formal evaluation was carried out using three expert participants and a heuristics-based evaluation tool [8].

In general, participants reported that the overall usability of the system was satisfactory, but problems were experienced when using the data glove to interact. This glove proved very awkward to use with some extreme arm movements required, resulting in arm elevation and extension away from the body when selecting menu icons. There were also problems relating to the resolution of the display – the text displayed was difficult to read when using either the stereo shutter glasses or the headset, although in general the resolution of the display images – i.e., not text – was felt to be sufficient. Overall, it was felt that the interface design was consistent with real world metaphors, but navigational errors did occur as it was possible to walk through walls due to no collisions being present. It was also suggested that the use of sound in the environment be enhanced and the inclusion of additional cues within the environment was requested.

In terms of presence, participants felt that there was a relatively high sense of presence produced by the display, but that the interaction methods detracted from the feeling of presence. In addition, some lag was noticeable – particularly where response speed was low due to high level of detail being rendered by the system. Participants did feel that the projected stereo added to the sense of presence, but if there were multiple viewers of the large screen display it was felt to be beneficial to use a mono display to avoid the slightly distorted stereo view experienced by the non-tracked viewers. There was also one instance of VR-induced sickness.

Key human factors issues identified

The key issues identified in this case study are

- Lag may still be an issue if system limits are pushed in terms of rendering of highly complex multi-faceted objects.
- There is a need for intuitive, natural interaction devices, to enhance usability, and minimise detraction from presence.
- Some sickness was experienced by one user.
- When this system was used in a co-located collaborative context, the best form of display was a mono projection screen.

CASE 2: REPRESENTATION OF MATERIAL PROPERTIES FOR PRODUCT DESIGN

Aim

The aim of this study was to evaluate the effectiveness of different methods of representing virtual objects. The particular application considered was the perception of material properties. This is important to consider if virtual prototypes are to reduce the number of costly physical prototypes that are used in manufacturing and industrial design – it is important that virtual objects represent the detail to an appropriate level [9]. In this case study, perfume bottles were chosen as example objects – these were considered to be of appropriate complexity in shape and form, required the representation of a variety of materials, including metal, plastic and glass, they were considered to be objects for which the weight and material properties were important in influencing user perceptions in terms of product quality, and it was possible to select items that did not have obvious or clear branding for participants. Participants were asked to make judgements regarding the object weight, volume and materials used; their responses were compared with the actual known values.

VR technology, VE and circumstances of use

Four visualisation methods were compared: photos, printed to scale on photographic paper, 2D wire frame models; 3D solid CAD models produced using AutoCAD and printed out and displayed on paper; and VR produced using Virtools software and displayed on a laptop computer screen, providing participants with the opportunity to have limited interaction, rotation and zooming with the object. Fig. 3 shows the VR representations of the different perfume bottles used during the experiment.

Summary of human factors evaluation

In order to avoid requiring participants to make absolute judgements, participants were asked to rank the bottles from lightest to heaviest, and largest to smallest volume. These estimated rankings were then compared with the actual correct rankings. For weight judgements, as expected, the real items yielded the most accurate estimate of weight. The VR and 2D CAD representations were the next most accurate. For volume, again the real items yielded the most accurate estimate of volume, and 2D CAD, 3D CAD and VR were all found to result in more accurate estimates of volume than photographs.

The consistency of the estimates was also considered; VR was found to produce the most consistent weight estimates, and the 2D CAD the most consistent volume estimates. In the case of the 2D CAD, it is possible that the omission of the information relating to material properties removed a distraction, and thus allowed more consistent judgements to be made. This is also cogent with the findings relating to estimates of item cost – in this case the 2D models yielded the least consistent rankings, and VR the most consistent – this finding represents the benefit of including the materials detail information in the VR representation.

FIGURE 3. VR model of perfume bottles displayed to represent materials.

When estimating the cost of manufacture of the object, the real objects, photographs and 3D CAD images all resulted in the same consistence of rankings, with the 2D model being least consistent.

Key human factors issues identified

The key issues identified in this case study are

- It is not necessarily the case that a more detailed representation will yield more accurate judgements in object weight, volume, item and manufacture cost.
- The poor performance of photos suggests that 'photorealism' may not be as important in representation of material properties as is often assumed.
- More research is required to evaluate the effects of increased interactivity, different levels of object detail – e.g., inclusion of sound and tactile feedback – or use of stereo vision on decision-making related to judgement of material properties.

CASE 3: SIMULATION FOR HUMAN FACTORS EVALUATION

Aim

The aim of this case study is to develop a low-cost simulator to allow decisions to be made regarding rail infrastructure. The priorities in the use of this simulator are to allow reliable and accurate decisions to be made regarding rail infrastructure, such as assessment of impact of lineside clutter.

VR technology, VE and circumstances of use

The technology used in this system is deliberately of a lower level of complexity and sophistication than is often seen in rail simulators. In this case, a Virtools model displaying simplified CAD objects is linked to a variety of types of input devices, ranging from a full-scale train cab to a computer keyboard. This allows a varying level of interaction with the system, either allowing the user to position themselves in a particular virtual track location, or to simulate the driving task in a reasonably simplistic manner – i.e., no moving base or haptic feedback and limited simulation of realistic resistance of rail controls.

Summary of human factors evaluation

For this case, a formal evaluation has not yet been conducted. However, a detailed user requirements exercise [10] was undertaken in order to determine the priorities when developing a system of this type. In this exercise, the priority was given by respondents to the need for accuracy in terms of the judgements made when using the system.

Key human factors issues identified

For this application, the key human factors issues identified are

- The need for comparable performance or perception with the real world.

- Identification of sufficient level of resolution of the display in order to yield reliable judgements.
- The limitations of the controls limit the extent to which judgements can be made regarding driver performance in a simplified system such as this.

CASE 4: INTERACTION METAPHORS – INPUT DEVICES

Aim

The aim of this application was to evaluate the design of a set of different input devices and menu designs for use within a VR design application. This work was conducted as part of the EU-funded project VIEW of the Future (IST-2000-26089) which was a project that had an overall goal to identify how visualisation technologies can be used to support automotive and aerospace design. Data from over 100 participants was collated to allow an overview analysis of the relationship between sickness, presence and usability to be performed.

VR technology, VE and circumstances of use

The system used was a stereo projection environment, where participants were wearing polarised shutter glasses. Participants were required to manipulate a virtual prototype object using one of three input devices (the Mike, Omni Control or Hornet) and one of three designs of menu – fan, linear or sphere [11]. Fig. 4 shows an illustration of use of the typical use of such input devices with a menu device such as this.

HF issues identified

A series of formal evaluations was applied to the system, including usability analysis, questionnaire and observation, measurement of presence and reports of simulator sickness from participants. Overall, one of the key findings was that it was not possible to evaluate either

FIGURE 4. Example of use of interaction devices with menu (input devices: mouse and dragonfly).

the interaction devices or the menu designs in isolation – in most cases there was an interaction, where a particular device would be more appropriate for a particular application. In general, a number of problems were experienced when using the input devices, and the key recommendations that emerged regarding device design were: to ensure that hand-held devices can be used by all hand sizes (at least 5 percentile female to 95 percentile male) as in the current form the devices were not easy to use for all hand sizes; that buttons on the interaction device should be easily accessible, differentiated from each other, e.g., by position, size or surface texture, easy to operate, and provide appropriate feedback when pressed; and that the interaction device should provide cues to indicate in which direction it is being pointed [11].

In addition the relationship between the different effects of VR use was examined. Participants completed questionnaires in which they were able to rate their symptoms of simulator sickness, using the Simulator Sickness Questionnaire [12], and presence, usability and enjoyment, using questionnaires specially developed for the study.

A positive association was found between presence and usability – i.e., that high usability scores were associated with high presence scores ($r = 0.49$, $p < 0.01$, $N = 71$). Presence and enjoyment were also positively correlated ($r = 0.30$, $p < 0.01$, $N = 80$) indicating that presence is a positive effect of VE use, and usability scores were also associated with high enjoyment scores ($r = 0.54$, $p < 0.01$, $N = 87$). In addition, presence was found to be positively correlated with the nausea dimension of sickness ($r = 0.27$, $p < 0.05$, $N = 80$). This is in contrast to previous results [13]. This may be due to a common causative factor, such as vection, or experience of self-motion, in the production of both presence and sickness [14]. In addition, a high level of oculomotor symptoms was associated with lower usability scores ($r = -0.21$, $p < 0.05$, $N = 87$), and lower enjoyment scores were consistently associated with higher symptom scores (Nausea: $r = -0.20$, $p < 0.05$, $N = 100$; Oculomotor: $r = -0.24$, $p < 0.05$, $N = 100$; Disorientation: $r = -0.22$, $p < 0.05$, $N = 100$).

Key human factors issues identified

For this application, the key human factors issues identified are

- The need to consider the interacting effects of design of the input devices and screen interaction menus.
- The continued evidence for some experience of sickness with projection screen displays.
- The need to consider anthropometric variations in design of interaction devices.
- The importance of presence and its potential impact on technology use.

CASE 5: INTERACTION METAPHORS – SPEECH

Aim

This study aimed to evaluate interaction issues using speech input in VR, where users were able to use free speech when instructing another person in order to complete a VR task. Twelve participants completed the study.

VR technology, VE and circumstances of use

The VE was displayed using Superscape software on an 800 MHz PC with a front projected monoscopic data projection system – 800*600 resolution. The VE used was a training tool for computer network-card replacement [15]. Participants were required to complete the task of changing a network card. The variables measured were task-completion time, total number of commands used, total number of words used, number of words per command and number of computer prompts required – which represented the number of illegal commands or incidents where a participant had to repeat a command.

HF issues identified

There was a large variation in the task time for users – ranging from 20 to 48 minutes. In addition, the number of commands used ranged from 89 – 403, and total number of words from 398 – 1057. In general users employed commands between two and five words long in order to complete tasks. The time taken to complete the task seemed to relate to navigation errors and orientation problems, rather than object manipulation requirements. Participants also tended to build up a usable command structure that they then re-applied throughout the task – this supports the concept of vocabulary habitability where the match between the language people employ when using a speech system and the language that the system can accept is a key issue in system usability [16].

By analysing the most and least common commands it is possible to begin to develop ideas regarding generic command structures. Movement in the VE was continuous, and so the most common command used was 'stop'. If movement in the VE was discrete then this command would hardly have been used, although actual navigation initiation commands would have increased as a result. The most common navigation commands were 'move', 'look', 'turn', 'walk' and 'go' – combined with supplementary commands such as 'forwards', 'backwards', 'right' or 'left'. The most common object manipulation commands were 'pick up', 'put down', 'open' – combined with object identifiers such as 'screwdriver', 'screws' or 'door'. Participants appeared to refrain from frequently using commands that used relative terms to other items in the VE. Barfield et al. [17] propose that input devices should account for the type of manipulations a user has to perform and be designed so that they adhere to natural mappings in the way the device is manipulated, as well as permit movements to coincide with a user's mental model of the type of movement in a VE.

This short trial support the idea that a task vocabulary could be developed based on the most common commands that users might therefore find most intuitive. Speech is highly context-specific, and therefore task vocabulary by its very nature will be application-specific and must be based on an understanding of the human factors issues surrounding speech input [18].

Key human factors issues identified

For this application, the key human factors issues identified are

- The potential for speech as an interaction metaphor for VEs should be investigated further
- Users are likely to develop a re-usable command set, either generally, or specific to each VE that they use.

CASE 6: COLLABORATION TECHNOLOGIES FOR INDUSTRIAL DESIGN

Aim

A new EU-funded project (CoSpaces IST-5-034245) aims to examine the ways in which technology can be used to support collaboration in distributed virtual manufacturing enterprises.

VR technology, VE and circumstances of use

The early system being evaluated is a Mixed Reality Architecture (MRA) cell [19] that provides the opportunity to 'share' virtual space and view each other's environments within a collaborative framework. Fig. 5 shows an illustration of the typical location of an MRA Cell within a standard office environment.

Key human factors issues identified

As this project is still in the early stages, no formal evaluations have yet been conducted, but the following issues have been identified as being of interest

- The impact of such technology on current working practice.
- The way in which the use of such technology can support collaborative thinking.
- How users understand and develop mental models of such a system.
- How collaboration when using such technologies can be measured.

FIGURE 5. Typical location of MRA cell within office environment.

DISCUSSION

Human factors impact of developments in VR

The most significant changes in VR from a human factors perspective that have occurred in recent years are improved processing power, developments in wireless technology and reduction in cost of projection technologies. Early research focussed on the use of head-mounted displays which tended to be linked to a powerful dedicated computer system using a set of wires enclosed in a large cable. The practical impact of hard-wired technology was a restriction on user movement and a safety hazard of tripping over wires. These systems have been replaced in general by either remote tracking technologies, such as magnetic or infra red technology, or have been addressed by having the participant wear a battery and computer console, in some cases on their backs, so that the wires are located close to their body and the risk of tripping is minimised.

When initial research into human factors issues associated with VR was carried out, system response rate was a particular concern – this led to low frame rate, and lag between user movement and display update. In general, this problem has been minimised due to the increased available computing power. Typical systems now run at 20 frames per second as a minimum, and so the impact of frame rate is much reduced, and whilst lag is still present in some systems it is much less noticeable. However, systems are still inevitably pushed to the limit, and in some cases this frame rate is not achieved, and lag is still noticeable.

However, arguably the most notable trend has been a move towards projection systems which have increased in power and declined in cost. The context of use of VR has changed from a single user enclosed in a headset system to a group of users collaboratively viewing a VE on a projection screen. Fig. 6 shows a prototype example of an interface designed for collaborative use [20]. One of the restrictions of some multi-user VR set-ups is that only one user is an active controller of interaction with the system – probably via head tracking using tracked shutter glasses and a hand-held input device, and all others are passive viewers, who may be provided with stereo viewing glasses. This can lead to a number of effects in terms of the way in which the display is used. Firstly, the active controller may be interrupted by

FIGURE 6. Example of co-located collaborative interface (Bayon et al., 2006).

instructions from other participants – this could potentially lead to a 'break in presence' [21]. Secondly, the other passive participants may have a distorted view depending on their position in front of the screen and relative to the active controller. In addition, it has been shown [22, 23] that participants who are passive viewers of a VE exhibit higher levels of sickness compared with those who are in active control.

In terms of VE design, the complexity of environments has increased as a result of increased computing capacity. This has the consequence of potentially increasing the number of objects in the environment, which could impact on optic flow. However, realism could also be increased, which will impact on the sense of presence. The relationship between presence, usability and sickness has yet to be firmly established.

The focus as far as environment design is concerned has moved towards interaction design. This was identified as an issue in the early 1990s, and the fundamental problem of a '2D view into a 3D world' still remains. No standard interaction metaphors have however emerged.

Finally, the issue of sickness still remains. Wilson et al. [24] wrote 'some participants, in some VR systems, with some virtual environments, do report or exhibit various side effects'. This statement still stands, and our research of over 300 participants appears to indicate that for about 5% of the population these effects can be so severe that they cause them to have to stop using the technology. Behr et al. [25] summarise the ethical issues associated with VR research and emphasise the importance of adopting a practical strategy to managing the ethical considerations of conducting VR research.

Human factors research priorities

On the basis of the case studies presented, the following research priorities can be identified:

Effects of use of VR

The set of effects outlined in 1995 identified a number of physiological effects that may result from VR use. To some extent the prevalence of these effects has been reduced with the increase in sophistication of the systems being used, but the issue of negative effects of using and viewing VR has not disappeared. It still appears that for a small percentage of the population, approximately 5%, these effects, particularly sickness, are experienced and are severe enough to prevent effective use of VR. Therefore the mediation of these effects, via system selection, environment design and training should be continued.

Of increasing importance is the issue of usability. There are still no standard metaphors or devices for interaction with VR applications, and the case for development of standardised guidelines has still not been firmly established. However, in the case studies presented in this paper, it was identified that poor usability of the input devices, and the interaction between device and interface design is critical if VR applications are to be successfully used. In addition, the topic of presence is still debated in the scientific literature, and should not be ignored, especially when considering its relationship with other effects such as usability and enjoyment, particularly when related to job satisfaction. This is also important in the light of increased incorporation of haptic and sound feedback with VE.

Effectiveness of VR applications

If VR technology is to be successfully implemented in industry, it must be shown to be effective at supporting task completion. Medical and educational applications are already appearing to succeed in this area, but there is still work to be done to ensure that industrial applications, such as virtual prototyping, design support and manufacturing analysis, do lead to more effective working practices. If this is considered, it is therefore also important to ensure that any evaluation methods incorporate metrics that indicate effectiveness, such as those that directly measure user performance or accuracy of perception of information from VR representations.

Impact of technological developments

The development of increased processing power, projection technologies and the introduction of wireless technologies are all technical developments that have greatly enhanced the experience of using a VE. However, it is important to remember that as processing power has increased, systems are still being pushed to the limit in order to allow display of increasingly complex visual displays, and to allow multiple users to view stereoscopic images simultaneously, so issues such as lag and image shadowing are still observed.

In addition to the developments in hardware, the previously distinct technologies of digital human modelling, computer-aided design and VR development toolkits are becoming more similar and compatible. There is still a challenge associated with integration of CAD models into VR viewing technologies – if any object simplification is required then it is hard to translate any modifications made in a VE back into the full, multi-faceted CAD model. It is important that these technical limitations are still acknowledged when evaluating the human factors issues associated with the use of such integrated systems.

Collaboration in use of visualisation technologies

Finally, the increased prevalence of projection-based technology, along with increased capacity of remote computer connections, means that the use of truly collaborative systems is becoming more realistic in a working, rather than laboratory, context. This collaboration presents interesting human factors challenges in terms of the way in which individual users, using individual interaction devices, may work together; how the use of such technologies affects existing team-working and co-located and distributed collaboration in different work and leisure contexts; and how we can measure collaboration in a manner that allows useful guidance to be produced for those involved in developing VR technologies and VE.

ACKNOWLEDGEMENTS

The work presented in this paper was funded by the EU projects VIEW of the Future (IST-2000-26089), IRMA (IMS 97007), Co-Spaces (IST-5-034245), Network Rail PLC and the Engineering and Physical Sciences Research Council (EPSRC).

REFERENCES

[1] Cobb, S.V.G., Nichols, S.C., Ramsey, A.R., and Wilson, J.R. (1999). Virtual Reality Induced Symptoms and Effects. *Presence: Teleoperators and Virtual Environments*, 8 (2), 169–86.

[2] Cobb, S., Nichols, S., and Wilson, J.R. (1995). *Health and Safety Implications of Virtual Reality: In search of an experimental methodology.* Proceedings of FIVE'95, 18–19 December, QMW University of London, UK, pp. 227–42.

[3] Nichols, S.C., Cobb, S.V.G., and Wilson, J.R. (1997). Health and Safety Implications of Virtual Environments: Measurement Issues. *Presence: Teleoperators and Virtual Environments*, 6 (6), 667–75.

[4] Witmer, B.G. and Singer, M.J. (1995). Measuring Immersion in Virtual Environments. *Technical Report 1014*, US Army Research Institute.

[5] Benford, S., Bowers, J., Fahlen, L., et al. (1995). *User Embodiment in Collaborative Virtual Environments*. Proceedings of ACM CHI'95 Conference on Human Factors in Computing Systems, Papers: Advanced Media for Collaboration, vol. 1, pp. 242–9.

[6] Sharples, S.C., D'Cruz, M.D., Patel, H., et al. (2006). *Human Factors of Virtual Reality: Where are We Now?* Proceedings of IEA.

[7] Modern, P., Chiabra, P., D'Cruz, M., and Cobb, S. (2004). *A Configurable Virutal Reality System for Multipurpose Industrial Manufactureing Applications*. Project IRMA IMS Confidential report D47.

[8] Tromp, J. and Nichols, S.C. (2003). *VIEW-IT: A Virtual Environment Usefulness Assessment Tool for Industry*. HCI international, 22–27 June. Proceedings of the 10th international conference on human computer interaction. Crete: Lawrence Erlbaum Associates.

[9] Sharples, S.C. and Saikayasit, R. (2006). *Effectiveness of Virtual Prototypes: Perception of Material Properties in CAD and VR*. Proceedings of IEA.

[10] Yates, T.K., Sharples, S.C., Morrisroe, G., and Clarke, T. (2007). Determining user requirements for a human factors research train driver simulator. In *People and Rail Systems: Human Factors at the Heart of the Railway* (J.R. Wilson, A. Mills, T. Clarke, B. Norris, eds) Aldershot: Ashgate.

[11] Patel, H., Stefani, O., Nichols, S., et al. (2006). Human centred design of 3D interaction devices to control virtual environments. *International Journal of Human Computer Studies*, 64 (3), 207–20.

[12] Kennedy, R.S., Lane, N.E., Berbaum, K.S., and Lilienthal, M.G. (1993). Simulator sickness questionnaire: An enhanced method for quantifying simulator sickness. *International Journal of Aviation Psychology*, 3 (3), 203–20.

[13] Nichols, S.C., Haldane, C., and Wilson, J.R. (2000a). Measurement of presence and side effects in virtual environments. *International Journal of Human-Computer Studies*, 52 (3), 471–91.

[14] Hettinger, L.J., Berbaum, K.S., Kennedy, R.S., et al. (1990). Vection and simulator sickness. *Military Psychology*, 2, 171–81.

[15] D'cruz, M.D. (1999). *Structured Training in Virtual Environments (STRIVE)*. PhD Thesis, University of Nottingham.

[16] Hone, K.S. and Baber, C. (2001). Designing habitable dialogues for speech-based interaction with computers. *International Journal of Human-Computer Studies*, 54 (4), 637–62.

[17] Barfield, W., Baird, K., and Bjorneseth, O. (1998). Presence in virtual environments as a function of type of input device and display update rate. *Displays*, 19, 91–8.

[18] Stedmon, A.W., Nichols, S.C., Patel, H., and Wilson, J.R. (2004). Free-speech in a virtual world: A user centred approach for developing speech commands. In *Contemporary Ergonomics* (P. McCabe, ed.) London: Taylor and Francis.

[19] Schnädelbach, H., Penn, A., Steadman, P., et al. (2006). *Moving Office: Inhabiting a Dynamic Building*. Proceedings of CSCW conference. Banff: Canada, pp. 313–22.

[20] Bayon, V., Griffiths, G. and Wilson, J.R. (2006). Multiple decoupled interaction: An interaction design approach for groupware interaction in co-located virtual environments. *International Journal of Human Computer Studies*, 64 (3), 192–206.

[21] Slater, M. (2002). Presence and the sixth sense. *Presence: Teleoperators and Virtual Environments*, 11 (4), 435–9.

[22] Stanney, K.M. and Hash, P.A.K. (1998). Locus of user-initiated control in virtual environments: Influences on cybersickness. *Presence: Teleoperators and Virtual Evironments*, 7 (5), 447–59.

[23] Nichols, S., Ramsey, A.D., Cobb, S., et al. (2000b). *Incidence of Virtual Reality Induced Symptoms and Effects (VRISE)*. HSE publication, p. 274.

[24] Wilson, J.R., Nichols, S., and Ramsey, A. (1995). Virtual reality health and safety: Facts, speculation and myths. *VR News*, 4 (9), 20–4.

[25] Behr, K-M., Nosper, A., Klimmt, C., and Hartmann, T. (2005). Some practical considerations of ethical issues in VR research. *Presence: Teleoperators and Virtual Environments*, 14(6), 668–76.

Meeting Diversity in Ergonomics (2007)
Pikaar, Koningsveld, and Settels (Eds)

12

RealPeople; Capturing the Emotions of Product Users

C. Samantha Porter, J. Mark Porter and Shayal Chhibber
Department of Design and Technology, Loughborough University, Loughborough, LE11 3TU, UK

Abstract. RealPeople is a DVD-based design resource, specified on the basis of interviews with designers. It is designed with the aim of making designers more aware of the specific characteristics of products that give pleasure to the people who own them. The context for RealPeople is a world of ergonomics that is becoming ever broader in terms of what designers are expected to understand about the users for whom they are designing. This goes beyond the physiological and psychological to the emotional. The resource contains information from an interview survey of 682 people concerning their attitudes towards functionality, usability, product pleasure and product preference. However, the main feature of RealPeople is the in-depth interview of an 100 people talking about their lifestyle, preferences for brands and style in general, and then describing three products that they own that give them the most pleasure. These descriptions are presented as 2–3 minute edited video clips, together with a breakdown of issues using the Four Pleasures framework. Each video clip is highly immersive, thought-provoking, and selected to encourage empathy between the designer and each individual in the database. The resource is currently being beta-tested by practising designers.

Keywords: designers, pleasure, product design, emotion, tools, methods

INTRODUCTION

The world of ergonomics is constantly changing, and the boundaries are being extended, as new technologies are developed and introduced into products, systems and services. A review of the papers, in the area of product design, presented at the International Ergonomics Association Triennial Congress, 2006 reveals two distinct themes; 'designing to accommodate different user groups' and 'tools and methods to facilitate understanding of, and empathy with, potential users'.

The first of these, 'designing to accommodate different user groups' (also known as universal design, inclusive design, design for all, design for the third age and design for more) is well described in the work of Johan Molenbroek and the Department of Industrial Design Engineering at Delft University [1]. It is further evidenced in the paper by Porter et al. [2] in their reporting of the HADRIAN system, a computer-based inclusive design tool developed, initially through the EPSRC 'EQUAL' (Extending Quality of Life) initiative, to support the design of kitchen and shopping based tasks. Goebel and Yoo [3], Oliveira et al. [4] and Wang and Chiou [5] also describe the development of products for the elderly and less able.

The second theme, 'tools and methods to facilitate understanding of, and empathy with, potential users' is also well represented, with a number of papers addressing this issue [2, 6–9]. It has been a topic of concern for ergonomists working in the area of product, system and service design for a number of years [10–12]. It is an obvious fact that designers should be primary users of ergonomics data and guidelines; applying an understanding of the consumers that they are designing for is imperative to the successful design of products. However, for many years it was evident that designers did not always use ergonomics data, guidelines and tools appropriately, if indeed they used them at all. Burns and Vicente [13],

Hasdogen [14], Pheasant [15] and Porter and Porter [16, 17] provide much evidence to support this and gave a number of reasons why this may indeed be the case:

- communication of ergonomics information at an inappropriate point in the design process
- communication difficulties between ergonomists and designers/engineering designers caused mainly by educational and practice differences
- communication of ergonomics information and data in an inappropriate fashion, by ergonomists to designers
- a number of 'fallacies' believed by designers, e.g., if a design is satisfactory for me it will be satisfactory for everybody else and the variability of human beings is so great that it cannot possibly be catered for in any design, but since people are so wonderfully adaptable it doesn't matter anyway [15].

In more recent years the use of ergonomics tools and methods in the design process has increased with ergonomists and designers alike realising the value of knowledge of the populations for whom they are designing in determining the success of the designed product. The development of tools such as the inclusive design tool, HADRIAN [18–19] and the website design tool, DENIM [20] that make ergonomics accessible to designers is evidence of this. Additionally, the increasing use of ethnographic methods by ergonomists, social scientists and design researchers [21, 22] is further substantiation of the recognition of the value of such tools and methods.

This increased awareness of the need of designers for ergonomics tools and methods tailored to their education and working processes has occurred against a backdrop of a considerable change in the relationship that consumers now expect to have with their products. There has been a shift, in the global market place of developed countries towards an 'experience' economy [23]. Young et al. [24] suggest that in the product design domain this shift has seen the contextual issues associated with products (their social/ideological context) become much more important than the physical ones, e.g., styling. Additionally, they advocate that the physical issues have evolved to have more emotional relevance, e.g., the semantic cues that a particular physical property may imply.

To satisfy these evolving consumer needs, the consideration of emotional satisfaction has to be integrated into the design process. Young et al. [24] highlight three key drivers that are accelerating the necessity for this change:

- Societal change; consumers are more knowledgeable and demanding and this places a greater pressure upon retailers and manufacturers. The designer is well placed to balance the consumer demand and the manufacturing budget.
- Technology driven by human pull; consumer preference, human factors and satisfaction will reshape the product creation process and override the current efficiency driven system.
- Emotional bonds; it is the contextual meaning of a product that creates a higher level of interaction and strategies have to change to meet this future demand.

Evidence suggests that consumers now desire products that they are able to engage with on an emotional level rather than products that are simply usable and functional [25–27]. Consumers have come to expect high quality products, with good functionality and usability, and it has been suggested that these factors no longer drive consumer choice. Consumers now look for more from the products that they buy; they are looking for pleasure and the fulfilment of their emotional needs.

Jordan [28] adapted Maslow's [29] hierarchy of human needs to create a hierarchy of 'consumer needs' that demonstrates the paradigm shift (see Fig. 1). As with Maslow's model,

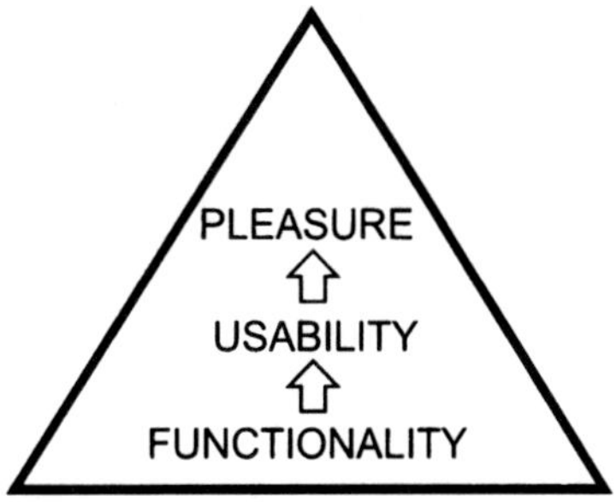

FIGURE 1. Jordan's 'hierarchy of consumer needs'.

the hierarchy follows the premise that once one level of need is met then the consumer seeks the level of need above. Jordan [26] argues that functionality and usability are at an acceptable level in most contemporary products and therefore consumers are now looking to their products to give them pleasure. Some researchers have questioned the validity of this model [30] but it serves to demonstrate the way in which that consumer–product relationship has changed.

Much work in the field of consumer research offers support to the argument that consumers have become more driven by pleasure, with literature from this field predominantly grounded in areas such as marketing, branding and consumer satisfaction in the service industries; ideologies and principles from this area of research also have much to offer design.

The literature shows that customers who are merely satisfied by the products of a brand have little 'brand loyalty' [31]. Differentiating a product from its competition, making it more desirable to the consumer, can encourage new consumers and strengthen brand loyalty [32]. Research has shown that today's consumers no longer make product choices by logical product comparisons; choice is increasingly driven by 'hedonic consumption' [33].

Burns and Evans [32] cite several examples of products that appear to owe their success to this change in consumer behaviour, such as the Apple iMac computer. It is a desirable and very successful product, and yet compromises in terms of both compatibility and upgradeability. Their explanation is a shift towards 'affective choice'; people now buy for pleasure making decisions based on the subjective feelings that are evoked by products.

Currently, designers tend to provide emotional input through intuitive techniques, lacking formal methodology. Additionally, the traditional usability approaches of ergonomists tend to view user satisfaction as the 'avoidance' of negative sensations rather than the promotion of positive ones [34]. Jordan argues [26] that in today's consumer market 'good' human factors and 'good' design are expected and usability is no longer a satisfier but rather a dissatisfier, i.e. where usability is poor, the user is dissatisfied.

However, tools and methods do exist that enable the formalisation of the design of pleasurable aspects of products. There are both generic tools and methods that are becoming more well known to the designer that can be used to reveal the ways in which consumers interact with products and the relationships that they have with them, e.g., ethnographic tools. There are others that are known in particular domains of product design e.g., Kansei engineering [35]. Additionally, there are also a number of other tools, techniques and methods available, from the design and related disciplines. However, many of these are relatively unknown to, and unused by, the design community. Hence there is a real need for research and development in this area.

THE ENGAGE PROJECT

In 2004, the European Commission funded the ENGAGE project; a Coordination Action [9]. The aim of the project was to bring the emerging area of research, described earlier, to the design and related disciplines, by bringing together individuals from research, design and industry to create an active knowledge community and knowledge base in the domain of emotional design. The consortium members are Instituto de Biomecánica de Valencia, University of Leeds, University of Palermo, Technische Universiteit Delft, Lund University, Linkoping University, Chalmers University, University Mariboru, Loughborough University, Packforsk, Fraunhofer-Technologie-Entwicklungsgruppe TEG, Interactive Institute, Volvo, Factory Design, Electrolux, Permas, Hergar, Proctor & Gamble, Middle Eastern Technical University, Gradient, Philips and the Design and Emotion Society. It is a European 6th framework coordination action, which is ran from September 2004 until February 2007.

ENGAGE had a number of objectives:

- to collect and classify emotional design tools, techniques and methods
- to identify best practice
- to promote a shared insight and common language
- to identify gaps in current tools, techniques and methods and to define directions for future research.

The main output of the consortium is its website, which is hosted by the Design and Emotion Society, based in the Netherlands. The Design and Emotion website (see Fig. 2,

FIGURE 2. Design and emotion society website 'homepage'.

FIGURE 3. ENGAGE website 'homepage'.

http://www.designandemotion.org/) comprises news, an events calendar, a knowledge base and the ENGAGE website. The ENGAGE website (see Fig. 3, http://www.designandemotion.org/engage) consists of newsletters, documents, both consortium-only and public domain, and a collection of published tools and methods for designing for emotion. The hosting of ENGAGE by the Design and Emotion Society will ensure the continuing support of the tools and methods collection and dissemination beyond the life of the ENGAGE project.

The members of the ENGAGE consortium have collected over 70 tools for designing for emotion and there are more than 50 tools in the collection. It has not been the role of the consortium to evaluate these tools. Individuals submitting tools fill in an online template where they provide the problem addressed, the solution provided, a description of the tool, its limitations, the theoretical background, images (if applicable), where the tool has been applied, in which stage(s) of the design process it can potentially be used, related tools and added value, publications and owner contact details. A review panel reviews and classifies the tools before they are published on the website.

Users are able to search the tools via two different search engines; these were included to ensure that all potential users (design researchers, industry and practicing designers) have a means of searching that is meaningful to them. The first is by the tool classification devised by the review panel; two broad categories, previously defined by Sanders [22]; generative tools for use in the earlier stages of the design process and evaluative tools for use in the later stages of the design process:

- generative tools and methods:
  - tools to COLLECT information (6 tools)
  - tools to REPRESENT/EXPLORE information (15 tools)
  - tools and methods to DEFINE product characteristics (11 tools)
- evaluative tools and methods:
  - tools to measure SENSORY characteristics (9 tools)
  - tools to measure the EXPRESSION/MEANING of products (6 tools)
  - tools to measure the EMOTIONAL reaction to products (10 tools)

The second method is by stage of the design process shown in Table 1 with tools also being classified for designing for radical innovation and/or incremental change.

TABLE 1. Stage of the design process

Type of design\Stage of process	Understand user/market	Explore ideas and concepts	Design specification	Test and evaluate	Market implementation

The ENGAGE website and collection of tools and methods was widely disseminated through the holding of a number of national and open events in a France, Germany, Holland, Italy, UK, Sicily, Spain, Sweden and Turkey.

Below two examples of typical tools are described; the first PrEmo is an evaluative tool that is described in brief and the second is RealPeople, a generative tool, the development of which is described in detail.

PREMO, PIETER DESMET, DELFT UNIVERSITY OF TECHNOLOGY

Emotional responses elicited by consumer products are difficult to measure because their nature is subtle (low intensity) and often mixed (i.e. more than one emotional response at the same time). PrEmo [36–38] allows respondents to report their feelings towards products with the use of expressive cartoon animations instead of relying on verbal expression. Fourteen emotions are portrayed by an animation of dynamic facial, bodily, and vocal expressions. These are shown in Fig. 4.

The unique strength of PrEmo is that it combines two qualities: it measures distinct emotions and it can be used cross-culturally because it does not ask respondents to verbalise their emotions. In addition, it can be used to measure mixed emotions. PrEmo can be useful as both a qualitative tool for creating insights into the relationship between product features and emotional impact that are valuable in an early design stage, or as a quantitative tool for evaluating the emotional impact of existing designs e.g., for creating an emotional benchmark. See Fig. 5 for one such example.

PrEmo has been used extensively by many different types of industry and recent applications include automotive design (Mitsubishi; Daimler-Chrysler), biscuit packaging (Bolletje), office chairs (Ahrend), mobile telephones (KPN Telecom), websites (Starchild foundation),

FIGURE 4. The fourteen PrEmo emotion animations.

gearlevers (Johnson Controls), cooking oil packages (Diamant), characteristics of city districts (city of Enschede), sport shoes (Nike), childrens wheelchairs (Havenkamp), fragrances (Procter & Gamble) and meals and snacks (KLM; Barilla).

REALPEOPLE: A 'PLEASURE' RESOURCE FOR DESIGNERS

RealPeople is a resource suitable for use at the beginning of the design process, to guide and focus the design direction and facilitate understanding of consumer groups by designers. In a previous study [30], data concerning peoples' attitudes towards the pleasure giving properties of the products they own were collected and classified against the 'four-pleasure' framework developed by Jordan [26, 34]: physio-pleasure, socio-pleasure, psycho-pleasure and ideo-pleasure. A rudimentary paper-based database of peoples' most pleasurable products was developed. Informal evaluation by designers indicated that a 'pleasure' resource would be of great value to them. This provided the premise for the further investigation of users' pleasure needs and the subsequent development of the RealPeople resource.

Development of the 'pleasure resource'

To ensure that the information collected will be accessible to designers it was necessary to consider carefully the design of the resource itself. The input of ergonomics information into the design process is immensely valuable [39]. However, as described earlier, there has traditionally been a communication 'gap' between the disciplines of ergonomics and design. Porter and Porter [12] suggest that the differences between them (and other related disciplines) may be due to consequences of innate ability, education and the real world practices of the different disciplines. Most ergonomics methods are quantitative or qualitative, and are

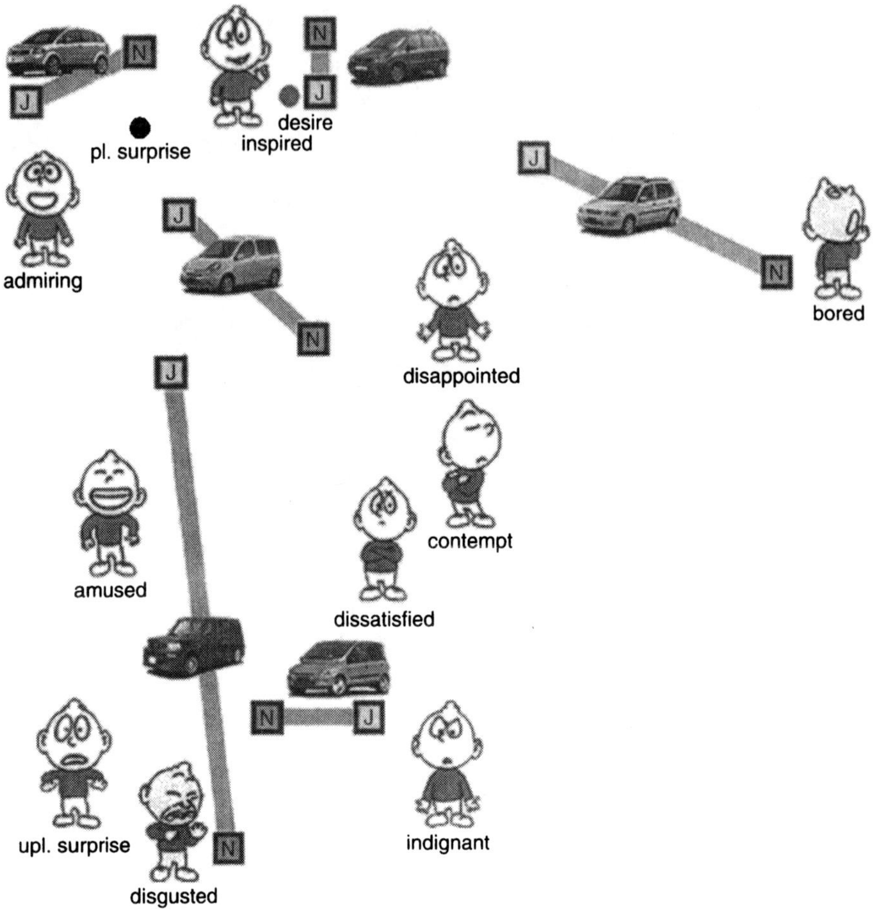

FIGURE 5. An 'emotional' benchmark.

essentially analytical tools. They provide data about people's capabilities and reactions to design variables, but do not generally lead directly to design solutions; this can lead to frustration for the designer [21]. The data are often in a scientific non-prescriptive form, which the designer finds hard to interpret [39]. This leads to human factors information frequently being left out of the design process [40, 41] or being used in an inappropriate way [10].

It is sometimes suggested that working in multi-disciplinary teams can overcome these problems and it is a solution that has reported successes [42]. However, Feeney and Bobjer [39] note that a multi-disciplinary approach can also be perceived by designers as intruding into their skill area, and that there is a general pre-occupation with technology and aesthetics within the design community. Ergonomics is also be seen as costly, and of low value and accessibility by designers Burns and Vicente [10]. Hasdogan [14] also supports this notion, concluding that much ergonomic data is geared towards workplace circumstances and the trained user, and that information is presented in forms more suited to ergonomists than designers.

Clearly the format of ergonomics information and its point of input into the design process are critical. Methods exist that enable the classification and evaluation of user 'pleasure', e.g., PrEmo, but little was known about what methods designers are aware of or are currently using. There was also a need to understand what types of information designers would like in a 'pleasure' resource and crucially, how they would like to access it.

Designer interviews

The aim of the designer interviews were:

- to investigate the current level of understanding in the design community of 'user pleasure' and related research methods, tools and resources
- to investigate designer requirements for a resource intended to enable a greater understanding of the pleasure needs of their target market
- to develop concepts of how a 'pleasure' resource may function.

The protocol for the designer interviews was developed following a series of informal interviews with 14 student and academic designers, 6 of whom had some industrial experience. This enabled the investigators to identify the issues of importance and to gain ideas for the resource. The protocol followed is described below.

A pre-interview information pack was developed (see Fig. 6) and sent to the 13 participants 1–2 weeks before their interview. The pack provided a comprehensive overview of the project and the interview they had agreed to take part in. It consisted of several different components: a covering letter; a brief overview document; 6 interview cards with an image and text relating to the 6 main topic areas of the interview, with space for annotations; and an A3 storyboard illustrating a sample concept of the resource. In conjunction with the structured interview protocol, the information pack facilitated effective interview sessions ideally suited to the time constraints of industry.

The major findings of the interviews are summarised below.

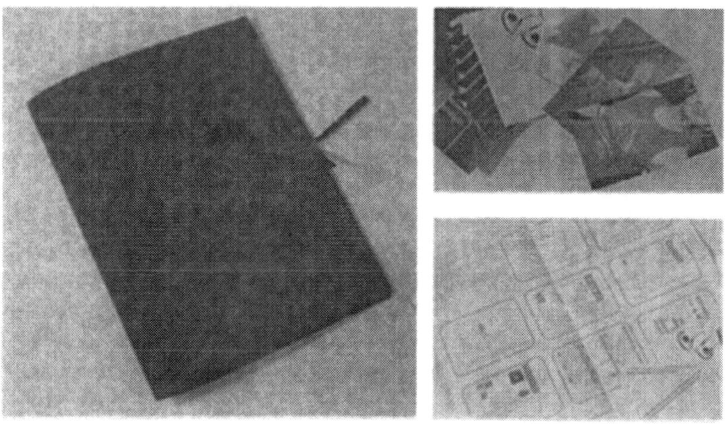

FIGURE 6. The pre-interview information pack.

How important is user pleasure in the design process?

There was a consensus that 'user pleasure' is of real importance from the 'designers' perspective. However, when placed in the context of an actual design project, it becomes one of a number of factors that designers have to consider and make compromises between. Four of the designers added that they often argue the case for certain features if they feel that the inclusion of them will benefit the user and, consequently, the client. However, there has tended to be a focus on providing pleasure through aesthetic appeal and physio-pleasure and a lack of appreciation of other ways of providing pleasure.

Which methods do you use to gain a more holistic understanding of the target user?

The majority of designers interviewed were strong advocates of user-centred research in the design process. However, the volume of research in the design process, and the emphasis placed upon it, varied. Nine participants were from companies that appeared to place a greater emphasis on 'design through research'. The other 5 design more intuitively, basing design decisions on their own feelings. The majority took an iterative approach to the design process, revisiting the research data or conducting further research to guide design decisions.

The majority of practicing designers used traditional design research methods, e.g. focus groups, to gain a more holistic view of the user. Eight of the designers also employed methods such as ethnography, role playing and brainstorming, and 'live the life' research methods (visiting consumers in their homes and spending extended periods of time with them to witness their day-to-day activities first hand) to give a more intimate picture of their target consumer. However, there did not appear to be any real scientific rigour in the research, which is understandable given the nature of design education, and the time and cost restrictions inherent in most projects. Few of the designers were aware of any pleasure-specific research tools.

Would you find a pleasure resource useful?

There was a very positive response to the potential of the pleasure resource; many raised the issue of updating the resource and the dangers of trends becoming dated.

There was concern from a couple of the interviewees about the role that the resource was intended to have in the design process and whether designers would misuse the resource by focusing too much on certain users in the database, or basing design decisions solely on information provided by the resource. This issue is strongly connected to the design education, and company design philosophy, of the designer using the resource; their willingness to design in an inclusive way and utilise their findings from the resource, to guide and inform further first hand research, is essential in the use of such methods.

What additional information would you like when designing?

The interviewees were satisfied with the suggested type, quantity and relevance of the data that would be collected; a mix of information, ranging from statistically tested general data to more intimate rich data about specific individuals. The majority of interviewees also felt that information about brands would be very useful; perhaps peoples' most pleasurable brands, as one participant suggested. Two of the designers also felt that information that allowed them to discover more about user groups' aspirations and attitudes would be useful.

How would you like the resource to function?

The designers wanted a flexible and intuitive resource that allowed access to information in different ways. There was consensus that access to information would have to be very quick; a consequence of this demand is that the resource layout and structure must be logical and easy to learn. Additionally, the resource should be a visually stimulating and interactive database. Concepts such as movie clips and sound clips, interactive user mood boards, and cut and paste workspaces where a designer could store information were also of interest.

How important is visual information during the design process?

The interviewees unanimously agreed that visual information is the quickest and most effective way to communicate ideas and themes to a range of different audiences, e.g., clients, senior management and internally between designers.

The interviews that formed this study led to several conclusions that have been used to inform the development of an accessible 'pleasure' resource for designers:

- User pleasure is of growing importance to designers but they are unaware of any tools/methods/resources that are available to them. They tend to incorporate only physio-pleasure and are less aware of other ways that they are able to bring pleasure to the user. The pleasure resource must be representative of all pleasure types and raise awareness in the designer population.
- Designers are increasingly employing user-centred design research methods to develop a more holistic view of the user, but methods are often 'quick and dirty', with an intuitive evaluation of data. Time is an issue for designers who are often working to very tight time constraints, so tools must be quick and easy to use.
- Designers would find a pleasure resource very useful but there were concerns of it being misused, e.g., focusing on particular users. The resource must be designed in such a way that accessing all users is encouraged/facilitated.
- The information in the resource must immerse the designer in peoples' lifestyles in a highly visual and engaging way. As much lifestyle information as practical would be appreciated.
- Designers require flexible and intuitive access to information in the 'pleasure' resource. They also emphasised the need for visual presentation as a means of communicating ideas in the design process.

While it is crucial to satisfy designers' needs, it is also important to develop mechanisms in the resource that promote inclusive design decisions, and reinforce the principle that the resource is a reference database aimed at inspiring and guiding design direction and research. The resource was never intended as a replacement for user-centred design and research, but as a tool that guides the designer into the key issues of their target consumer earlier in the design process. The above findings were used to guide the design of the RealPeople DVD-based design resource.

Collection of the data to inhabit the resource

Two separate research activities were undertaken concurrently, to provide the different types of information requested by the designers.

General trend data

The first was the collection of general trend data concerning attitudes towards functionality, usability, product pleasure and product preference from 582 participants, in 11 UK locations (including London, Cardiff, Edinburgh and Nottingham). Participants were asked to select from a number of statements the one which best described their feelings towards a particular aspect of product functionality and usability, and to identify any of the statements concerning pleasure that represented their own feelings. These are shown in Table 2. A quota sampling method was used to ensure an even spread across gender and age (equal numbers of participants in the ranges 18–25, 26–35, 36–45, 46–55, 56–65, 66–75).

Nominal data were collected from the trend data questionnaire and analysed to identify trends. A non-parametric χ^2 test was employed in two ways: as a test of fit, and as a test of independence, to identify population trends and gender and age differences. The pleasure data from the smaller sample was also included. A number of trends were identified at $p < 0.05$ but only those where $p < 0.01$ are presented below. They are presented in the categories functionality, usability, physio-pleasure, socio-pleasure, psycho-pleasure and ideo-pleasure. Each category includes population trends, age trends and gender trends.

Functionality

There are a number of significant trends in the functionality category.

Population Trends The sample population showed a preference towards products that have the exact functionality required (41%), or slightly higher levels of functionality that can be explored (42%). There was a low appreciation of products with more functionality than will be used (9%), or maximum functionality (8%).

Age Trends The older age groups prefer products that display exact functionality. This is mirrored by a trend of younger age groups preferring products with more functionality that they can explore, with increasingly older age groups finding this less appealing.

Gender Trends 45% of females, as opposed to 36% of males, have a preference for products that have the exact functionality required; the difference approaches significance. This would is mirrored by a slight preference in males, 45% as opposed to 39% of females, towards products that have additional functionality that can be explored.

Usability

Population trends There was a preference across the sample population, 60% for products that are simple and easy to use. 27% of the sample prefer products which are initially a challenge to use but then easy.

Age trends 60% of the subjects preferred products that are simple and easy to use; a high proportion of these are from the older age groups, 83% of the 56–65 age group, compared with lower percentages in the younger age groups, e.g., 43% of the 18–25 year old age group and 50% of the 26–35 year old age group. Approximately 27% of the sample preferred products that were initially challenging to use, but after a period of ownership were easy to use. A higher proportion of that 27% are from the younger age groups, 31% of 18–25 and 36% of 26–35, indicating that younger age groups are more willing to use products that require a level of learning or are a challenge to use. The percentage of the 56+ age range seeking a similar level of usability was markedly less at 17%. This highlights the fact that younger age ranges tend to be more comfortable with technology and with learning to use it.

TABLE 2. Atttitude statements

Product characteristic	Attitude statement
Functionality	I like products that: – have exactly the functionality I know I will use – have exactly the functionality I know I will use, plus some other functions that I am interested to explore and evaluate whether they will be useful – have more functionality than I will probably use – have the maximum available functionality, despite the fact that I will not use many of these functions
Usability	I like products that: – are simple and easy to use first time – are challenging to learn, but once learned they are easy to use – that are challenging to use even after I have owned them for a while e.g., a computer game with increasingly more challenging levels – have 'secrets' and hidden features that I have to discover over time, whilst learning to use the product
Physio-pleasure	– the colour(s) of a product are important to me – the touch and feel of a product when I interact with it are important to me – the 'right' sound of a product in use is important to me e.g., the clunk of a car door – the way materials are used in a product is important to me – the shape and form of a product is important to me
Socio-pleasure	I like products that: – demonstrate I have a discerning taste to other people – demonstrate to other people that I am successful – tell other people something about me e.g., sports watch, ethnic clothing – are a talking point amongst my friends and/or my family. – are a talking point amongst any group of people – allow me to socialise – that fit into any social context e.g., a wristwatch that can suit a formal or casual situation – that are understated
Psycho-pleasure	I like products that: – express an aspect of my personality – allow me to complete tasks easily – operate in a meaningful way to me e.g. desktop metaphor on a computer – have some level of personal significance to me e.g., a gift from a loved one
Ideo-pleasure	I like products: – that represent an ideology that I believe in e.g., eco-friendly, fair trade, materialism – Where the overall aesthetics (the combination of form, textures, colour, etc. that create the aesthetic) of a product are important to me – from a particular brand – by a particular designer

Gender trends Responses demonstrate that males and females have similar attitudes towards product usability; a desire for products that are either, simple and easy to use, or perhaps challenging when initially used.

Physio-pleasure

Population trends The shape and form of a product was important to 66% of the sample population. Additionally, it was the most valued of all of the physio characteristics for the sample. Colour and tactility of the product are very important to 60% of the sample and 39% think that the sound that a product makes is important.

Age trends The younger age groups value the colour of products to a greater extent than older age groups, e.g., 72% of 18–25 year olds compared to just 47% of the 56+ age group. A similar age effect is seen with the tactile aspects of products, e.g., 71% of 26–35 year olds compared to 49% of 56+ year olds value the way a product feels to touch. There were not the same significant effects when considering the sound that a product makes and the overall form.

Gender trends 70.2% of the female sample valued colour as a product characteristic, compared with just 48.9% of the male population. Although product sound was valued by only 39% of the total sample, the data when viewed across gender shows an effect. 45% of the males in the sample population valued this characteristic, compared to just 33.8% of the females.

Socio-pleasure

Population trends Only 33% were interested in products that demonstrate that they have discerning taste to others and 19% were interested in products that demonstrate that they are successful. Few were interested in products that tell other people something about themselves, e.g., a sports watch. 38% like products that are a talking point amongst their friends and/or family, with less people being interested in products that are a talking point amongst any group of people. 62% like products that fit into any social context and 40% like products that allow them to socialise

Age trends The 26–35 age group, 46%, are significantly more interested in products that demonstrate discerning taste to others in contrast to the other age groups in the sample, e.g., 36–45, 27.8%, and 56+, 26%. The younger age group are also more interested in products that tell other people something about themselves: 18–25, 40%, compared to 46–55, 12%. There is a general trend of a reduced appreciation of products that tell people something about them with increasing age. The younger age groups are also more interested in products that are a talking point to family and friends – 18–25, 50% and 46–55, 20% – and in products that are a talking point amongst any group of people. Evaluation of this product characteristic across the entire sample population showed that it did not appear to be of great only 20% valued it. However, it was the younger age groups that found this aspect more appealing than the older groups. They are also more interested in products that allow them to socialise, e.g., 18–25, 49%, 56+, 27%. The middle age groups show a greater preference for products that are understated, e.g., 63% of 36–45 year olds, compared to 46% of 18–25 year old, or 43% of 56+ year olds.

Gender trends More males, 38.2%, show a preference for products that demonstrate a discerning taste than females, 26.5%. Females, however, find more pleasure in products that allow them to socialise, 47% compared with 34% of males.

Psycho-pleasure

Population trends The results for this product characteristic correspond closely with the results for usability. However, this psycho-pleasure attribute relates more to the sense of satisfaction from task completion and the ability to achieve this easily; 82% desired this characteristic in products. 71% found products with personal significance to them to be appealing.

Age trends Younger age groups tend to find greater pleasure from products that express their personality more than older age groups, e.g., 73% of 18–25 year olds, compared to 27% of the 56+ age range.

Gender trends Females gain significantly more pleasure from products that express an aspect of their personality than males: 52% as opposed to 45%. They also experience more pleasure from gifts that have a personal significance; 77% versus 64%.

Ideo-pleasure

Population trends 63% like products that represent an ideology e.g., eco-friendly. 72% value the overall aesthetic of a product and a majority do not like designer/branded products.

Age trends It is evident from looking at the data that overall aesthetic is an important characteristic to the sample population as a whole. There is a steadily reducing appreciation of this characteristic as the age groups get older e.g., 18–25 year olds 82.7%, compared to 60% of 56+ age group. Similarly, younger age groups seem to gain more pleasure from products by a particular designer than older age groups, for example, 26.5% of 26–35 year olds compared to 10% of 56+ year olds stated that they found such products pleasurable. There are no apparent age differences in the appeal of products that represent an ideology or loyalty to a particular brand.

Gender trends 68.9% of females like products that represent an ideology they believe in, compared to 57.1% of males.

In-depth data collection

The second activity was in-depth interviewing of an additional 100 people, providing more intimate and richer data from people discussing their three most pleasurable products, again across genders and a wide age range. An 'inclusive' design approach was taken to facilitate accessibility, by designers, to the broadest ranger of users possible. An interview questionnaire was developed through several pilot studies; the content was strongly influenced by the views of practising designers [13]. The questionnaire is subdivided into several sections:

- Lifestyle – a series of open-ended questions regarding different aspects of a person's life, e.g., leisure activities, favourite music and career aspirations. The aim of including this information is to immerse the designer in an individual's lifestyle.
- Pleasure attitude questionnaire – this was the same as used in the general survey; this pool of data were added to the general data set. It also gives the designer a brief overview of each person's general opinions towards pleasure in products.
- Brand choice – participants list several brands that they like, aspire to, or feel reflect their personality. These data are included to satisfy the need, specified by designers, to understand the aspirations of different individuals
- Style choice – participants were presented with images showing a sample of products from four product categories: mobile phones, chairs, fonts and wristwatches. Each category had 5 examples selected through a focus group with designers, to represent the style possibilities in that category. The participants selected which one of the 5 samples in each category they like the most, and briefly explained why. This was designed enhance the designer's understanding of particular users and their aesthetic preferences.

After completing the lifestyle questionnaire the participants took part in an informal video-taped interview where they presented the 3 products they own that bring them the most pleasure, explaining specifically what it is that gives them pleasure. The footage was edited into 2–3 minute high quality movie clips that encapsulate the pleasure the chosen product brings them. Each video clip is highly immersive, thought-provoking, and is selected to encourage empathy between the designer and each individual in the database. In addition to this, a table of bullet points of the most pertinent reasons why the product brings them pleasure, analysed and presented with respect to the 4 pleasure framework, is provided as a quick reference for the designer.

Each interview took place in the participant's home or a personalised work space; showing their chosen products in, as far as practically possible, their natural environment. It also allowed the interview session to be conducted in as relaxed and informal manner as possible, to facilitate the collection of such an intimate type of data.

RealPeople; the resource

Using the conclusions of the designer interviews as the design specification for the resource, the trend data in which general population preferences and gender and age differences were identified and the rich intimate data from the 100 individuals who were interviewed, the 'resource' was developed. RealPeople is a 'stand alone' DVD-based resource with pages being developed in Macromedia Director MX and collected data being stored in a database accessed via the scripting language 'Lingo'. The 'look' and 'feel' of the resource have been developed through an iterative design process using team members, other colleagues and designers for evaluation.

On the search page (see Fig. 7) users are able to select individuals or product types, choosing gender, age category and income bracket, giving the flexibility the designers asked for. As the categories are defined, all those individuals not included fade and the designer is left with the images of the relevant individuals/products to look at in more detail.

Figures 8–12 show the type of information available for each of the participants. Each page has a constant homepage for the participant, on the left, through which it is possible to access all of the other pages for that participant.

Additionally, users are able to:

- make and save notes
- have a slideshow of the videos as a screensaver selected via the search categories of interest at a particular time
- save items of interest (including notes) in project portfolios; these will be editable and can also be shown as slideshows
- export notes and portfolios to external media for use elsewhere
- search history and open previous searches.

Evaluation of the resource

A beta version of the resource has been evaluated by practising designers and views are extremely positive. The commercial version of RealPeople should be available for purchase in Summer 2007.

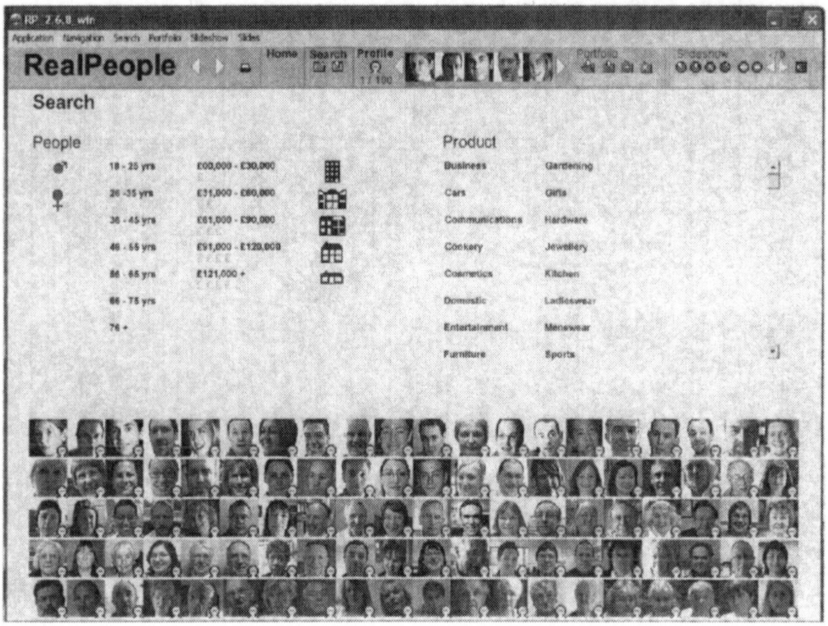

FIGURE 7. Search page showing selection categories and photographs of the 100 participants interviewed.

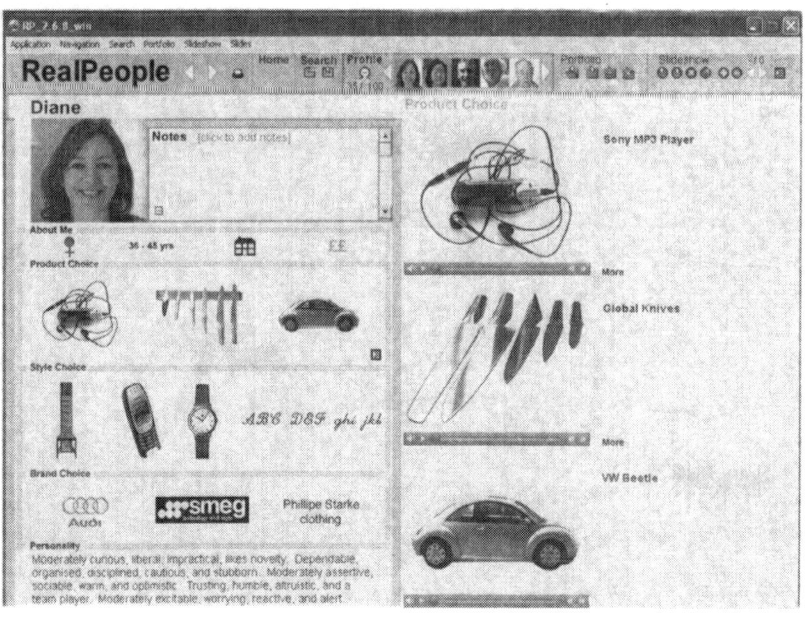

FIGURE 8. Individual 'Home' page and product videos.

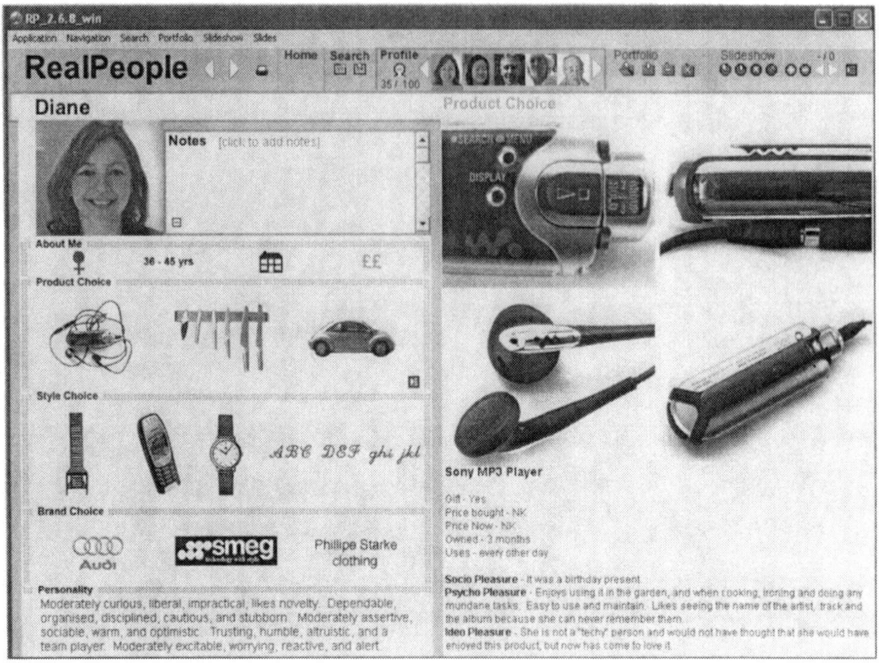

FIGURE 9. Individual 'Home' page and further product information.

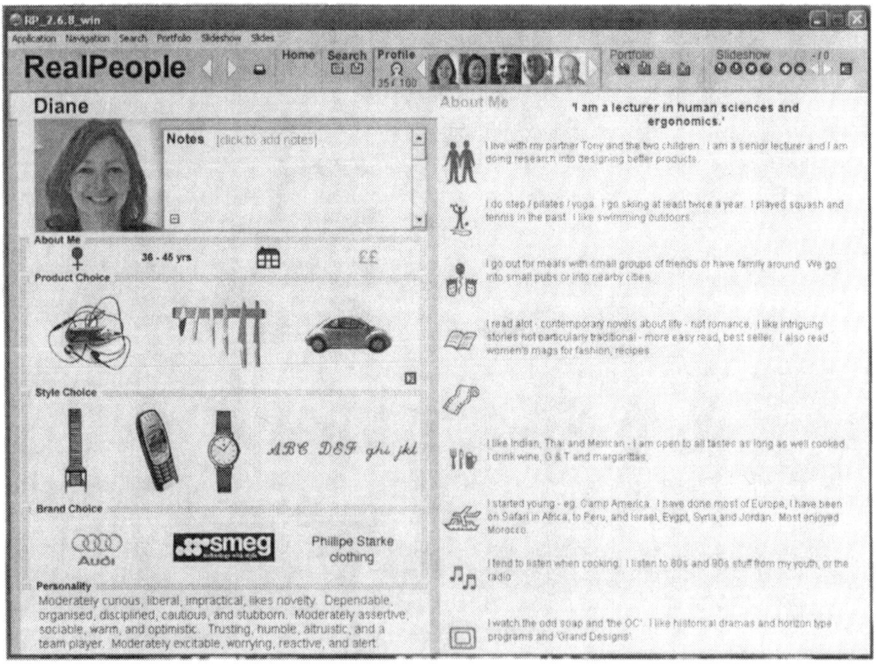

FIGURE 10. Individual 'Home' page and lifestyle information.

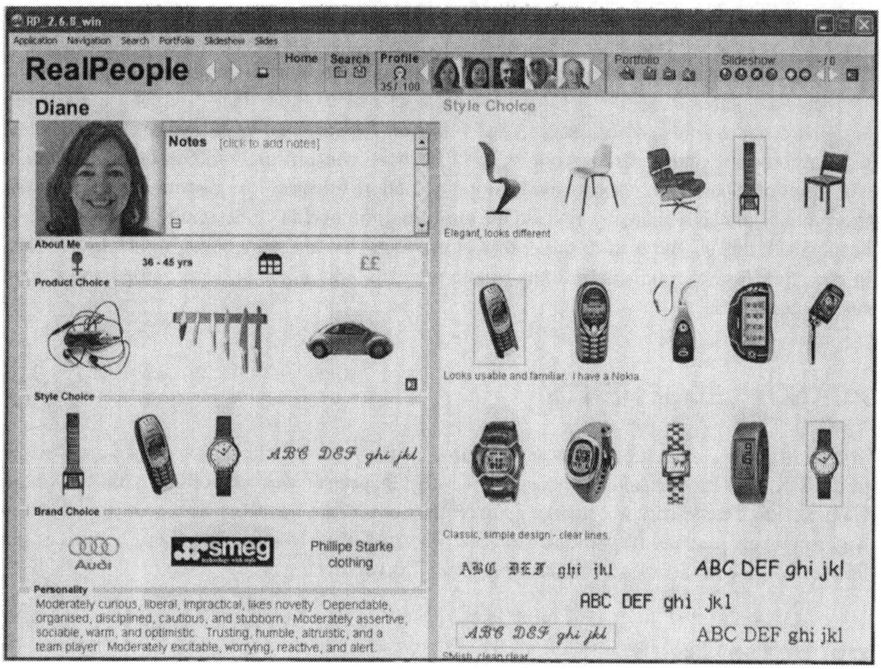

FIGURE 11. Individual 'Home' page and style choice.

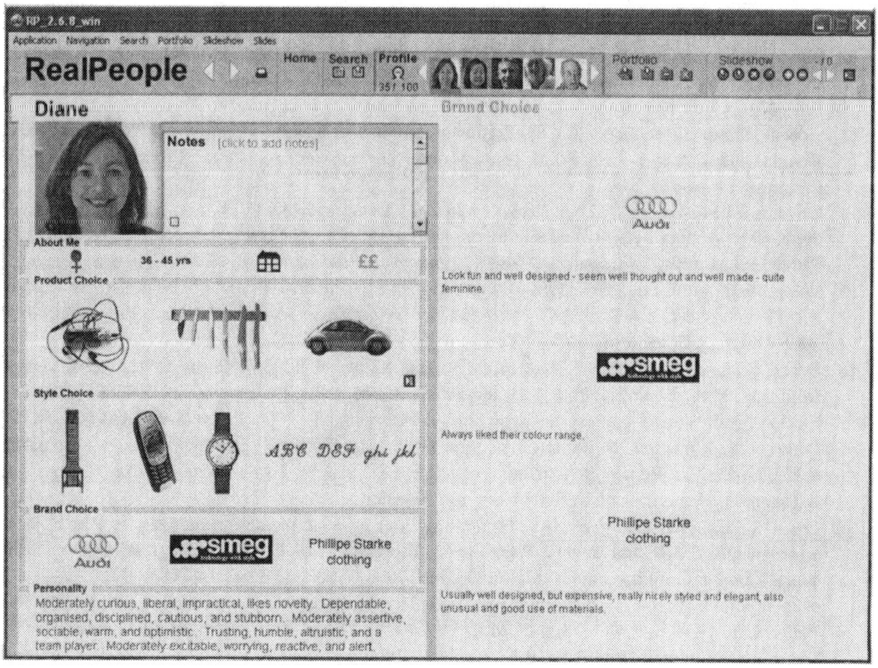

FIGURE 12. Individual 'Home' page and brand choice.

THE FUTURE FOR DESIGNING FOR EMOTION

The case for designing for users' emotional needs is now well established and there is an increasing interest in the development of tools and methods that facilitate and support designer's capabilities in this area. The collection, review and classification of tools and methods, currently being carried out by the ENGAGE consortium, will continue as a Design and Emotion Society activity, once the funding period finishes. The dissemination activities that have been carried out, by ENGAGE, mean that the profile of designing for emotion has been raised; this should lead to the design of products that are more engaging for users. This, in turn, may lead to products becoming more precious, less likely to be discarded and a more sustainable future.

ACKNOWLEDGEMENTS

The RealPeople research project was supported by the Arts and Humanities Research Council in the UK, and Procter and Gamble. The ENGAGE project was supported by the European Commission, Framework VI; thanks go to all the consortium members who contributed in any way to this chapter and in particular to Pieter Desmet, Paul Hekkert and Gael Laurans of the Faculty of Industrial Design Engineering, Delft University.

REFERENCES

[1] Molenbroek, J.F.M. and Dekker, M.C. (2006). Exploring the boundaries of design for all. In: Pikaar, R.N., Koningsveld, E.A.P., Settels, P.J.M. (eds), *Meeting Diversity in Ergonomics*. Proceedings IEA 2006 Congress, Elsevier.

[2] Porter, J.M., Marshall, R., Sims, R., Gyi, D.E., and Case, K. (2006b). HADRIAN gets streetwise. In: Pikaar, R.N., Koningsveld, E.A.P., Settels, P.J.M. (eds), *Meeting Diversity in Ergonomics*. Proceedings IEA 2006 Congress, Elsevier.

[3] Goebel, M. and Yoo, J.-W. (2006). Ergonomic product design for elderly users. In: Pikaar, R.N., Koningsveld, E.A.P., Settels, P.J.M. (eds), *Meeting Diversity in Ergonomics*. Proceedings IEA 2006 Congress, Elsevier.

[4] Oliveira, M.M., Campos, C.F., Viana, S.S.M., and Camara, J.J.D. (2006). Applied Ergonomics to the Kitchen Aiming Security and Well-being for the Elderly. In: Pikaar, R.N., Koningsveld, E.A.P., Settels, P.J.M. (eds), *Meeting Diversity in Ergonomics*. Proceedings IEA 2006 Congress, Elsevier.

[5] Wang, M.H. and Chiou, W.K. (2006). From user trip to universal design of the peelers. In: Pikaar, R.N., Koningsveld, E.A.P., Settels, P.J.M. (eds), *Meeting Diversity in Ergonomics*. Proceedings IEA 2006 Congress, Elsevier.

[6] Boess, S.U., de Jong, A.M., Rooden, M.J., and Kanis, H. (2006). Inside investigative design. In: Pikaar, R.N., Koningsveld, E.A.P. and Settels, P.J.M. (eds), Proceedings of IEA2006 Congress, Elsevier, 16th World Congress on Ergonomics, Maastricht, NL, 2006, ISSN: 0003-6870, [CD-ROM]

[7] Porter, C.S., Chhibber, S., Porter, J.M., and Healey, L. (2006a). RealPeople: Designing pleasurable products. In: Pikaar, R.N., Koningsveld, E.A.P., Settels, P.J.M. (eds), *Meeting Diversity in Ergonomics*. Proceedings IEA 2006 Congress, Elsevier.

[8] Gyi, D.E., Cain, R., Campbell, R.I., Facilitating 'user push' in the design process. In: Pikaar, R.N., Koningsveld, E.A.P. and Settels, P.J.M. (eds), Proceedings of IEA2006 Congress, Elsevier, 16th World Congress on Ergonomics, Maastricht, NL, 2006, ISSN: 0003-6870, [CD-ROM]
of incorporating ergonomics analyses into workplace design. *Applied Ergonomics*, 31 (3), 291–300.

[9] Porter, C.S., Desmet, P.M.A., Signes, M.J., Sperling, L., Ozcan, E., and Tito, M. (2006b). ENGAGE: Designing for emotion. In: Pikaar, R.N., Koningsveld, E.A.P., Settels, P.J.M. (eds), *Meeting Diversity in Ergonomics*. Proceedings IEA 2006 Congress, Elsevier.

[10] Burns, C.M. and Vicente, K.J. (1994). *Designer Evaluations of Human Factors Reference Information*. Proceedings of 12th Triennial Congress of the International Ergonomics Association. vol. 4, Ergonomics and Design, Toronto.

[11] Helander, M.G. (2000). Seven common reasons to not implement ergonomic. *International Journal of Industrial Ergonomics*, 25 (1) (January), 97–101.

[12] Porter, C.S. and Porter, J.M. (1997). 'The interface between ergonomists and product designers'. In: Seppala, P., Loupajarvi, T., Nygard, C.H. and Mattila, M., (eds), *Proceedings of the 13th Triennial Congress of the International Ergonomics Association*, Finnish Institute of Occupational Health, Tampere, Finland, pp. 240–242, ISBN 951-802-189-9

[13] Burns, C.M., Vicente, K.J., Christofferson, K., Pawlak W.S., Burns, C.M., Vicente, K.J., Christofferson, K., and Pawlak W.S., (1997). Towards viable, useful and usable human factors guidance. *Applied Ergonomics*, 28 (5/6), 311–322.

[14] Hasdogen, G. (1995). The nature and limitations of user models in the household product design process. *Design Studies*, 17 (1), 19–33.

[15] Pheasant, S. (1988). *Bodyspace*. London: Taylor and Francis.

[16] Porter, C.S. and Porter, J.M. (2000). Designing for usability: Input of ergonomics information at an appropriate point, in an appropriate form, in the design process. In: P.W. Jordan and W.S. Green, (eds), *Human Factors in Product Design*, London: Taylor and Francis, pp. 15–25.

[17] Porter, C.S. and Porter, J.M., (2000), CoDesigning: Designers and Ergonomics. In: Scrivener, S.A.R., Ball, L.J. and Woodcock, A. (eds), *Collaborative Design*, Springer-Verlag, London, 2000, pp 27–36, ISBN 1 85233 341 3

[18] Porter, J.M., Case, K., Marshall, R., Gyi, D.E., and Sims, R.E. (2004). 'Beyond Jack and Jill': Designing for individuals using HADRIAN. *International Journal of Industrial Ergonomics*, 33, 249–64.

[19] Marshall, R., Case, K., Oliver R., Gyi, D.E., and Porter, J.M. (2002). A task based 'design for all' support tool. *Robotics and Computer-Integrated Manufacturing*, 18 (3–4) (June–August), 297–303.

[20] Newman, M.W., Lin, J., Hong, J.I., and Landay, J. (2003) DENIM: An informal web site design tool inspired by observations of practice. *Human-Computer Interaction*, 18 (3), 259–324.

[21] Suri, J.F. and Marsh, M. (2000). Scenario building as an ergonomics method in consumer product design. *Applied Ergonomics*, 31 (2), 151–7.

[22] Sanders, E.B.-N. (2004). *Ethnography and the Empowerment of Everyday People*. A white paper written for Microsoft Corporation.

[23] Pine, B.J.II and Gilmore, J.H. (1998). Welcome to the Experience Economy, Harvard Business Review July-August, 76 (4), 97–105.

[24] Young, R.A., Van der Veen, G.J., Illman, M.E., and Rowley, F.J.B. (2000). *Creating Enhanced User Experiences: The Designer as a Co-operator by Facilitating Communication*. In: Scrivener, S.A.R., Ball, L.J., Woodcock, A. (eds), Proceedings of Co-designing, CollaborativeDesign, Springer-Verlag.

[25] Hummels, C. (1999). *Engaging Contexts to Evoke Experiences*. Proceedings of the first Design and Emotion conference, November 3–5, Delft University of Technology, pp. 39–45.

[26] Jordan, P.W. (2000). *Designing Pleasurable Products*. London: Taylor and Francis.

[27] Jordan, P.W. and Servaes, M. (1995). Pleasure in product use: Beyond usability. In *Contemporary Ergonomics* (S. Robertson, ed.) London: Taylor and Francis, pp. 341–6.

[28] Jordan, P.W. (1997). Products as Personalitie. In: M.A. Hanson, (ed.), *Contemporary Ergonomics* London: Taylor and Francis, pp. 73–8.

[29] Maslow, A. (1970). *Motivation and Personality*. 2nd ed. New York: Harper and Row.

[30] Porter, C.S., Chhibber, S, and Porter, J.M. (2002). Towards an understanding of pleasure in product design. In: D. McDonagh, P. Hekkert, J. van Erp, D. Gyi, (eds), *Design and Emotion, the Third Episode: The Experience of Everyday Things*, 2004 Loughborough, UK: Taylor and Francis, pp. 291–5.

[31] Schneider, B. and Bowen, D.E. (1999). Understanding customer delight and outrage. *Sloan Management Review*, 41 (1) (Fall), 35–45.

[32] Burns, A.D. and Evans, S. (2001). *Beyond Satisfaction: An Investigation of the Delighted Consumer and Implications for the Design of Desirable Products*. Proceedings of the International Conference of Engineering Design, Glasgow, August 21–23.

[33] Hirschman, E.C. and Holbrook, M.B. (1982). Hedonic consumption: Emerging concepts, methods and propositions. *Journal of Marketing*, 46, 92–101.

[34] Jordan, P.W. (1999). Pleasure with products: Human factors for body, mind, and soul. In: P.W. Jordan and W.S. Green, (eds), *Human Factors in Product Design*, Taylor and Francis, pp. 206–18.

[35] Nagamachi, N. (2002). Kansei engineering as a powerful consumer-oriented technology for product development. *Applied Ergonomics*, 33 (3), 289–94.

[36] Desmet, P.M.A. (2002). *Measuring Emotions: Development and Application of an Instrument to Measure Emotional Responses to Products*, PhD Thesis, Department of Industrial Design, Delft University of Technology, Dordrecht: Kluwer Academic Publishers.

[37] Desmet, P.M.A. (2003). Measuring emotion: Development and application of an instrument to measure emotional responses to products. In: M.A. Blythe, A.F. Monk, K. Overbeeke, P.C. Wright, (eds), *Funology: From Usability to Enjoyment*, pp. 111–23.

[38] Desmet, P.M.A., Hekkert, P., and Jacobs, J.J. (2000). When a car makes you smile: Development and application of an instrument to measure product emotions. In: S.J. Hoch and R.J. Meyer, (eds.), *Advances in Consumer Research*, vol. 27, pp. 111–17.

[39] Feeney, R. and Bobjer, O. (2000). Communicating ergonomics data and principles to other professions. Proceedings of IEA 2000/HFES 2000 congress.

[40] Bruder R. (2000). Ergonomics as mediator within the product design process. Proceedings of the IEA 2000/HFES 2000 congress.

[41] Vicente, K.J., Burns, C.M., and Pawlak, W.S. (1998). Better handbooks, better design. *Ergonomics in Design*, pp 21–27.

[42] Haslegrave, C.M. and Holmes, K. (1994). Integrating ergonomics and engineering in the technical design process. *Applied Ergonomics*, 25 (4), 211–20.

[43] Chhibber, S., Porter, C.S., Porter, J.M. and Healey, L. (2004) Designing pleasure; designers needs. In: A. Kurtgozu, (ed.), *Proceedings of the Fourth International Conference on Design and Emotion*, Middle East Technical University, Ankara, July, [CD-ROM].

Part IV
Research and everyday practice

Meeting Diversity in Ergonomics (2007)
Pikaar, Koningsveld, and Settels (Eds)

13

An Integral Approach to make Software Work

E. Mulder

ErgoS Engineering and Ergonomics, 7500 AG Enschede, The Netherlands

Abstract. An approach to the systematic diagnosis and improvement of software in use is described in this Chapter. 'Integral' implies not just dealing with the software as such, but also with the way in which it is used in its full context, in task and work organization. This holds for the diagnosis as well as for the resulting advised improvements: not just dealing with software changes, but also with changes in, for example, work organization, user instruction or hardware. The framework of the approach is projected on engineering steps in ergonomics projects. The diagnosing of software in this approach is compared with more common usability evaluations, such asexpert walkthroughs and user tests. An application of the approach is conducted in the area of health insurance administration in the Netherlands. The approach detects usability problems in several levels of software design (like functional design, dialogue design and presentation design) and several domains (like user authorisation, software configuration, application flow, user instruction and hardware).

Keywords: usability evaluation, human computer interaction, best practice, software design, ULD/CTD/RSI.

OVERVIEW

An approach to the systematic diagnosis and improvement of software in use is proposed. The approach in this chapter is addressed as '*integral*' to emphasize:

- integral diagnosis: dealing with the software and the way it is used in its full context;
- integral improvements: not just dealing with software changes, but also with changes in, for example, work organization.

History

The integral approach has evolved from the way the author contributes to software design projects. Typically, the emphasis is on how users fulfil tasks. Users are consulted at their place of work, carrying out their task.

Since the year 2000, there has been an increasing demand to have software evaluated and improved. Sometimes, line managers ask: 'Can you come over and have a look? I think we're losing production due to bad software.' Other times the initiative comes from 'Human Resources' or 'Health and Safety': 'I wonder whether this software contributes to ULD?' (Upper Limb Disorders, or also referred to as 'RSI' or 'CTD'.)

Over the years an approach has been developed to diagnose and improve software. ErgoS [1] published this approach in 2005 together with a concise set of software guidelines.

How to make software work better

Is software not working? Some examples

Software is not working optimally when a user has to perform actions that could have been done by the automated system. In a call centre, users often had to request financial details of a client's account. Whereas 99% of the requests (of course) concerned the most recent changes on the account, the automated system answered the user's request simply with a list of financial transactions, from which the user each time had to choose the details of the most recent transaction. Even worse,, when the system finally honoured this second request with recent details, the user had to scroll to the bottom of the information, because the system systematically added the last requested information at the end of the list, but always displayed the top of the list after an information refresh. Explained like this, it seems easy to detect this flaw. But in reality, the flaw is obscured because the system behaves very consistently in answering requests. So everybody became used to it, and even to like the way the system behaved so predictably.

Software is not working optimally when it presents users with cluttered screens, overloaded with functionality and eye-catching attributes. Often too 'eye catchy'; in other words, distracting. Compare the 'head aching' Fig. 1 with the simplicity offered by Fig. 2.

Software is not working optimally when it does not immediately show options from which one frequently has to choose. Software becomes nerve-wracking when a choice has to be made from several tens of options, which have to be remembered because they cannot be made visible in one go. Compare the way a user may choose from many items in Figs. 3 and 4. Also notice that the list in Fig. 4 is not obscuring any information.

Software is not working when users' authorization does not match task demands and users' expertise. In one case, this defect forced call centre employees to direct some requests to be handled by the back office, while they knew very well what to do and had all the necessary information at their hands.

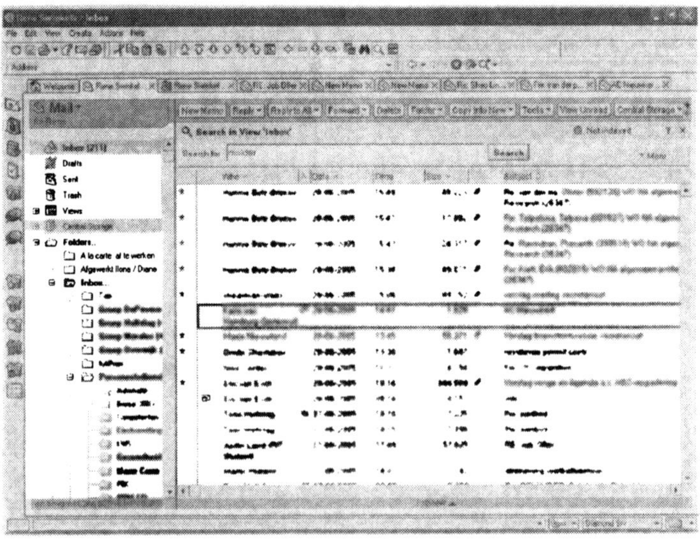

FIGURE 1. Cluttered mail application (blurred for privacy).

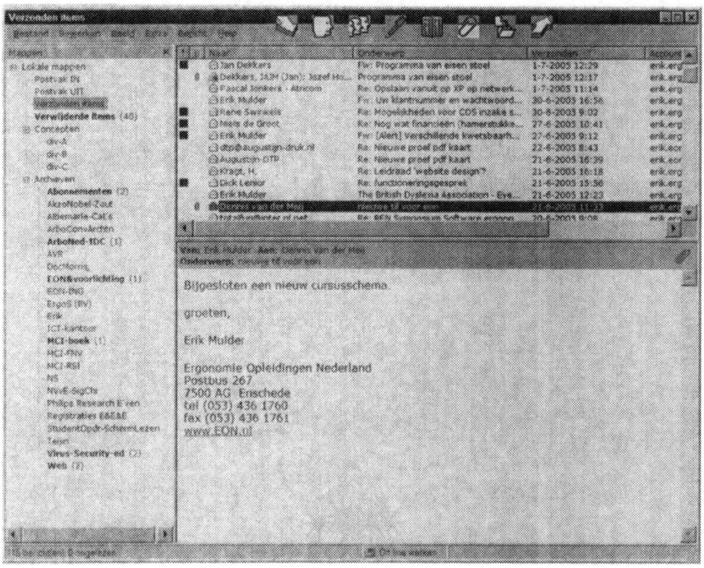

FIGURE 2. Simple mail application (partly fictitious).

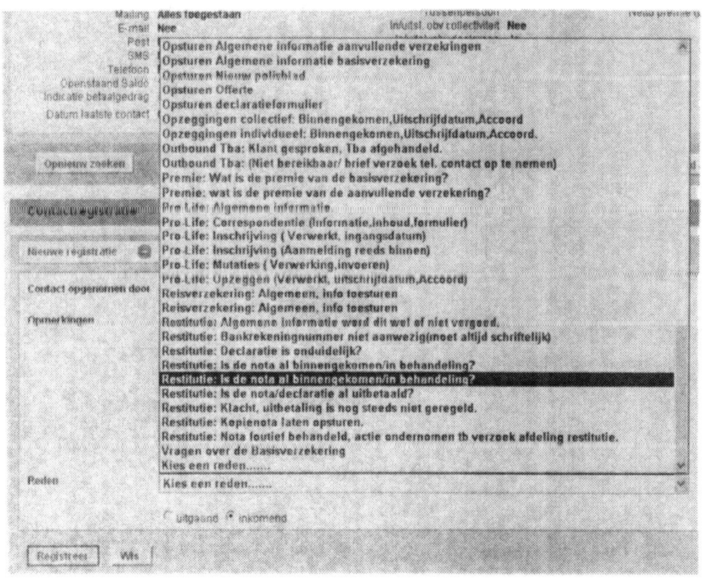

FIGURE 3. Estimate the amount of items in the huge dropdown list by the length of the mover in the scroll bar.

This implied a threefold penalty for the company:

1) Lots of unnecessary work in the organization.
2) Poor task quality because employees were under-loaded and not completing the 'product'.
3) Clients not being served to the full at the time.

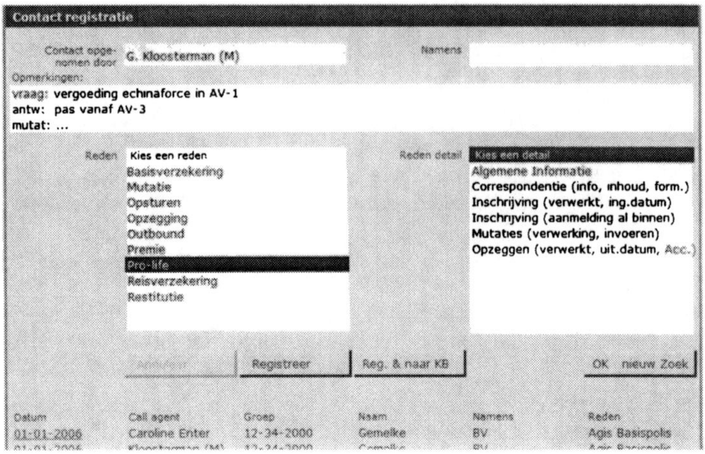

FIGURE 4. Improved fully visible list, split in two hierarchical levels.

Working software

Software that works well facilitates productivity without health risks for the users. The relationship between software, productivity and health is described in the Appendix.

To summarize: diagnosis and improvement is needed at all design levels of software, among which are:

- task allocation and task flow;
- information design;
- dialogue design;
- amount and kind of control actions needed.

Integral

Integral diagnosis

The word 'integral' in the title refers to integral diagnosis. The core evaluation is done by a usability expert consulting a user during work. Software is diagnosed together with the method of use in its full context. This differs from most software usability evaluations [3, 4], which separate one or more of: expert, user, task and context.

Integral improvement

The word 'integral' also refers to integral improvements, i.e. is not limited to the software itself. In many projects we find that changing the software itself is beyond budget or time schedule, or that software simply is not to be controlled because it is compulsory standard software.

Nevertheless, there may be more cost-effective solutions 'outside' the software than inside it. These types of improvement may be found in:

- user instruction on break schedules, efficient action sequences, keyboard short-cuts, etc.;
- other sources, destinations or formats of data;

- organization of work: alternative work flows, task enrichment, job rotation, user authorizations, etc.;
- providing different hardware such as displays and input devices.

Changes to the software itself also have several levels, including, but not limited to (expensive) redesign:

- designing new software;
- selecting alternative software;
- adjusting software in future releases;
- configuration by system administration;
- configuration by user.

Future developments

The integral approach should lead to derived 'best practices', dedicated to certain business domains or types of software. Very recently an approach was developed by the author for the Dutch health insurers; more about this in Simplifying diagnosis on page 225.

Additional research on the effectiveness of the approach and on the operational usability of ergonomic software guidelines is planned. Unfortunately these guidelines, like a lot of standards, tend to be difficult to use according to their own metrics.

THE INTEGRAL APPROACH

Step 1. 'Promoting and selling' the project

In most organizations, software use is not considered to benefit from guidelines for occupational health, nor is it common to expect advice for software productivity improvement from an ergonomist. Often, ergonomists will be asked for guidance on physical aspects of production systems, like seating or manual materials handling. In other domains, such as building architecture or software, ergonomists are often out of sight.

Therefore, the first step of a project to improve software is 'promoting and selling the project', getting the support of company management or line management or other stakeholders. Important in this step is to show the possible productivity gains or, putting it the other way around: show the implicit costs of leaving software the way it is.

Analysing projects in the last five years, the author has found that productivity gains are often more readily accepted than health gains. Avoiding 60% of the mouse clicks or non-productive 'windows' handling on the screen is apparently easier to grasp as improvement than relieving the user by avoiding cluttered and overcoloured screen images.

There is also a strong prejudice that health is just connected with seating, armrests, screen height, 'ergonomic' pointer devices and the like. Apparently, because it is so much easier to conceive the effects of physical objects, the invisible mental aspects are overwhelmed. More discussion of the relationship between software, health and productivity can be found in the Appendix.

Therefore, the emphasis of this first step is on 'selling' the project (in reality, a short consideration on feasibility, project scope and planning). So in step 1 the project is planned. Decisions are taken about goal, means, time, budget, scope etc. The integral approach explicitly demands decisions about which software and which jobs are covered by the project.

Ways to improve software or its use

An important decision in this step is about which type of improvements will be dealt with in the project. All types mentioned before in 'integral improvement' on page 215 may be part of the project, but mostly there are limitations.

Including or excluding certain types of improvement have substantial implications on the criteria used for evaluation and on what deficiencies one should concentrate.

For example when software itself may not be changed, it is fine to detect awkward dialogue boxes, because there may be ways to avoid them or to reduce their impact. But it is inefficient in this case to focus on potential changes in these very dialogue boxes.

This first step results in

- Project scope: including which software, which tasks, in what domains may improvements be suggested?
- Support of project partners, such as (line) management, ict and users.
- Project structure and planning.

Step 2. Stakeholders and desk research

The goal of this step is to roughly investigate the main issues, which may include task flows, complaints of users and line managers, health risks and expected usability issues.

Actions taken are:

- Consulting one or more stakeholders involved in production with this software (line managers).
- Researching materials related to the software: instructions, manuals, types of hardware, screen prints and the like.
- Analysing screen prints to identify whether legibility is an important issue in the project. The integral approach possesses an easy technique to gather and analyse screen prints and characters on the level of pixels.

Step 3. Consulting users

This step of user participation may be considered as the core of integral evaluation in the approach. The user is not asked to directly pinpoint problems in the software, but rather an efficient source for leading the evaluator through the tasks and through the software.

A good and simple opening question would be: 'Please show me what your most common activity is with this software'. This simple question will be the start of getting insight into:

- the start, flow and end of tasks;
- task frequencies, duration and criticality;
- which information items on the screen are important and frequently read;
- which screens, windows, dialogues, controls give rise to workload (cognitive or physical) and are awkward to use.

Note: besides focusing on the most common user activity it is important to ask for rare but critical activities.

Best practice: from task to details

Consulting just a few users will suffice in most cases. The concern is not on investigating user opinions about the usability, but rather on task execution combined with 'on the fly' expert reviewing.

The amount of users to be consulted, therefore, is determined by task-variation and user diversity, not by statistical confidence. In most situations it suffices to consult two to four users for one to two hours each. Try to involve user and task diversity; novice and expert users, mouse movers and key pressers and so on.

The first consultation or interview has an emphasis on task structure and task execution, as the ergonomist needs to get a good working knowledge of these. In the next interview more attention may be given to details of software dialogues. Observations should be postponed to the last interviews; by that time the ergonomist knows the ins and outs of task execution without a need to interrupt the user.

Despite these different accents, during all the interviews the ergonomist should be attentive on all levels, tasks, dialogue and display details. During the first interview a bizarre display detail may show up while investigating the global task execution. And in the last interview the user still has to be asked: 'Will you tell me in short what your task involves?' Although the ergonomist probably already knows, the last user may have a particular, valuable view on task executing.

In between the user consultations the ergonomist needs to use his or her skills for task analysis and expert review to pinpoint bottlenecks, which need more attention in the next interview.

The 'integral' aspect of this diagnosis is described in more detail in differences with other methods on page 220.

Step 4. Sorting out results and checking with users

Structuring the gathered results

Researching the materials in step two and consulting the users in step three gives a lot of information, which at first has little structure. A first division in the findings can be made by distinguishing the items related to the 'look and feel' from the items related to task flow.

The 'look and feel' items may be well structured by connecting each item to a type of information or type of control. The task flow items may be structured along the task flow.

Having a meeting with users and stakeholders

It is important to get feedback on the structured results from the users and consulted stakeholders. Presenting the (anonymous) results goes so far. A meeting with users and stakeholders widens the coverage and makes the results more reliable:

- Users will react – often enthusiastically – and thereby give a clear indication whether a resulting item was an accidental problem for a particular user or a structural usability problem.
- Hearing each other, users often come up with more details or examples of usability items.
- Users will help in prioritizing the items.

Assigning priority

The priority of each item is determined by two independent factors:

- quantity (task frequency, duration, criticality) of occurrence of the usability problem;
- quality in negative terms (severity, inconvenience) of the usability problem.

Step 5. Agree to have problems solved

In this step, others (project partners, stakeholders, ict, ...), need to give their commitment to solving the prioritized items according to agreed requirements. This is not just deciding which problems will be solved, but also how well, and when.

Economic solutions

Besides priority, there is another deciding factor as to whether or not to go for certain improvements: the amount of cost and time needed to implement a solution. For example, there may be items with low priority but which are easy to implement, like distributing a 'post-it' note with the ten most used keyboard short cuts.

Example of choosing in costs and time limits

Suppose a high priority has been given to the fact that users have to duplicate information by hand from one application to the other (which occurs more often than not):

- A good solution would be integrating the functionality of one application into the other, which probably is very expensive and not to be expected in the next few years.
- A second best solution would be adding an automated data connection. which still may require quite an investment.
- A cost-effective, temporary solution may be to provide keyboard macros or easy buttons on the screen for copying the most wanted information items to the clipboard, which will free the user from laboriously mousing to select and copy these items.

Step 6 Tracking improvements

Once improvements are agreed on, there is a need to guide and support the designer, because the designers probably are confronted with new ergonomic requirements.

In a number of projects the ergonomist finishes his or her part of the project after step four or five. This should be avoided by incorporating this step six in the project structure. So when the project is sold and planned in step one, time and budget have to be allocated to stay connected to the project as a guide for designers implementing the improvements.

Finally, if possible, a short evaluation should be carried out. It would be better to compare formal quantitative measurements at the beginning and the end of the project.

PROJECT FRAMEWORK

Common engineering projects and this integral approach differ in quite some characteristics. It is interesting to compare the steps of the integral approach with common steps in ergonomic engineering projects. In Table 1 this comparison is made with steps mentioned in Chapter 3, New challenges: ergonomics in Engineering Projects by Pikaar [2].

A major difference is the size of the projects.

- The investments according to Pikaar in ergonomic projects are between €0.1 to 10 million, including an ergonomic participation for an estimated 0.1 to 5% of the total investment.
- Projects for this integral approach sometimes are too small to be called a 'project'. Investments will be between €5 and 100 thousand, of which the ergonomic participation is relatively high, from five to even 50% in small projects.

TABLE 1. Comparing steps of integral approach with common project steps.

Ergonomic Engineering (Pikaar)	Integral approach of software (Mulder)
1 Feasibility 2 Problem definition	1 Promoting the project
3 Analysis 4 Functional design 5 Detailed engineering	2 Stakeholders and desk research 3 Consulting users 4 Sorting out results and checking with users 5 Agree to have problems solved
6 Implementation 7 Commissioning 8 Evaluation	6 Tracking improvements

Another difference is the role of functional design. As it is all about improvement, there is little design of new functionality. Functionality is not so much changed, but rather improved by detailed design and implementation

Quickly visible is that the start of the integral approach is rather a small step: simply promote and 'sell' the project and spend some hours, or at most one day, considering feasibility and planning of the project.

On the other hand there is quite an emphasis on what in most projects is called analysis. Steps two, three and four of the approach are generally seen as the core of software improvement projects: getting to know what should be improved. Sometimes we even see that the participation of an ergonomist in software projects basically is limited to step three of the approach. No need to say that such a limitation easily leads to missing cost-effective opportunities for improvements.

Concerning the functional design, step four of Pikaar; one could say that there hardly is functional design, for most is about improving existing functionality. But the table still shows that the major decisions in the approach are taken in checking the results with the users and in getting agreement on which problems are going to be solved.

DIFFERENCES WITH OTHER METHODS TESTING USABILITY

Typical for this integral approach

User + Task + Expert = Integral approach

The integral approach differs from other evaluation methods in some aspects:

- The integral approach has a solid base in the users' expertise in executing tasks supported by the software.
- The integral approach uses the surplus value, which arises from the 'real time' combination of usability expert and user. The expert lacks task knowledge, and has difficulties imagining a user's mind. The user lacks discrimination between effective and ineffective interaction, and has blind spots due to being used to the way of working. Deficiencies of both will be compensated for simultaneously.

- The integral approach is directed to several types of improvements, not limited to adjusting software.
- As the integral approach needs access to users performing their tasks, it less effective for designing new software from scratch.

Seen from the perspective of Jakob Nielsen, the integral approach seems the ultimate evaluation. Nielsen [3] advises applying alternately the expert review (heuristic) and the user tests in iterative design phases, in order to increase the chance to identify usability issues. Bias [4] proposes a pluralistic usability walkthrough, combining experts and users as well but this is a group meeting, not dealing with 'real life' task execution.

Although not mentioned as such by Nielsen and Bias, they implicitly adhere to quite a lot of the analysis pattern in the integral approach:

- Analysis starts off with a global expert review in step 2: *stakeholders and desktop research*
- Alternate user interviews and expert considerations in step 3: *consulting users.*
- Getting results with users and stakeholders in a meeting in step 4: *sorting out results and checking with users.*

Production tasks versus public software

To judge the differences between the integral approach and more common methods, one should be aware of the different goals of each.

Notice that the integral approach aims to identify the biggest problems in order to make a limited number of effective improvements towards a more efficient and healthier use of software. The approach works best with software in use, for bounded production tasks.

This differs from Nielsen, who aims to identify as many usability problems as possible in order to produce usable new software, often for a large public, perhaps through web sites.

Compared with expert walkthroughs

Expert walkthrough (e.g., cognitive walkthrough and heuristic evaluation [3]) is often carried out without a working knowledge of 'real life' tasks.

- [illegible]
- [illegible] mismatch [illegible] real tasks: Is [illegible] information and control [illegible]

This latter mismatch occurs frequently due to the fact that software developers get simple demands like: 'design screens which support all these actions'. No one tells them that just 6 of the 30 database fields are involved in 95% of the tasks. In other words, when no

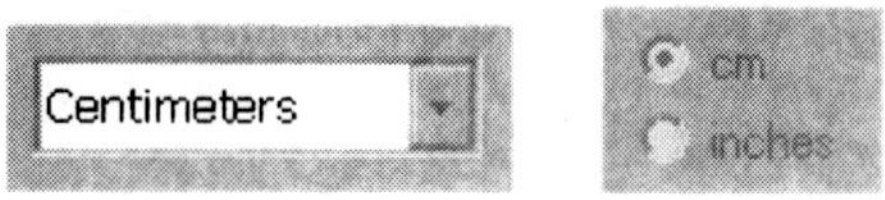

FIGURE 5. Dropdown list and radio buttons.

dedicated screens are designed for the 95% of simple tasks, users are loaded with a far too crowded interface most of the time. (Hence: requirements for software often lack adequate task information.)

Compared with user tests

Tests carried out by users basically fill the gap caused by the fact that designers are not like users, and do not perform real life tasks with the software. Therefore user tests are essential in designing software. But this is not enough. Users often do not complain about inefficiencies as long as they understand the system and know how to carry on executing their task.

Example

When users have to use the dropdown list in Fig. 5 they know perfectly well what to do: (1) click on the little down arrow, (2) find the value, (3) move the mouse to this value down in the list and (4) click again.

The user may click anywhere in the list control to get it unfolded, but as this does not work with all lists most users tend to click on the 'far too small for frequent clicking' down arrow.

For frequent use the dropdown list is too laborious. The radio buttons on the right of Fig. 5 only require one click and that may be anywhere in the imaginary rectangle surrounding button and text label, offering an easier goal for pointing with the mouse.

A user might easily fail to detect that the dropdown list is not efficient and comfortable for very frequent use. Nevertheless, an expert in combination with a user will quickly find out whether the control is used frequently and give an adequate priority to get it improved.

(Strangely enough, we quite often see these dropdown lists with just two items: Yes and No; which actually represents the functionality of one simple checkbox.)

SOME CASES: APPLICATION OF THE INTEGRAL APPROACH

The approach was applied to a diversity of usages, like administrative applications, CAD software and intranet portals; all in occupational context with no public use. Applying this approach to architectural CAD software has led to a guideline for selecting and configuring CAD software used in building design.

Three examples are described below.

- The first case is a quick screening (about 3 hours per application).
- The second is a detailed evaluation of one application including some redesign (30 hours).
- The third case is somewhat different because this had, beside the goal for 'making software work', the goal of being a pilot to improve the approach (54 hours)

Case 1: Screening of six software applications

Rationale

Users complain about small characters, lots of 'mousing' and small screens. The applications, mostly database-oriented, are used by about 200 workers. Part of the workerforce deals with clients on the phone.

The goal of the project is to answer these questions:

- Are the applications effective?
- Is there a health risk in using these applications and hardware?
- What types of solutions are there for the user complaints and the problems found in the screening?

Actions and time needed

- step 1 Promoting
  3 hrs: contact by telephone, writing quotation and plan.
- step 2 Stakeholders and desk research
  4 hrs: surveying manuals and reviewing screen prints, identifying issues.
- step 3 Consulting users
  4 hrs: six interviews of 1/2 hr per application, each time with one or two users.
- step 4 Sorting out and checking with users
  4 hrs: sorting out results and preparing presentation.
  2 hrs: presenting and discussing results so far.
- step 5 Agree to solve problems
  2 hrs: adjusting results to get final evaluation report.
- step 6 Tracking improvements
  This was not the scope of the project.

Some of the findings and advice

- For a certain critical reporting task every two weeks: clean up Excel from unused toolbars, borders, etc. to free up space for actual data. Together with a bigger screen, this takes away the need to reduce the zoom factor from 100% to 70%, causing the user to peer at the screen.
- Reduce network reaction times for most tasks by reading data from decentralized (mirror) servers. The few tasks needing to write and update data still use the central server. This improvement aimed to reduce the stress caused by waiting; especially while users had clients on the phone.
- Rearrange data tables to avoid frequent switching to another window to copy one value from it.
- Assemble a task-dedicated screen to avoid skipping 10 fields not needed for an ordinary financial booking (40/hour).
- Implement a running total while booking. In the old situation one could only check the correctness of bookings after completing the whole day, hoping no error was made.
- Invest in new screens, slightly bigger and much better focused.

Case 2: Detailed evaluation of a database application

Rationale

A Human Resource manager expects health risks and workload may be reduced while keeping up or increasing productivity. The application is used by about 100 people with rather monotonous database tasks and dealing with paper forms.

Goal of the project:

- Lowering mental and physical workload.

Actions and time needed

- step 1 Promoting
  4 hrs: contact by email, writing quotation and plan.
- step 2 Stakeholders and desk research
  6 hrs: visit stakeholders and department using the application, reviewing screen prints, identifying issues.
- step 3 Consulting users
  4 hrs: interviews with three separate users.
- step 4 Sorting out and checking with users
  4 hrs: sorting out results and preparing presentation.
  3 hrs: presenting and discussing results so far.
- step 5 Agree to solve problems and
- step 6 Tracking improvements
  8 hrs: to assemble a report for redesigning, including examples of redesigned screens and alternative task flows.

Some of the findings and advice

- Add dedicated screens for 99% of the tasks, because about 80% of the database fields for these tasks stay empty or hold standard values.
- Rearrange task flows and involved screens in order to facilitate copying data from another application in one go per paper form. In the original situation one had to switch about 2–20 times between two applications for handling one paper form. Users made a solution to this awkward situation by jotting down about 20 values on a bit of paper, which actually was forbidden because of quality regulations.
- Clean up the screens from lines, superfluous labels, borders etc.

Case 3: Detailed evaluation of a database application

Rationale

Due to a law change in the Netherlands, Dutch health insurers need to re-design their administrative software, and temporarily their call centres need to process three times as many client questions. The approach is applied because the line manager of the call centre expected benefits for productivity (quantity and quality) and the health and comfort of employees. About five of the 200 employees have, or did have, CTD-related complaints.

Goal of the project

First of all, there is a need to increase productivity. In this case, this means dealing with client contact in a shorter time, better client satisfaction and fewer errors. At the same time health risk and discomfort for the employees should possibly decrease.

Beside this, the project is a pilot to improve the approach. A simplified approach had to be developed, which should be usable for occupational health specialist after only a short (six hours) training.

Actions and time needed

- step 1 Promoting
  8 hrs: contact by email, meeting, writing quotation and plan.
- step 2 Stakeholders and desk research
  8 hrs: visit stakeholders, gathering and reviewing screen prints, identifying issues.
- step 3 Consulting users
  8 hrs: interviews with three separate users.
- step 4 Sorting out and checking with users
  8 hrs: analysing and sorting out results.
  4 hrs: preparing presentation and meeting.
  4 hrs: presenting and discussing results so far.
- step 5 Agree solving problems
  4 hrs: preparing a meeting with software designers. This is a fairly delicate meeting as the decision takers at the same time were the designers of the existing software.
  4 hrs: meeting with designers.
  2 hrs: reporting results
- step 6 Tracking improvements
  4 hrs: keeping in contact with redesigners and stakeholders. Due to reorganizations about half of the improvements were postponed to beyond the project deadline.

Some of the findings and advice

Basically, the advice was a shortlist of 23 items, five with the highest priority, 12 with a high priority and six with a medium priority. Attached to the two page list was a six page explication, giving some background and an estimate of costs of the existing problems.

The five items with highest priority:

- Communication between the call centre software application and some 'side' applications should be supported better. The best and most costly solution would be to fully integrate the side applications into the main application. A second best solution would be to have an automated transfer of the main items, or at least the clientIDnr, to these applications. A cheap, but still quite effective solution would be to copy the clientIDnr in the clipboard automatically, so a simple paste action <Ctrl+V> would suffice to transfer the most important data in other applications.

- Authorize (some of) the call centre employees to complete some tasks that are at the moment transferred to the back office.
- Improve the logging of client contacts at the end of each call. This had a lot to do with the list in Figures 3 and 4 (page 214).
- Automate or standardize the communication with the back office. In the existing work organization several methods were in use; most of these required the employees to print out data and attach this hard copy to a written note.
- Avoid some annoying cases of secondary windows. These windows not only disturb the information presentation, but also require non-productive overhead handling, like closing them.

SIMPLIFYING DIAGNOSIS WITH 15 ITEMS

Step three (page 217) is a major component of the integral approach. This step efficiently integrates several ways of usability testing in one go. At the same time this pinpoints a weak spot in the approach because the evaluator also in one go has to carry out:

- task analysis,
- heuristic evaluation/cognitive walkthrough and
- user test.

All this with guidelines or standards such as ISO9241 [5] in mind.

To help evaluators in this difficult task, it is possible for restricted domains to give them a more restricted framework, which is easier to handle.

The author developed for the Dutch health insurers a better-to-handle approach. Especially the diagnosis is simplified by replacing the general HCI-guidelines like ISO 9241 with 15 most occurring deficiencies found in administrative software. These 15 items resulted from many projects in administrative software improvement carried out by the author and his colleagues.

The 15 items, arranged from global issues to detailed issues:

Concerning task

- incomplete
- boring or unnecessary
- 80/20 rule: 80% of tasks needs just 20% of user interface

Concerning dialogue

- wrong ordering
- user not in control
- confirmations
- pop-ups
- intolerance, bringing in errors

Concerning information

- limited (scanned) image handling
- not easily readable
- coding strong or messy

Concerning control

- inefficient selection
- scrolling
- no (short cuts) keyboard control
- laborious mouse control

All these items are described on two pages, one page to assist the expert in detecting the deficiency and helping him or her to estimate the seriousness of the occurrence of the item in the particular situation. The other page holds information to convince others that they have to solve the problem. This is done, firstly, by giving examples ofr what the problem might cost in the existing situation. Secondly, some solutions are shown in different domains and 'pricing categories'.

The author found that transforming the guidelines as 15 most occurring deficiencies makes it much easier for project partners with less expertise to grasp and handle ergonomic issues related to software. Surprisingly this holds for software designers as well. A general strategy to get designers working according to ergonomic software guidelines is by giving them a workshop. And although the 15 items do not cover all the guidelines, the workshops based on 15 items are more effective. Designers get fewer guidelines, but have a stronger recognition and motivation for the most important issues.

CONCLUSIONS

The approach detects usability problems on all design levels of the software, from global task design down to cleaning up screens from graphic frills. The approach also leads towards solutions in different domains like work organization, adjusting software, changing hardware and instructing users.

It is hard to get project partners really involved in solving the identified problems. It should be clear from the beginning of a project that the ergonomist stays connected to the project in the phase of implementation of the improvements.

DISCUSSION

A more difficult question is: does the approach detect all the important usability problems? This cannot be clearly concluded from the applications of the approach so far. A positive indication is the fact that users, stakeholders and other usability experts in the feedback meetings hardly ever add new issues with a high priority to the list of problems found when a detailed evaluation had been done. In the range of issues with little priority this often is the case. Therefore it is important not to present the issues with less priority as being a complete list of deficiencies.

APPENDIX: SOFTWARE, HEALTH AND PRODUCTIVITY

SOFTWARE INFLUENCES HEALTH (ULD/CTD)

Physical influence

Physically this looks quite straightforward for most people. Presumably, health risks increase and productivity decreases with:

- more mouse and keyboard actions;
- smaller mouse click areas (demands longer and more precise muscle control);
- unfavourable mouse actions like dragging (pressing a button while moving increases muscle contraction);
- smaller, less legible, characters (tense posture due to peering at the screen).

At the congress (IEA-2006) a special symposium was dedicated to the health risks of computer use: 'Unravelling the causes of Upper Extremity Disorders among computer users'.

- Lingen [6] presents some software aspects, which can also be found in [1] and are part of the basis of this approach.
- Kraker [7] investigates quantitatively the relation between software and musculoskeletal disorders. She also gives some references for important mental aspects influencing musculoskeletal disorders from the use of computers.

Mental load

Another major factor is mental state. It is proven that personality and stress are important factors in developing ULD/CTD for PC users; see references in [7].

Simple cognitive loads are also important; these increase significantly (co)contraction of muscles, as Van Galen [8] shows. So, for both physical health and mental workload it is important to avoid, e.g., memorizing and selecting data from crowded screens.

To summarize, software ergonomics is important at all design levels, among which are:

- task allocation and task flow;
- information design;
- dialogue design;
- amount and kind of control actions needed.

REFERENCES

[1] Ergo, S.E. (in Dutch) (2005). *Mens Computer Interactie: Arbothemacahier nr.16*. Den Haag, The Netherlands: Sdu Uitgevers.

[2] Pikaar, R.N. (2007). New Challenges: Ergonomics in Engineering Projects. This volume.

[3] Nielsen, J. (1994). *Usability Engineering*. Morgan Kaufmann: San Francisco.

[4] Bias, R.G. (1994). The pluralistic usability walkthrough: Coordinated empathies. In *Usability Inspection Methods* (J. Nielsen and R.L. Mack, eds) New York, NY: John Wiley and Sons.

[5] International Standards Organisation. (1998). ISO 9241: Ergonomic requirements for office work with visual display terminals. Geneva: Switzerland.

[6] van Lingen, P. (2006). How to design software user interfaces to prevent musculoskeletal symptoms. In: *Meeting Diversity in Ergonomics*. IEA 2006. Maastricht: Elsevier, Oxford.
[7] de Kraker, H. van Lingen, and P. Blatter, B.M. (2006). Are software characteristics related to musculoskeletal disorders in computer workers? In: *Meeting Diversity in Ergonomics*. IEA 2006. Maastricht: Elsevier, Oxford.
[8] Van Galen, G.P. and Müller, M. (in Dutch, English summary) (1995). Repetitive Strain Injury, Stress, Muscle Tension and the Neuromotor Noise Concept. *Tijdschrift voor Ergonomie, nr.2* pp. 3–17, NVvE.

Meeting Diversity in Ergonomics (2007)
Pikaar, Koningsveld, and Settels (Eds)

14

Surgical Workflow Analysis: Identifying User Requirements for Surgical Information Systems

A. Jalote-Parmar

Faculty of Industrial Design and Engineering, Delft University of Technology, Landbergstraat 15, 2628 CE Delft, The Netherlands

P.M.T. Pattynama

Department of Intervention Radiology, Erasmus MC, Dr Molewaterplein 40/50, Rotterdam, The Netherlands

H. de Ridder, R.H.M. Goossens and A. Freudenthal

Faculty of Industrial Design and Engineering, Delft University of Technology, Landbergstraat 15, 2628 CE Delft, The Netherlands

E. Samset

The Interventional Centre, University of Oslo, Rikshospitalet, Sognsvannsveien 20, 0027, Oslo, Norway

Abstract. Developing surgical information systems to support time-critical processes with significant internalised domain knowledge requires a systematic approach to analysing the surgical workflow. Such an approach is necessary if the information needs that govern surgical tasks are to be identified and subsequently communicated within multidisciplinary development teams. Addressing this need, this chapter proposes a design framework, known as a *workflow integration matrix*. This aids ergonomists, system developers and designers to create a knowledge base of information requirements within the surgical workflow to define the user requirements.

Keywords: Design method, multidisciplinary team, surgical workflow, user requirements, surgical information systems

INTRODUCTION

Operating theatres in hospitals are currently changing at a rapid pace as a result of the introduction of innovative surgical techniques such as minimally invasive procedures and robotic surgery. Technological innovations lie at the core of the application of such surgical techniques, which are expected to become widespread in the near future. At the same time these surgical techniques are constantly giving rise to new research and development activities, especially in the area of surgical information systems responsible for data acquisition and organisation, image processing and display. Some examples of recent technological developments are a planning tool for radio frequency ablation (RFA) [1], augmented reality solutions to provide additional information support for surgeons [2], and intraoperative monitoring for anaesthesia as described by Sanderson [3]. Contrary to the commonly held assumption that new technology will equate to a more efficient workplace, it appears to increase the information load on surgeons by increasing the number of information sources at their disposal [4, 5]. Information overload is becoming one of the contributory factors in human error, affecting the quality and efficiency of surgical procedures, hence patient safety [6]. One way to tackle this problem is to develop surgical information systems that deliver appropriate information to the surgeon at the right time [7] and in a surgeon-friendly way – 'surgeon-friendly' in the sense that the content and presentation of the surgical system meet the surgeon's cognitive and visual requirements in the surgical workflow, i.e. the boundaries governing the surgical tasks and decision-making during the three stages of a surgical procedure, before (preoperative), during (intraoperative) and after (postoperative).

Note that all three stages of the surgical procedure have to be included, as the system should not only assist surgeons by providing information support while they are performing surgical tasks, but also promote consistency in patient treatment strategies.

Developing surgical information systems to support time-critical processes with significant internalised domain knowledge requires a systematic approach to analysing the surgical workflow. Such an approach is necessary if the information needs that govern surgical tasks are to be identified and communicated systematically within multidisciplinary development teams. Addressing this need, this chapter proposes a design framework, known as a *workflow integration matrix* (WIM), for creating a knowledge base of information requirements within the surgical workflow to help ergonomists, designers and system developers define the user requirements. This is in line with the user-centred approach to healthcare design as advocated by Buckle et al. [8]. The framework assists the team members, in particular the designer, to create and document an overview of information needs and recommend appropriate design methods for investigating the surgical workflow. It enables the user requirements and technological possibilities to be integrated, thus facilitating the co-design process within a multidisciplinary team. Recently, it has been applied to support development in the domain of minimally invasive surgery (MIS). This research is being carried out by a multidisciplinary pan-European research consortium known as ARIS*ER (ARIS – augmented reality in surgery), which aims to explore and build augmented reality-based solutions to improve information support, focusing on task visualisation and navigation support during MIS. The research team includes clinicians, a human computer interaction designer and technologists in the fields of augmented reality, imaging and robotics. The next part of the introduction briefly explains the need for surgical information systems in MIS and the design challenges related to identifying the information needs within the surgical workflow.

Test case: MIS

MIS is characterised by limited site entry ports and an indirect view of the surgical space. It is beneficial for the patient, as it causes less tissue damage, but at the same time it is challenging for doctors, owing to limited access and reduced visualisation and haptic feedback. This compels surgeons to rely on advancements in medical imaging technology to guide task visualisation and navigation during operations [9]. Inadequate information support in MIS has undoubtedly been a factor in the limited acceptance of this surgical technique among the surgical community at large. Lack of acceptance and limited visualisation, among other things, have led several technological research labs to develop computer-aided surgical systems that integrate multiple technological solutions, such as those found in stereotactic surgery and augmented reality surgical systems. Stereotactic surgery is a minimally invasive form of surgical intervention that makes use of a 3D coordinate system to locate small targets inside the body and perform actions such as ablation (removal) and biopsy [10]. Various technological possibilities for presenting the patient data to the surgeons are being explored in augmented reality, using live video imagery which is digitally processed and augmented with computer-generated graphics [11]. This also includes modelling real-time 3D computer-generated images of human anatomy based on preoperative patient-imaging modalities such as computerised tomography. In his paper, Shuhaiber [12] points out that advancing user-friendliness in augmented reality has revived interest in real-time surgical anatomy as a way to maximise the number of safe surgical hands in the coming century. In our view, Shuhaiber's suggestion of technological innovations such as '... augmented reality to become user friendly ....' is valid if the requisite information is presented to surgeons in real time when they really need it. A deep understanding of the information needs of surgeons, corresponding to their surgical tasks, is therefore required before the technology is developed.

Design challenges

Designing and developing surgical information systems poses a number of problems for the designer. First, the complexity of the surgical domain makes it difficult to investigate and identify the information needs. This complexity is due to the multiplicity of information sources, unpredictability of events and the highly time-critical nature of task performance. Further factors are uncertainties and surprises regarding the patient's condition, which are quite common in the surgical domain [13]. Second, the information needs and the overview of the surgical workflow have to be communicated to the development team systematically. Here the challenges are as follows:

- *To create a knowledge base on surgeons' information needs within the surgical workflow.* This requires an understanding of the physical and cognitive parameters involved in setting the boundaries of surgical tasks. The challenge is to identify and represent the important elements in the surgical workflow that influence surgical tasks, and finally to identify the need for exchange of information between surgeons and the system such that it does not obstruct their current task and at the same time provides information support for the next task.
- *To understand the timing of surgeons' information needs.* The challenge is to identify surgeons' information needs from moment to moment (when and which information is required) in the context of the surgical workflow. We know that the patient's medical condition can change during the surgical workflow, forcing surgeons to revise their original plans while performing surgery. It is very important, therefore, to consider how much time is required to perform each task, and to consider the time flows in the three stages of a surgical procedure (before, during and after).
- *To document and present the complexity of the surgical workflow.* In a multidisciplinary design and development team comprising technologists, surgeons and designers it is often tacitly assumed that all the members have the same insight into the complexity of the processes in the surgical work environment at the product planning stage. In reality this is often not the case, resulting in a gulf between the team members in their understanding of surgeons' requirements and design specifications. The challenge is to communicate the overview of the surgical workflow so that the technologists and surgeons have a shared understanding of their respective needs and possibilities.
- *To select design methods.* Analysing the surgical workflow involves selecting and integrating appropriate user observation and task analysis methods [14, 15] so as to generate richer results. In this context there are a number of well-established methods at the disposal of designers, e.g. contextual inquiry [16], group interviewing techniques such as focus groups [17] and task analysis techniques such as hierarchical task analysis (HTA) [18].

- *To create an observation framework.* Integrating methods generates a rich data set, but to optimise efforts and generate focused results an observation framework needs to be built that facilitates systematic observation and helps to document and present the findings in a structured way.

The chapter (a) explains the role of the designer in a multidisciplinary consortium, (b) describes the selection of the method and the creation of the proposed framework, known as WIM and (c) outlines how WIM can be applied to identify the user requirements for the development of a surgical system for a specific minimally invasive procedure, RFA. It concludes with a broader look at future challenges in the development of surgical information systems.

THE ROLE OF THE DESIGNER

In order to develop a co-design approach to the design of medical systems it is important to understand the role of the designer in a multidisciplinary consortium. The following conclusions were drawn, based on discussions with people from medical and engineering organisations:

- There is often a disconnection between clinicians and technologists in large medical engineering organisations. At present, when it comes to understanding surgeons' requirements it is the product manager who is responsible for communicating with them, usually through informal meetings or focus group sessions. These sessions generate mainly qualitative data, which is difficult to formalise [19]. As there is no formal framework for recording data, much of the data gathered is lost or randomly documented, making it difficult to access or communicate to the technologists so as to produce well-informed decision-making.
- The product manager's approach to product development is business and technology-driven rather than user-driven, with the result that technologists have a biased conceptual model

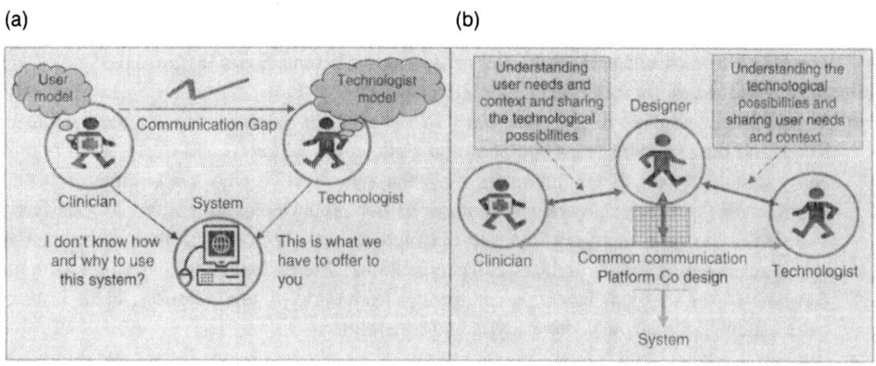

FIGURE 1. (a) The current scenario: the gap between the technologist and the clinician. Conceptual model adapted from designers' model by Norman (1998). (b) A suggestion of the role the human computer interaction designer can play in bridging the communication gap between the clinician and the technologist.

of user needs and the work environment. Conversely, when the clinicians receive the system they are unable to relate to the technologists' model and therefore have difficulty adapting the system optimally to the medical context. At present, most of the communication between technologists and clinicians takes place indirectly, through the final system (Fig. 1a).

- To streamline the process of co-design at the product development stage we propose that designers play a role in organising communication between clinicians and technologists (Fig. 1b). To do this they will need to understand the surgical needs and connect them up with the technological possibilities through a common communication platform.

THE CREATION OF WIM

Based on the design challenges, this section describes the WIM requirements, the integration of user observation and task analysis methods and the creation of the WIM components.

Requirements for WIM

The key requirements for a framework that facilitates systematic observation and helps to document and present the findings in a structured manner are as follows:

- What should be observed, and how, in order to identify surgeons' visual and cognitive information requirements in the surgical workflow?
- What are the key factors as regards information needs in the surgical workflow that influence the surgical task and hence determine the system requirements?
- How should the relationships between time, tasks and information needs within the surgical workflow be identified and related to future workflow within a common framework?

Integrating observation methods

Given the project requirements, various user observations and task analysis methods were selected and integrated. They were then used to analyse the surgical workflow and identify the user requirements. Note that some of the methods chosen are elaborate, consisting of multiple techniques: we have only selected those components of these methods that are appropriate when it comes to identifying information needs within the surgical workflow. We touch upon this topic very briefly in this subsection.

- *Surgical observations.* First, focus groups were conducted with surgeons to identify the specific problems involved in MIS. This helped us to understand surgeons' wishes, problems and requirements as regards the system. Focused interviews and observations were then carried out with the surgeons and the medical staff in the surgical context at all three stages of surgery (before, during and after). To gain insight into the critical decision-making factors and medical treatment strategies the four principles of contextual inquiry (as proposed by Beyer and Holtzblatt [16]: context, partnership, interpretation and focus) were applied during the observations. To analyse task flows in detail, selected surgical procedures were observed by means of video documentation, as mentioned by Fetterman [21] in his ethnographic study. Surgical milestones (steps) and the visual and cognitive information requirements within the surgical workflow were identified, based on the results of analysing the focus group sessions and the contextual inquiry. To identify the medical goals of the surgical milestones and the details of the tasks and procedures involved, the video clips were analysed using HTA [18]. Note that the application of HTA was restricted to decomposing the surgical milestones to a single layer of tasks. Further breakdown of surgical tasks at procedural level was not required at this stage, since the aim of the surgical system was to provide information support during surgical tasks, not to automate them. The surgical milestones were further decomposed on the basis of surgery-specific parameters (see Fig. 2, y-axis) which the surgeons identified as being the ones most important to their physical and cognitive tasks and system needs, also as recommended by Rasmussen and Pejtersen [22] in their works on the virtual ecology of work. The choice of parameters is explained in section on 'WIM Components'.

- *Interactions with the technologist.* Focus group sessions were conducted so as to provide a better understanding of the domain expertise of the technical partners in the research consortium.

WIM components

Based on the requirements mentioned above, the selected methods were applied in the surgical domain to generate an understanding of the key factors that surgeons identified as being the most important ones in supporting and influencing surgical tasks. These findings enabled the components of the framework to be defined and aided in decomposing the surgical milestones. The study was conducted on two minimally invasive procedures that are aids to treating liver cancer, RFA and endoscopic liver surgery. It included two focus group sessions with a total of 15 clinicians, 10 surgeons and 5 intervention radiologists at each session. Twenty unstructured interviews were conducted at the three stages of surgery (before, during and after). Additionally, 15 surgical procedures were observed by taking notes and 10 procedures were video-recorded. All the observations from the study were converted into key insights, as proposed by Owen [23]. An affinity diagram, as used in contextual design, was created to group all the insights semantically.

The WIM components, which are structured along two axes (x and y), were formulated, based on these groups (see Fig. 2). The x-axis represents the sequential breakdown of the surgical tasks into surgical milestones. These are presented sequentially on a time axis connecting the three stages in the surgical workflow. Note that the number of grids is flexible, depending on the nature of the problems identified at each stage.

The y-axis is in two parts, representing the information requirements corresponding to each surgical task: (a) current workflow, consisting of a breakdown of physical and cognitive parameters and (b) future workflow, consisting of user requirements, technology trends and possibilities. Note that the relationship between the current workflow and future workflow is included in order to permit tracking of surgical tasks and needs in relation to technological possibilities.

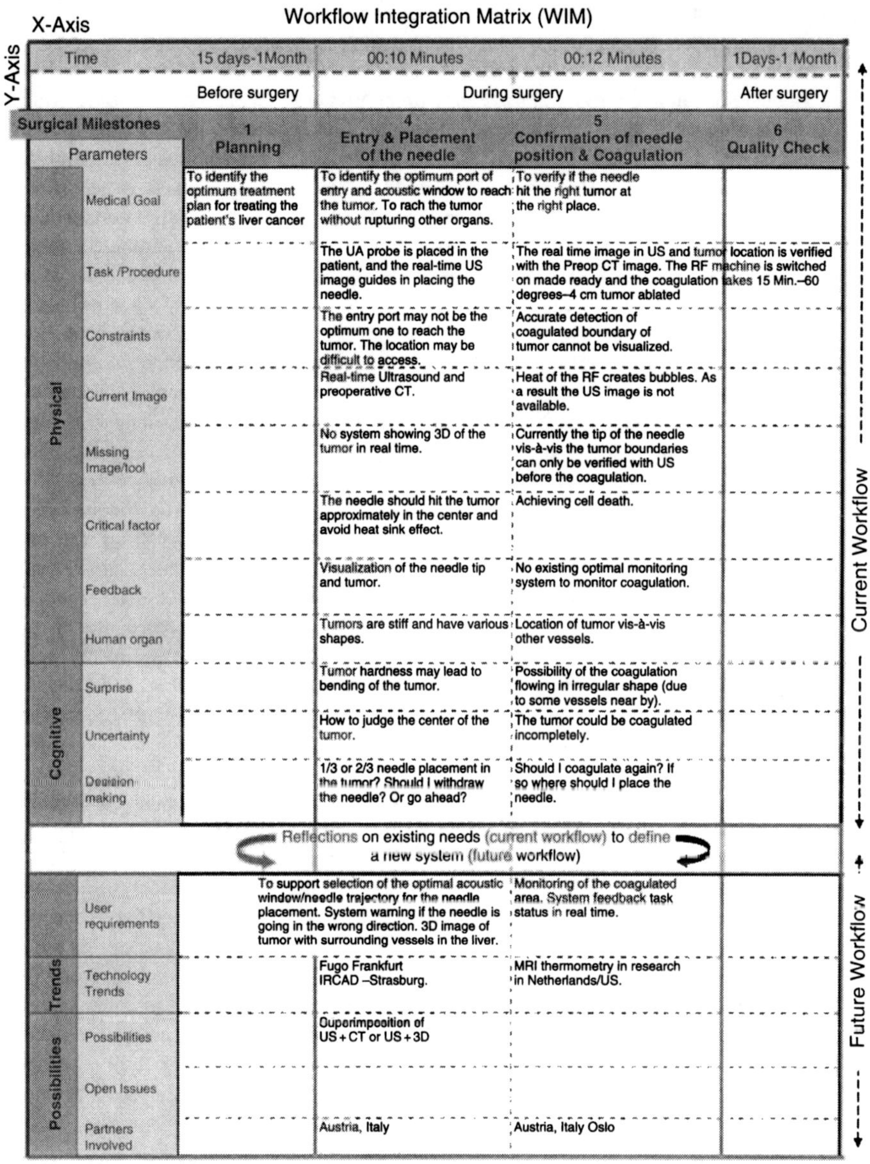

FIGURE 2. Workflow integration matrix: the *x*-axis is the task and time axis; the *y*-axis is the parameter that corresponds to carrying out these tasks. A real-world example: two columns of the analysis of the radio frequency ablation procedure as recorded by the pan-European consortium are displayed in Tasks 4 ('entry/placement of the needle') and 5 ('confirmation of needle position & coagulation').

- *y-Axis, current workflow*. This represents surgeons' physical and cognitive processes that specify the information needs during the surgical task. An understanding of these needs is necessary before the user and system requirements can be outlined.
  1. *Physical parameters*. These include the following sub-parameters: *Medical goals*: the medical goal at the start of the surgical task. *Procedure:* a description of the tasks the surgical procedure involves. *Constraint:* current limitations on the surgical task. A system may be developed to overcome these limitations. *Current image:* the currently available imaging modality used for the task. *Missing image/tool:* a missing image or tool related to a particular task, e.g. to define the content of the image to support task visualisation or navigation for the surgical task. *Critical factor:* each surgical task has a corresponding critical factor which has to be achieved, or in some cases avoided, to complete the task successfully. *Feedback:* the response received during the task from either the system or the patient's body. In case of MIS, the requirement is to identify what feedback is missing for which surgical task. *Human organ:* the organs of the patient undergoing surgery respond differently to each surgical task, so it is also important to specify the nature of the organ. This understanding is required to select the technological approach.
  2. *Cognitive parameters*. These include the following sub-parameters: *Surprises*: often, while performing a surgical procedure, there are unexpected findings in the patient body, or a new situation arises owing to error in carrying out the surgical task. These findings/events often come as a 'surprise' as they were not known at the start of the procedure, by means of any preoperative patient data. In some cases they can result in breakdown of the surgical tasks. They must be documented. *Uncertainty:* surprise leads to uncertainty regarding whether to continue the current strategy for performing the surgical task. This requires a new surgical strategy to be formulated, which could be decided for a particular task or an entire surgical procedure, taking overall patient conditions and new 'surprise findings' into account. *Decision-making:* the decision-making has to be consistent in each task, as well as in the overall patient treatment strategy corresponding to the three stages of the surgical workflow.
- *y-Axis, future workflow*
  3. *User requirements*. The findings from the cognitive and physical factors corresponding to each task can be summarised in the form of user requirements. The technological possibilities can be explored, based on these user requirements.
  4. *Trends and possibilities. Technological trends:* seeking solutions for each surgical issue requires technology to be developed, which is a complex, expensive and time-consuming process. A worldwide search for findings of current research needs to be carried out and these findings have to be integrated. If we are aware of existing global advancements and available technology solutions we can avoid reinventing the wheel. In some cases we should avoid selecting possibilities which are too futuristic, being excessively dependent on distant technology breakthroughs. *Possibilities:* each team member will have his or her own ideas on future possibilities of the system, and for the optimum design to emerge it is important to capture these ideas.

APPLICATION OF WIM

The research in ARIS*ER involves developing a medical system which assists clinicians by providing task visualisation and navigation support during the MIS procedure known as RFA. This is a treatment for retarding liver cancer by means of radio frequency emissions from the tip of a needle-like probe that heats the cancerous cells and causes cell death [24]. WIM has been applied (a) to analyse the surgical workflow and identify the key user requirements, (b) to incorporate the technological possibilities and (c) to communicate these findings to the partners in the consortium.

WIM as an observation and communication tool

WIM has been applied by the consortium to incorporate the viewpoints of all the team members. It has enabled the designer to structure the investigations and integrate the findings from the user observation and task analysis methods as mentioned in sections on 'Integrating Observation Methods' and 'WIM Components'. An example of the data gathered from the observations can be seen in two columns as described in Tasks 4 and 5 (Fig. 2). The WIM content was shown to one of the expert interventional radiologists, who approved the data on its medical merits. WIM was then used to present the surgical workflow and user requirements to the technologists and clinicians at a focus group session. This made for a streamlined discussion, and the observations from this session revealed several technological possibilities. These findings were noted on the WIM y-axis corresponding to the parameter 'technological trends/possibilities'. WIM enabled the clinicians to communicate their needs and wishes, and at the same time to participate in suggesting solutions. It enabled the technologists to have an overview of the surgical workflow. Even while engrossed in defining the technicalities of system design, the team members did not lose sight of the key medical requirements.

To facilitate group discussion, the WIM was printed on A1 size paper. It was uploaded in digital form to the consortium web site, where each team member could access it and add data. Interaction with the data and the communication of needs and possibilities remain possible at a later stage.

WIM as a system design tool

- *Identifying user requirements.* Each parameter on the WIM y-axis acts as a condition, influencing the information needs related to the surgical task. This makes it easier to ascertain what information is required corresponding to which task at what point in the surgical workflow. This in turn gives the designers the flexibility to reflect on several dimensions of information requirements and thus to provide rich data on user requirements. For a brief example illustrating the process of identifying user requirements using the various WIM parameters, see Fig. 2, x-axis, Task 4, 'entry and placement of the needle', and the corresponding cell in the *human organ* parameter on the y-axis, 'tumours are stiff and of varying shapes and sizes'. This can be understood as saying that the physical characteristics and shape of the human organ are important parameters when it comes to selecting the best possible position to place the needle. This is due to the fact that the RFA needle ablates the tumour over an area approximately 4 cm in diameter, so if the tumour is elongated the needle may have to be reinserted to treat the remaining area of the tumour. The placement of the needle is also dependent on another parameter on the y-axis, *critical factor*, which refers to the 'heat sink effect'. This effect, which results in a difference in temperature in the blood flowing in the vessels very close to the tumour, is caused by the anatomical location of large vessels near the tumour. It prevents the heat from the RFA needle spreading evenly in the tumour, hence affecting the completeness of the treatment. The technological solution to make the placement of the radio frequency needle more accurate can be based on the following inferred user requirements: (a) a warning if the needle is going in the wrong direction, (b) a 3D visualisation of the shape of the tumour and the nearest vessel so that the clinician can monitor all needle placement angles and (c) a specification of the size of the closest vessel. This example illustrates the process of identifying the task-specific information on surgical anatomy that surgeons require, which can be augmented in real time and presented to them. It also enables the critical factors to be specified that have to be incorporated when subsequently identifying the design requirements.

Our experience is that the combination of methods suggested provides rich enough data for the user needs to be understood, and the technological requirements for the system identified, at the product planning stage. Once the technological approach has been selected and the project proceeds towards the product design stage, the methods mentioned above can be applied elaborately, if necessary, to generate multiple levels of analysis, e.g. an elaborate task analysis or a combination of cognitive work domain analysis and HTA. An example of applying HTA to the hierarchical decomposition of a laparoscopic procedure can be found in Mackenzie et al. [25].

- *Connecting the three surgical stages.* Another advantage of WIM is that it enables the information flow through the three stages of the surgical workflow (before, during and after) to be connected. This provides an overview of the user requirements at all three stages. This overview enabled one of the consortium partners to successfully generate a framework for defining and optimising augmented reality solutions in liver surgery using WIM [26].

Incorporating the needs of other team members in the operating theatre

WIM is currently being used to analyse surgeons' information needs, but the same framework can be used to investigate the user requirements of the other team members in the operating theatre. There are two ways of approaching this. (a) Where information feedback from another team member or system needs to be incorporated into the surgical system, the WIM structure remains the same, apart from the addition of a new parameter on the y-axis, for example, *Team 1/System 1*. This may imply: 'user observations indicate that the surgeon needs feedback from Team 1 or System 1 during Task 4'. This information could be entered in the corresponding cell. (b) Where a system is being developed specifically for anaesthesiologists or other medical staff, the structure of the WIM remains the same and the investigations have to be conducted in the same order to gather the user data, but the parameter on the y-axis (*human organ*) needs to be adapted slightly in line with the domain expertise. For example, an anaesthesiologist might require information on blood pressure rather than the human organ. These parameter adaptations and additions can be made based on initial interviews with the anaesthesiologist.

THE NEXT STEPS IN THE DEVELOPMENT OF WIM

The following steps are currently in progress to extend WIM.

- *Storyboards.* The visualisation support provided by the new medical system may change the current surgical workflow, so we need to speculate on possible changes in the surgical workflow. Based on the user requirements, a number of optional storyboards representing scenarios of technological possibilities will be created using an adaptation of storyboard technique (a technique for projecting future scenarios of product usage graphically). These will be subjected to a technological evaluation cycle with the surgeons and the technologists. The advantage of storyboards is that they show the sequence of actions on a time scale, making it easier for surgeons and technologists to select the solution with the maximum potential in future.

For example, if computerised tomography is selected as the imaging modality to assist the clinician in the current procedure, the workflow will change, as this requires either the patient to be taken to and from a CT room or a portable CT machine to be brought into the operating theatre/intervention room.

- *User interface and design specifications.* The user interface and design specifications can be drawn up after selecting the key technological approach, based on the storyboards. These specifications will be based on combining the user requirements and the inputs from the technology partners on the system specifications. Finally, a number of alternative interface designs will be simulated for each task. These alternatives will undergo an iterative design cycle based on heuristics evaluations until they are finally approved by the surgeons.

FUTURE CHALLENGES IN THE DESIGN OF SURGICAL SYSTEMS

The surgical domain is a fragile and complex web of experts, with constant decision-making and uncertainties linked to patient safety. Any unwanted technological interference in key treatment strategies or surgical tasks can lead to fatalities. Future medical systems will need to support not only task visualisation and navigation but also decision-making. This could perhaps lead to bio-intelligent systems that understand the patient's tissue quality and aid decision-making in real time. In this regard the following challenges are worth considering when introducing new systems in the surgical domain.

- *Individuality.* In order to perform surgery, surgeons develop expert skills and techniques, either individually or in compliance with standard hospital protocols. The question is how far does a system consider these individual skill sets? Or does it assist in advocating or teaching a standard protocol for carrying out the procedure even in different hospitals?
- *Uncertainty.* While performing surgery, surprise medical findings during the procedure often create a critical moment of uncertainty in decision-making. In this case the surgeons invite other specialists to the operating theatre to ask their advice. For example, while performing surgery to remove tumours from a liver, many new tumours are discovered half-way during the procedure. As they were not identified in the preoperative CT, they were not scheduled for resection. The challenge is how far can the surgical system respond to such uncertainty and instant decision-making in the surgical environment? How far can it aid the various experts present in the operating theatre? Should it only allow linking to in-house experts, or should it also connect with global experts to share in the decision-making?
- *Multimodality.* Given that multimodality (systems including audio, video and haptic feedback) will be a critical feature of these surgical systems, it will be important to identify the relationship between the multimodes and the dynamic context of surgical tasks. For example, there are already a number of context-related sounds and sources of visual information in the operating theatre: what will be the effect of adding yet more information? The first question is to what degree will multimodality support or interfere with surgical tasks? Second, is multimodality required, and if so, which modalities should be involved? In this context there is a need to document attention time [3].
- *Modularity.* Surgical workflow-driven information systems are still under development. Most of them focus on specific types of surgery, such as RFA or liver resection. The issue, however, is to what extent will these surgical systems be adaptable to other types of surgery, or will they be modular in nature? How will they incorporate future technological breakthroughs?

CONCLUSION

As regards planning and developing surgical systems for MIS, this chapter has focused on investigating surgeons' information needs in the surgical workflow. WIM has aided the analysis of the surgical workflow and the identification of user requirements. WIM is currently evolving and it is to be tested for various other medical cases before arriving at a generic solution, leaving ample scope for further iterations and improvements. Given the project requirements, it has been applied to generate information requirements for augmented reality solutions in surgery, but we would argue that, as a framework, WIM is not technology-governed; rather, the technology is to be determined from the WIM findings. The next focus will be on investigating whether a generic model of WIM could be applied to non-surgical environments. In future, based on how applicable WIM proves to be in the medical domain, various other applications could be considered, driving the planning and development of new ideas and products not only in the academic world but also in industry.

ACKNOWLEDGEMENTS

ARIS*ER is funded by the European Union as part of the Sixth Framework Programme for Research, under the Marie Curie Actions for Human Resources and Mobility. The authors would like to thank the medical team at Erasmus MC hospital for granting permission to observe a number of surgical procedures and interventions, and also for sharing their thoughts and ideas. Special thanks are due to all the technology researchers and members of the board of the ARIS*ER consortium for their input.

REFERENCES

[1] Villard, C., Soler, L., Papier, N., Agnus, V., et al. (2003). *RF-Sim: A Treatment Planning Tool for Radiofrequency Ablation of Hepatic Tumors*. Proceedings of the Seventh International Conference on Information Visualisation (IV'03), IEEE Computer Society Press, pp. 561–67.

[2] Freudenthal, A., Samset, E., Gersak, B., et al. (2005). Augmented reality in surgery (ARIS*ER). *Endosc. Rev.*, 10 (23), 5–10.

[3] Sanderson, P. (2006). The multimodal world of medical monitoring displays. *J. Appl. Ergon.*, 37 (4), 501–12.

[4] Albayrak, A., Meijer, D.W., and Bonjer, H.J. (2004). Current state of ergonomics of operating rooms of Dutch hospitals in the endoscopic era. *Minim. Invasive Ther. Allied Technol.*, 13 (3), 156–60.

[5] Taylor, R.H., Lavallée, S., Burdea, G.S., and Mösges, R. (1996). *Computer-Integrated Surgery: Technology and Clinical Applications*. Cambridge, MA: MIT Press.

[6] Bogner, S.M. (1994). *Human Error in Medicine*. Hillsdale, NJ: Lawrence Erlbaum.

[7] Woods, D.D. (1986). Paradigms for intelligent decision support. In *Intelligent Decision Support in Process Environments* (E. Hollnagel, G. Mancini, D.D. Woods, eds) New York: Springer, pp. 153–74.

[8] Buckle, P., Clarkson, P.J., Coleman, R., and Ward, J. (2006). Patient safety, system design and ergonomics. *J. Appl. Ergon.*, 37 (4), 491–500.

[9] Veelen, M.A. van and Goossens, R.H.M. (2005). Ergonomics in surgery. In *Human Factors in Design, Safety, and Management* (D. de Waard, K.A. Brookhuis, R. van Egmond, T.H. Boersema, eds) Maastricht: Shaker Publishing, pp. 387–95.

[10] Benardete, E.A., Leonard, M.A., Weiner, H.L., et al. (2001). Comparison of frameless stereotactic systems: Accuracy, precision, and applications. *Neurosurgery*, 49 (6), 1409–15.

[11] Azuma, R., Baillot, Y., Behringer, R., et al. (2001). Recent advances in augmented reality. *IEEE Comput. Graph. Appl.*, 21 (6), 34–47.
[12] Shuhaiber, J.H. (2004). Augmented reality in surgery. *Arch. Surg.*, 139, 170–4.
[13] Hunink, M.M.G. (2005). Decision making in the face of uncertainty and resource constraints: Examples from trauma imaging. *Radiology*, 235 (2), 375–83.
[14] Annett, J. and Stanton, N. (2000). *Task Analysis*. London: Taylor & Francis.
[15] Crystal, A. and Ellington, B. (2004). *Task Analysis and Human Computer Interaction: Approaches, Techniques, and Levels of Analysis*. Proceedings of the 10th Americas Conference of Information Systems, New York.
[16] Beyer, H. and Holtzblatt, K. (1998). *Contextual Design*. Los Altos, CA: Morgan Kaufmann.
[17] Sanders, E.B.N. and Williams, C. (2001). Harnessing people's creativity: Ideation and expression through visual communication. In *Focus Groups: Supporting Effective Product Development* (J. Langford and D. McDonagh-Philip, eds) London: Taylor & Francis.
[18] Annett, J. and Duncan, K. (1967). Task analysis and training in design. *Occup. Psychol.*, 41, 211–21.
[19] Langford, J. and McDonagh, D. (2003). *Focus Groups – Supporting Effective Product Development*. London and New York: Taylor & Francis.
[20] Norman, D.A. (1998). *The Design of Everyday Things*. London: MIT Press.
[21] Fetterman, D.M. (1998). *Ethnography*, vol. 17. USA: Sage
[22] Rasmussen, J. and Pejtersen, A.M. (1995). Virtual ecology of work. In *Global Perspectives on the Ecology of Human–Machine Systems* (J. Flach, P. Hancock, J. Caird, K.J. Vicente, eds), vol. 1. Hillsdale, NJ: Lawrence Erlbaum, pp. 121–56.
[23] Owen, C.L. (1992). Context for creativity. *Design Studies*, 13 (3), 216–28.
[24] Yan, C., Li-Yan, Z., Man-Ku, D., et al. (2003). Ultrasonography guided percutaneous radiofrequency ablation for hepatic cavernous hemangioma. *World J. Gastroenterol.*, 9, 2132–4.
[25] MacKenzie, C.L, Ibbotson, J.A., Cao, C.G.L., and Lomax, A.J. (2001). Hierarchical decomposition of laparoscopic surgery: A human factors approach to investigating the operating room environment. *Minim. Invasive Theory Allied Technol.*, 10 (3), 121–7.
[26] Kalkofen, D., Reitinger, B., Rishlom, P., et al. (2006). *Integrated Medical Workflow for Augmented Reality applications*. Proceedings of AMI Workshop, Copenhangen.

Meeting Diversity in Ergonomics (2007)
Pikaar, Koningsveld, and Settels (Eds)

15

Ergonomic Intervention in Hospital Architecture

J. Villeneuve

Association paritaire pour la santé et la sécurité du travail, secteur affaires Sociales (ASSTSAS), Montréal, Canada

S.L.M. Remijn

ErgoS Engineering and Ergonomics, Enschede, The Netherlands

J. Lu, S. Hignett

Healthcare Ergonomics and Patient Safety Research Unit, Department of Human Sciences, Loughborough University, Leicestershire, UK

A.E. Duffy

Ontario Safety Association for Community and Healthcare (OSACH), Toronto, Canada

Abstract. The reorganization of healthcare services in western countries has produced a considerable number of hospital renovation and construction projects. Ergonomic intervention in architectural projects is a relatively recent phenomenon, and the Triennial IEA Congress 2006 was an exceptional opportunity to observe the progress being made in ergonomic practice in various countries. This chapter presents a summary of the Symposium: Healthcare Ergonomics – Architecture. A surprising kinship of approaches and methods has emerged from these papers.

Participatory ergonomics is a particularly effective approach to the design phase of architectural plans. Several methods involving direct users of the future installations are presented: dynamic simulation, site visits, process and flow analysis, field observation, etc. Case studies illustrate the applications and the results achieved.

Due to the scope and complexity of architectural projects, workstation ergonomics (microergonomics) has given way to the analysis of much larger production systems and services (macroergonomics).

Despite definite progress, the potential of the discipline is still undervalued in this field. Building professionals (architects, engineers, etc.) and government agencies should be made aware of the added value that ergonomics can provide.

Keywords: Hospital architecture, design process, participatory ergonomics, simulation, macroergonomics, standards, multidisciplinarity

INTRODUCTION

In recent decades, western countries have made extensive reforms to health and social services which have led to major changes in building stock. The phenomenal increase in healthcare costs associated with improvements in medical technologies and ageing of populations has forced government agencies to optimize health service efficiency by developing home care and ambulatory services and by merging healthcare institutions into integrated networks of community services. Parallel to this situation, the concentration of hyperspecialized care has resulted in the construction of mega teaching hospitals in major urban centres.

Following post-occupancy evaluation in a number of renovated or new healthcare facilities, certain problems in workplace design have been identified, both in patient care units and in other related areas such as laboratories, food services, administrative offices and so on. Ergonomic interventions were required to correct these situations, when in fact the problems could easily have been avoided at the planning stage.

Many design projects are carried out with insufficient consultation between direct users, management and professional designers (architects, engineers, etc.). Generally speaking, construction standards do not consider ergonomics or health and safety criteria.

Serious design faults, which could be avoided by a more rigorous analysis of users' activities and processes, may occur, and may generate discontent and affect operational efficiency. Sometimes new buildings have to be altered immediately after construction. This is both costly and unproductive.

This chapter outlines the results of professional practice and research in the field of ergonomics and hospital architecture presented at the Triennial IEA Congress 2006. This is a relatively recent and innovative practice which has led to some interesting breakthroughs in a number of countries. The following developments are presented and discussed in this chapter.

- In French Canada (Québec), the implementation of an ergonomic intervention program in architectural projects – the PARC [1, 2] program - has been surprisingly successful, with several hundred interventions. This program has spread to the Daycare sector [3], where it has also produced some remarkable outcomes. Discussions with our colleagues in English Canada have allowed us to make some exciting breakthroughs in other provinces [4, 5].
- In The Netherlands, the design approach was originally developed in 1985 in process control rooms [6]. This engineering approach was further developed by Remijn into architecture with a focus on the end-user. It uses a step-by-step guide from process and task analysis to a set of requirements for the architectural design [7]. The method is applied particularly in healthcare buildings [8].
- In the UK, there are national construction guidelines published by the NHS. Research has been carried out on ergonomics and architecture to improve these standards [9].
- In France, the *Laboratoire d'ergonomie des systèmes complexes – LESC* [10], has produced keynote work on the theory of ergonomic practice within architecture [11].
- In Brazil, a multidisciplinary ergodesign team is part of the University Hospital's regular services [12].

The Maastricht congress showed surprising similarities in the approaches and methods used by Symposium speakers. The principles of participatory ergonomics have been particularly well demonstrated. This concept invites users, especially hospital staff and management, to participate in the design process. An overview of the methods used by the co-authors is presented in the form of case studies. Despite some variants, there is an unquestionable convergence of approaches among the co-authors.

The impact of ergonomic interventions in hospital architectural projects is beginning to bear fruit, particularly when it comes to dealing with standard-setting bodies. However, it seems that building professionals and project managers are largely unaware of the potential of ergonomics in environmental design to shape improved interfaces for future building occupants.

ERGONOMIC APPROACH TO ARCHITECTURAL PROJECTS

The approach presented in this section is the result of 15 years of experience with the PARC program, and a number of interventions in small and large architecture projects in healthcare facilities [13].

The approach of the PARC program

The approach is *participatory* in that, to succeed, it requires the direct participation of end-users in a well-structured consultation process.

The approach is *prospective*, and concentrates on anticipating future activities and on ensuring that the design concept layout provides suitable conditions for the comfort, safety and performance of users in the new facilities.

The approach also gives *decision-making support* to institutional players and designers by helping them to understand the demands of the work and thus to make appropriate design choices. It therefore complements well the expertise of the architect, which is concerned more with the construction of the building itself.

Levels of ergonomic analysis

In a functional analysis of users' activities in existing or future buildings, it is useful to distinguish three levels (see Fig. 1):

- a *macroscopic level*, involving the building as a whole in relation with its site, and the location of the main activity zones within the building;
- a *mesoscopic level*, involving the functional relationships between the different facilities within each activity zone, i.e. a department or related departments;
- a *microscopic level*, involving workstations and the performance of activities on the premises.

From this standpoint, the building as a whole is seen as a broad system comprising a set of subsystems (the departments) which themselves comprise [illegible] subsystems (the

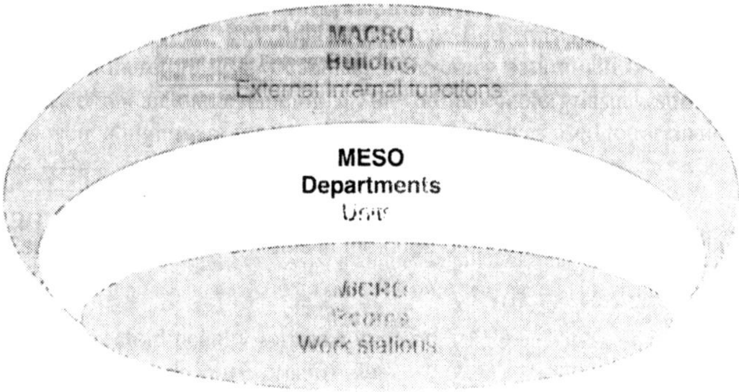

FIGURE 1. Level of ergonomic analysis.

workstations). All the systems are open, and close functional relationships exist between the different levels.

This level is concerned with the relationship between the position of the building and the characteristics of the site: its size, position in relation to the cardinal points and neighbouring streets, future easements and the type of soil and vegetation.

The limitations and possibilities of the site with respect to the *external functions* of the building and required surface areas significantly affect the design concept in terms of the volume of the building – basically, its shape and the number of floors. The external functions of a building include pedestrian and vehicular entrances and exits, access for service and goods delivery, terraces, gardens and parking. The access points must be positioned in compliance with safety and entrance/exit surveillance requirements.

The other major aspect to be considered concerns the position of the main activity zones in relation to the *internal functions* of the building: reception, administration, care units, dietary services, specialized facilities, community facilities, staff facilities and technical facilities. The positions of these zones must enable the best possible proximal relationship among them and with the outside of the building, to facilitate internal circulation of people and materials.

The proximity of the various zones to one other will depend on their respective functions and the compatibility or incompatibility of their activities. For example, contact between soiled and clean materials should be avoided.

Analysis of horizontal (floor by floor) and vertical (floor to floor) circulation of people and materials will enable the activity zones, stairwells, lifts, laundry chutes and technical channels to be located in the best possible position in the building.

Mesoscopic level

The mesoscopic level is concerned with the activity zones themselves. Each zone may contain one or more related services, and it is important to assess the functional relationship of each facility with the others.

If we take a care unit as an example, the layout of the facilities will depend on the exterior shape of the building and on the internal circulation of people and materials. Whatever the shape, the indoor space will almost always be arranged on either side of a long corridor or at the junction of two or three shorter corridors.

At this level, the layout should minimize staff movement and promote communication between the facilities in the activity zone.

Microscopic level

The microscopic level is concerned with the organization of workstations within a service area. A workstation is always related to a set of other workstations. It is important to identify these relationships and ensure that the physical layout takes account of chronological work sequences so that the physical and informational links are as efficient as possible. The layout of workstations in a laboratory, for example, should be designed to avoid unnecessary handling of samples, and should reflect the chronology of the operations performed on the most common types of samples. This obviously requires prior analysis of flow process during handling of samples by laboratory technicians. The specific constraints of each workstation must also be identified, especially if employment injuries or significant incidents have been declared in the past.

Environmental considerations, such as the location of windows according to the orientation of the building, potential sources of noise and ventilation requirements, must be considered when designing workstations according to the type of activity to be performed. A computer-based activity in an administrative office does not involve the same design constraints as a workshop activity requiring the use of noisy mechanical tools or toxic products.

Equipment and tools are generally selected at the microscopic level, on the basis of ergonomic criteria that take account of user safety and comfort as well as production goals. The choice of a specific production system sometimes has an impact on the design as a whole. Such choices are made at the macroscopic level, because they affect the initial design concept. The decision to use a tunnel washer rather than separated washing machines and dryers in a laundry, for example, will determine the whole design of the laundry area and also the upstream areas – dirty laundry reception area, weighing area and tunnel feed area – and the downstream areas – sorting, drying and folding.

An iterative process

At first glance, the design process appears to be linear, moving from the general to the specific, from the building as a whole (macro) to specific workplaces or living spaces (micro). In reality, however, it is more iterative in nature, and is characterized by continuous switches back and forth from the general to the specific. For instance, to be able to design a hospital properly it is absolutely vital to begin with an ergonomic study of the activities of room users in the different care units. The concept layout depends to a very large extent on the format of basic units such as patient rooms, nurse's stations, examination rooms, etc., according to the architectural programming. This exercise is rather like a LEGO session, where the shape of the blocks determines the end result.

An initial evaluation at the micro level ensures a smoother passage to the meso and the macro levels. In the process of developing the drawings from the concept-layout stage to the execution stage, it is always necessary to switch back and forth from one level to another, in order to make sure that the results are both coherent and harmonious [14]. This method is not popular among architects, who have yet to be convinced of the need to make more detailed analyses of user activities at the concept-layout stage before making final decisions about the architecture of the building as a whole. Yet this type of analysis helps prevent costly reviews when the plans have reached a more advanced stage (preliminary or execution plans) when it is much more difficult to rethink the original design layout concept (Fig. 2).

Participatory process

The main concern of the ergonomist is to ensure that all the players who have information relevant to the project are involved in the design phases. The way people are involved is a key element to success. The method proposed here is based on positive experiences in a number of institutions that truly took charge of their projects.

In a large-scale project, three types of committees are required. Their mandates are separate but complementary, as shown in Fig. 3 The Steering Committee is composed of representatives of all the social players concerned, and is led by the general management. Its mandate is to define the project orientation and make strategic decisions. The Technical Committee, composed of technical professionals and led by the project leader, is responsible for designing and carrying out the project in compliance with the Steering Committee's orientations. Finally, the User Committees are formed by department or by theme, and are composed of representatives of the executives and workers concerned by the project. Their mandate is to help define requirements and then provide an informed opinion on the design proposals, based on their expertise in the field.

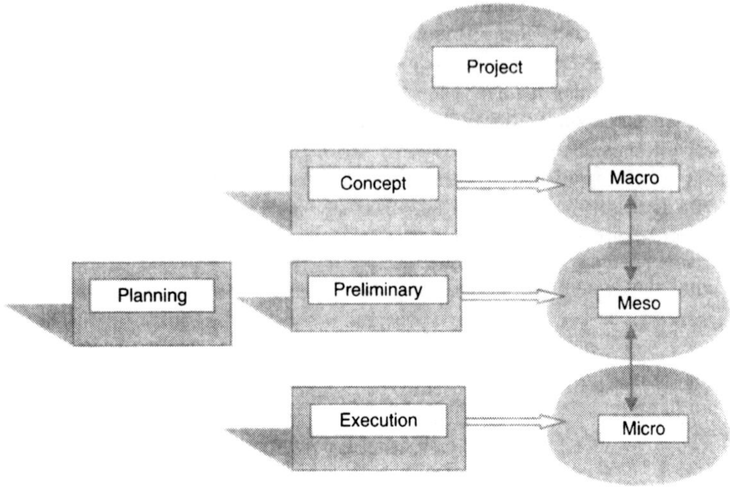

FIGURE 2. Phases of a project and levels of analysis.

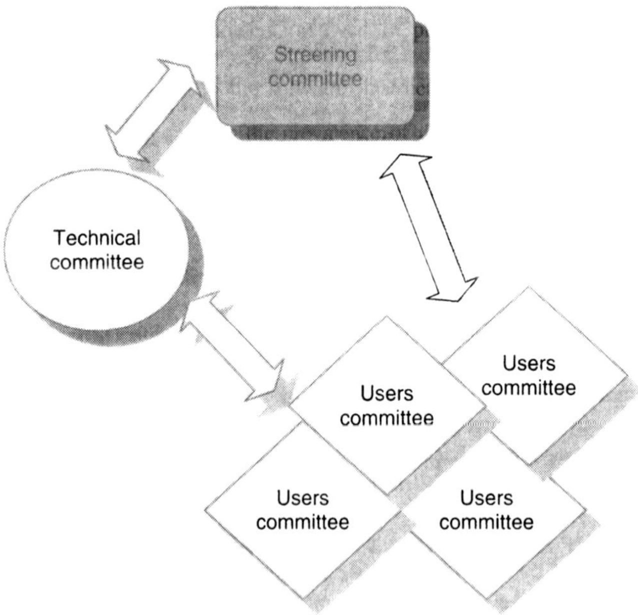

Close and constant contact must be maintained between the committees. The project leader plays a key role in managing these contacts.

FIGURE 3. Participatory structure

Close and constant contact should be maintained between the committees. In particular, each committee should delegate a representative to the Steering Committee, which is in fact the decision-making body. The project leader plays a key role in managing the contacts between the Steering Committee, User Committees and the professionals on the Technical Committee.

In a smaller project, one Steering Committee with representation from all stakeholders could work well as long as there is clear support from senior management.

Case Study No. 1: Participatory ergonomics

Based on: "Moving to a New Emergency Room: A Major Challenge!", by Villeneuve, ASSTSAS[1].

Study

In preparation for moving to a new emergency room, management and the unions agreed to undertake an allocation process, which took approximately 2 years. The role of the ergonomist in this project was to provide support for the project team and to propose the appropriate methodology. The goals were firstly to ensure that the new facilities would meet operational needs and secondly to ensure that the completely new and radically different workspace, approximately three times larger than the old space, would meet the expectations of medical and support staff.

Nearly a year has passed since the opening of the new emergency department. Did everything go as planned?

Project data

Emergency department	New	Old
Staff	27 full time + on call	24 full time + on call
Physicians	43	23
Stretchers	43–52	23
Radiology room	1 larger	1
Total area (m^2)	4447	1316
Total cost (M$ can.)	14.3	

Participatory management

The project team consisted of 12 employees and managers representing all job categories and all three shifts.

Several activities were organized: simulation of on-the-job activities, analysis and correction of the architectural plans, manpower planning, redefinition of duties for all three work shifts, review of the roles of the teams, guided tours of the facilities before the move, and training of resource people to provide support for the teams during the start-up phase.

Staff members involved in the consultation process were given time off so that representatives could participate actively in the project group. Close monitoring of the architectural plans proved to be extremely important throughout the design process. A number of changes were made along the way, thus minimizing post-construction adjustments.

[1] The ASSTSAS is a joint sector-based association dedicated exclusively to occupational health and safety in the health and social services sector.

Simulation

Large-scale simulations (e.g. step-by-step role play of a patient arriving by ambulance) and simulations on a smaller scale (e.g. models of the nursing stations) made it possible to see how effectively staff would be divided up into the various sectors of the emergency room and to review certain operations (e.g. procedure following the death of a patient). This process led to some adjustments in the staffing plan, specifically the addition of staffing hours to replace employees at mealtimes and breaks.

Teams' work

The importance of flexibility and versatility of the teams soon became apparent, given the surface area to be covered and the variations in traffic in the different sectors. This entailed continuing education for medical staff, with a nurse educator to provide support in day, evening and night shifts, and staff rotation, except in psychiatry.

Review of job descriptions

A detailed review of job descriptions for each position was also carried out. A chronological list of activities was drawn up for each job, to identify any grey areas (responsibilities belonging to everyone and no one) and to clarify the roles and professional prerogatives of each category of care giver. The ultimate aim was to harmonize practices and ensure adequate coverage of every aspect of the patient care continuum.

Start-up and opening

As the opening of the new department approached, guided tours conducted with simulation of real patient case scenarios were organized for the entire staff. This exercise relieved some of their apprehensions about the radical change in their work environment.

The presence of resource people from different job categories proved to be a decisive factor in the critical start-up phase of the operations. These individuals had received training in problem-solving techniques and had reviewed the facilities, equipment and standards for the new emergency room. Their role was to support staff on site and to help the head nurses with their supervisory tasks.

Eight months after the opening – How are things going?

Performance of the Team

Overall, the start-up went well. The head nurse, assistant head nurses and nurse educators listened carefully to staff concerns and were able to solve problems quickly. Because of the large surface area, staff versatility was maintained. The nursing shortage, particularly in the first few months, was very difficult to manage.

The Facilities

Of course, nothing is perfect, but overall the facilities have met expectations. As projected, the large surface area has made it difficult to achieve an overview of activities in the ER. On the other hand, patient comfort and confidentiality are far superior. Staff have also benefited from improved work spaces and rest areas, and they are not tripping over each other as they were before.

Improved information technologies compensate for the inconvenience of longer distances: electronic notice boards, portable telephones and an increased number of computer stations facilitate access to crucial information throughout the emergency room.

There are still Patients in the Hallways...

The bad news is that there are still sometimes patients in the hallways. Overcrowding of the emergency room is attributed to the opening of the new mother/child health centre which takes up the top two floors. New patients, particularly obstetrics patients, are now pouring into the emergency room. The psychiatry sector is also constantly busy. Several steps have been taken by top management to deal with this situation.

Winning conditions to participation

Here are some of the best practice conditions for participatory ergonomics in an architectural project.

Where does the project leadership come from?

To quote an architect on this key issue, "I don't believe the architect is the main reason a project is or isn't successful. The drive and direction should come from the people who deliver the care: the physicians, nurses, technicians, and so on." [15].

In our view, this is the only constructive way to manage a project, and the ideal opportunity to apply the principles and methods of participatory ergonomics. Nevertheless, meeting the requirements for strong internal leadership is not an easy matter. The major role of the ergonomist is to support the people working in the hospital involved in this complex undertaking [16].

What are the winning conditions?

The most important conditions are commitment and consistent support from upper management. The project manager should be skilled at participatory management. Listening to users' needs with respect to professional prerogatives should be a major concern. The following conditions should therefore be met by hospital management, design professionals, employees and their trade union representatives.

Hospital Management

Transparency and sharing of information are essential conditions, with the exception of certain information which must of course remain confidential. It is impossible to encourage staff participation without ensuring the transparency of major project-related decisions.

The transition and start-up of the new facilities is much easier if staff are involved from the beginning, especially if there are major changes in work organization, technology and staffing.

In the early phases of the process, there are generally no cost overruns or scheduling delays. Considerable savings can also be achieved by significantly reducing any modifications made after the construction work has been completed.

Physicians

Time constraints generally make it difficult for doctors to participate actively. Medical staff representatives should be appointed and asked to plan the schedule accordingly. It is desirable that physicians are regular members of the users' group, so they can better understand everyone's needs.

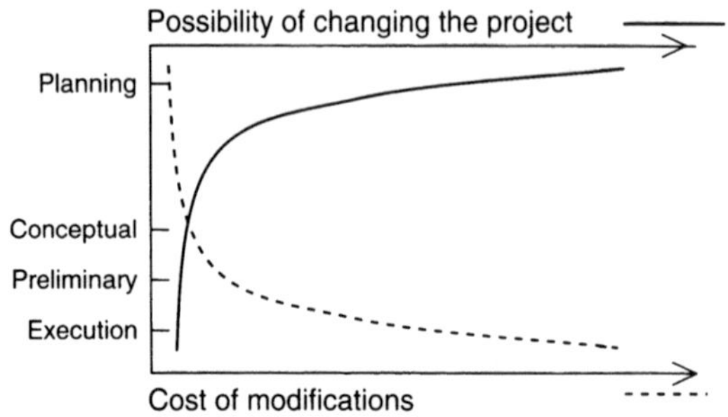

FIGURE 4. Possibility of changing the project.

Employees and Trade Union Representatives

These stakeholders have a real interest in participating as long as management listens to their opinions and comments, which is why the mandate and powers of the joint employee–management committees must be clearly defined. Recommendations should be followed up and it is important to explain why some have not been implemented or have been changed.

Provision for staff paid time off, or time off in lieu, should be included in the budget to avoid overburdening employees who remain on duty.

Designers, architects and engineers

Open-mindedness and listening to staff comments are of the utmost importance. Designers, architects an engineers should also strive to provide comprehensible answers to technical questions and should be able to manage seemingly inconsistent demands. The ability of these professionals to interpret these demands and propose "creative compromises" is a major asset. The ergonomist often has to act as intermediary between the designers and staff by providing additional documentation for the needs and criteria to be met.

Aesthetic and technical considerations should not overshadow users' needs and the functionality of the facilities. All too often these arguments do not prove to be justified after analysis. As the team leader of an emergency department expansion project put it, "The department design proposed by the architectural consultant should fit your needs and functional requirements even in a renovation project. It should not be your needs that change to fit a design." [17].

Last-minute interventions at the final preliminary or detailed planning stage should be avoided, because any significant change means a major revision of the plans. As the project advances, the possibility of introducing changes becomes increasingly limited and more expensive (see Fig. 4).

METHODS TO ANTICIPATE FUTURE ACTIVITY

The primary requirement of a design project is to produce a design that is consistent with the future activities of users. Consequently, the activities to be performed in the new facilities must be correctly anticipated.

This is a significant challenge, since the level of uncertainty is always high, and particularly so in an environment as complex as a hospital. It is an even greater challenge for users, who are not skilled at designing new facilities. Indeed, the mental representations of users are founded on past experience, derived mainly from the practices of the institution in which they are currently working. However, a design project involves major changes of practice, because the premises themselves change, as do the technologies, operating methods, care philosophy and organization of work. Reference to what currently exists is therefore ineffective to some extent, and can even be an obstacle to the development of the project. In addition, if the participation of users is to be productive it requires increased support, enabling them to guide the professionals correctly in their design choices.

Some methods are useful in predicting future activities: dynamic simulation, process and flow analysis, organized site visits, observation and links analysis.

Dynamic simulation

Simulation is undoubtedly one of the most effective ways of testing design concepts. Obviously, not all situations can be simulated, and priorities should be established on the basis of precise criteria, including the ease with which situations can be reproduced and the risks to human health and safety.

The proposed *dynamic simulation procedure* can be applied to all steps of the design phase – programming, design draft, preliminary plans and specifications, and detailed plans and specifications. It can also be reproduced at the three functional analysis levels – in other words, at the level of the building itself (macroscopic), the various departments (mesoscopic) and the individual workstations (microscopic).

The general logic of the simulation process (Fig. 5) is to develop future activity scenarios on the basis of the data used to define the design project. Placing real users in a layout representing the proposed design concept simulates the scenarios. The suitability of the concept for the predetermined future activity scenarios is then assessed. This gives rise to proposals for changes to the concept or the scenarios, and finally to a design that fully satisfies users' expectations.

It also provides an opportunity to confront different viewpoints in a positive way. Architects, engineers, departmental heads, employees and the ergonomist all have very different views of the work. This diversity of viewpoints is not an obstacle to project development. On the contrary, it provides an overall vision of the projected situation without which the design exercise may be defective. Simulation is thus an excellent way of confronting viewpoints and reaching a creative compromise.

Simulation props

The major props used to create dynamic future activity simulations are *the enlarged plan, the three-dimensional representations and the full-scale simulation with mock-up*.

Any of these simulation props can be used, depending on the circumstances. All simulations begin with a plan. However, users may not be used to reading design plans, and their natural ability to do so may vary tremendously between individuals. Some people will find it easy to situate themselves in a given space and to project themselves into a two-dimensional plan. However, the vast majority find this task difficult, especially when the plan is on a small scale. For this reason, inexperienced users must receive guidance to help them read and interpret the plan, and it is better to work with larger-scale plans (1:50 rather than 1:100).

Three-dimensional simulation props are much better in providing users with a meaningful representation. The closer the simulation context is to the proposed reality, the better the results will be in terms of the reliability of the reference points used for the design.

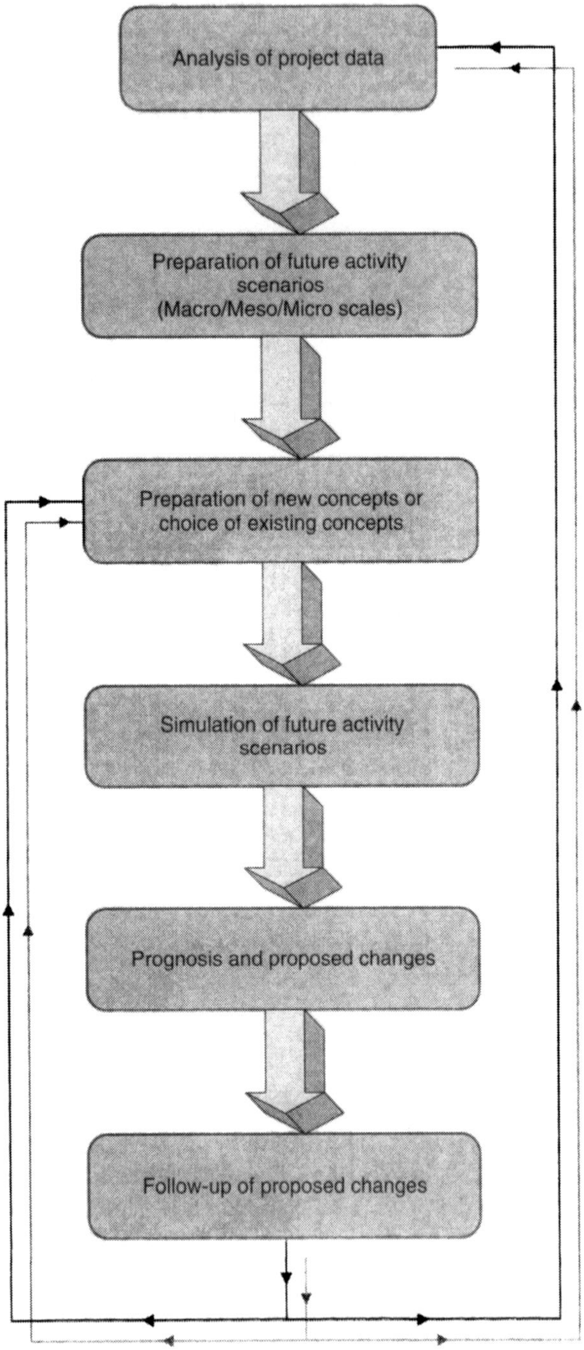

FIGURE 5. Dynamic simulation process.

Three-dimensional simulation software is now available at reasonable cost. On request, architectural firms can easily provide three-dimensional representations produced with Auto-CAD that are tremendously useful in visualizing the building as a whole, or specific areas inside the building.

One excellent type of simulation, which has the advantage of involving real users in action, is the full-scale simulation. It is relatively simple to organize, either in comparable existing facilities or in a room that is large enough to reproduce the situation with a mock-up of all fixed and movable equipment. Simulations such as this are appropriate when copies of the same layout model will be reproduced a number of times in the building. This is often the case in hospitals, where different floors all follow the same model. They are also appropriate for testing operations of a critical nature, where errors may have serious human or financial consequences.

Process and flow analysis

Analysing traffic patterns of people and equipment is a basic exercise in validating proposed layouts at a building level. To do so, the traffic flow patterns must be examined at each step of the process:

- Identify traffic flow for people and equipment in the future building, from the entrance to the exit.
- Select priority traffic patterns.
- Describe activity scenarios associated with the process being analysed: functions, physical premises, people involved, equipment and tools used.
- Identify potential hazards (physical, biological and chemical).
- Identify the level of sterility required (soiled, clean and sterile).
- Make recommendations to eliminate hazards and improve operating efficiency.

Certain basic principles must be respected in this analysis:

- The building shall be completely wheelchair accessible.
- Traffic flow patterns for soiled and clean supplies shall be clearly separated to prevent transmission of nosocomial diseases.
- Mechanical devices shall be provided to avoid manual lifting of people or objects.
- The forward movement principle should guide the layout of the facilities to avoid backflow in operations, particularly in the production sectors (e.g. kitchen, laundry, laboratory, central sterilization, etc.). This makes the facilities more efficient.
- The proposed layouts should respect the natural chronological sequence of operations so that staff and clients can avoid unnecessary steps, have a comfortable space to live in, and have everything they need to perform their duties close at hand.
- As far as possible, traffic patterns for equipment should be kept separate, making maximum use of the basement and reserved service elevators. These elevators should be well positioned in the building and be dedicated to either clean or soiled supplies.
- Public, semi-public and staff-only areas should be clearly defined and controlled by a simple, effective system (e.g. magnetic card).
- Finding locations in the building should be as natural as possible. There should be explicit visual cues inside and outside the building.
- Signs should support the natural visual cues, but people do not always read signs.

The following section provides an excellent example of process analysis and simulation applied in an operating room design project.

Case Study No. 2: Tools for users' participation

Based on: "Integrating Ergonomics into the Architectural Design Processes: Tools for User Participation in Hospital Design", by Remijn, ErgoS Ergonomics and Engineering.

In the pre-design and schematic design phases of the operating suite, the emphasis must be on the requirements concerning the routing, relationship between areas, logistics and planning. The ergonomic consultant has to derive these requirements from user information. A basis tool for this user participation is an ergonomic analysis in which users are interviewed and their task performance is observed.

The user information can be structured in process flow diagrams as a model of the functions in the operating suite. By allocating features of work activity in the existing situation, both the users and the project team have a good overview of the way these functions need to be realized in the design. Furthermore, the process flow diagrams can be used to simulate possible future changes in the activities and work organization, to optimize logistics and planning for example.

Relationship of functions or areas

At the start of the preliminary design phase a diagram that illustrates the relationships between functions or areas is a helpful tool for users to discuss the global layout aspects. Figure 6 shows an example of such a diagram (in Dutch). The diagram is derived from the process flow diagrams and additional features of the future situation.

It is the architects' task to provide such a diagram in a number of different layouts for discussion by the user groups. To support users in the review of the design proposals, the ergonomic consultant should generate a list of criteria to provide structure for the process. These criteria are derived from the results of the ergonomic analysis. Additional decision techniques can support an objective choice in layouts.

Layout on magnetic board

To investigate the demands for the layout of the operating rooms, a magnetic board is used, with scale representations of all persons and objects found in the room (see Fig. 15.7).

In a number of sessions with surgeons and surgical personnel the various layouts were discussed (What is the existing layout? What should be changed in the new OR?), and the use of a magnetic board made it easy to modify the layout. The results were used to determine the required dimensions of the operating rooms and were also used as input for mock-up sessions.

Mock-up evaluation

A mock-up is a full-size model, using simple (mostly wooden or cardboard) materials. This is a very useful tool in the design phase, used for instance to examine the layout of an operating room. The mock-up must be constructed in a flexible way to enable testing of different layouts. Figure 8 shows an impression of the operating room mock-up. Based on the magnetic board sessions, the end-users were asked to review a layout by performing predefined work scenarios. This resulted in requirements for the design of the operating rooms. A particular advantage

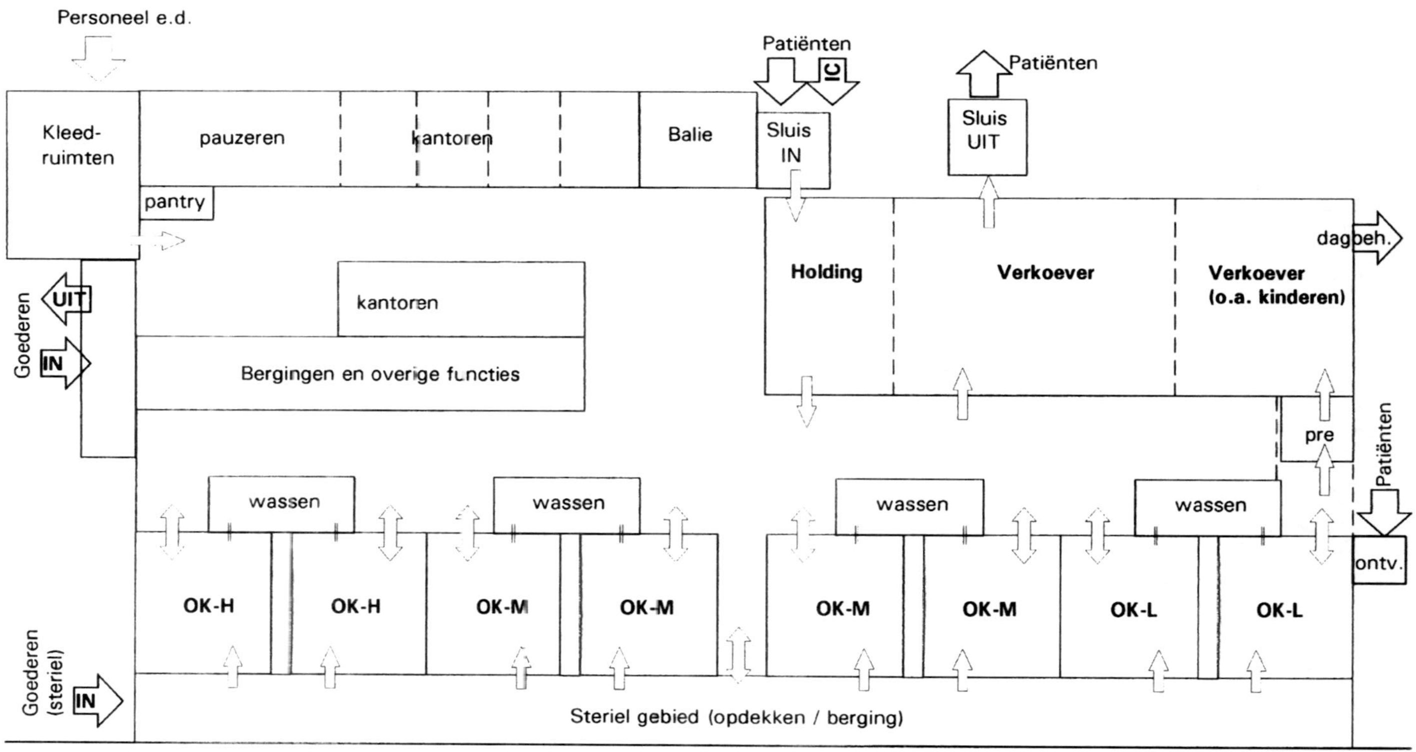

FIGURE 6. Example of a diagram which shows the relationship of areas in an operating suite (in dutch).

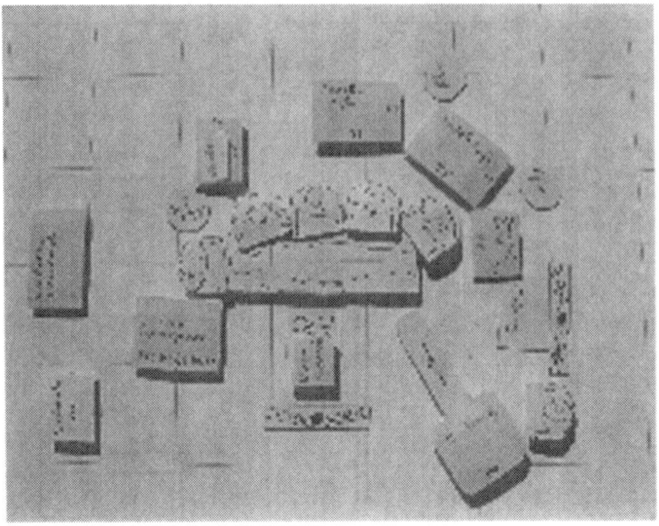

FIGURE 7. Layout on a magnetic board.

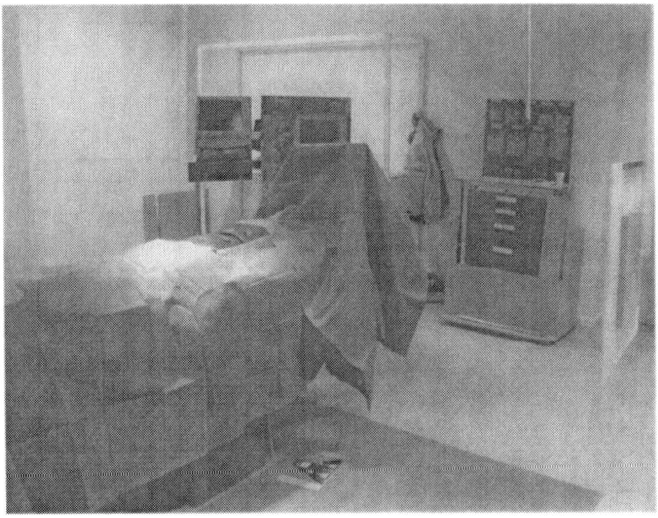

FIGURE 8. Mock-up of the operating room.

of the mock-up sessions was that the end-users could experience the actual dimensions of the proposed operating room.

The above-mentioned tools proved to be very helpful in building design, but the success of a design procedure lies not only in the organization of the process itself, but is also largely dependent on the way architects and design teams deal with the user input.

Visits to reference sites

Visits to reference sites are an excellent way of creating a more open attitude to the new operating methods that are often derived from new technology. Such visits encourage those involved to think about changes in practice and move away from what presently exists. They are therefore a vital component in the project definition process. They tend to be organized spontaneously, but all too often are improvised and achieve only part of their goal.

Here are some basic conditions required to ensure that visits to reference sites are as useful as possible [18].

Forming the project group

Whatever the type of visit, the project group should include a decision-maker, the project manager, direct user representatives, a workplace health and safety representative and the professional designers. Each person has a different viewpoint and the questions he or she raises will help the group understand how the site works, how it is organized and how all these issues are relevant to their own project.

Establishing objectives

Wherever possible, the project group should meet beforehand to prepare the visit and establish precise objectives. What specific aspects does the group wish to consider? What information is needed on operations, care approaches, technologies, layout designs and ergonomic, and health and safety questions? The group members should draw up questions in advance, although there should also be room for spontaneous questions that arise during the visit. There is nothing to prevent the questions from being sent to the site beforehand.

Selecting sites

Preference should be given to sites that meet the project's general objectives. It is best to visit newly constructed or newly renovated buildings, because the technologies and layouts will reflect recent changes in the medical field. The number of sites visited will depend on the complexity of the project and the expertise available within the group. Sometimes new functions may be added with which users do not have much experience. In such cases, it is essential to make several site visits, so that users are able to give clear opinions on the department's future orientations and on the layout proposals submitted by the architects.

Informing the host team of the visitors' expectations

The host team should be told about the visiting group's expectations, so that it can ensure that the right people are available. Indeed, the choice of host representative is very important. "Tourist" type visits, led by the institution's public relations officer, should be avoided, since they will give a very superficial idea of the premises and tend to emphasise the positive aspects of the building. It is just as important to know the negative aspects, and it is therefore important to meet with the people who are most familiar with operations.

During the visit

An introduction to the department and the background to the project should normally precede the site visit. A classical way of conducting a visit is to follow the chronological order of the department's operations, beginning, for example, with the patient arriving in the waiting room and moving through the various stages to discharge from the hospital. Members of the group should feel free to ask questions.

The following are some examples of key questions that might be asked:

- What are the department's main goals and preferred approach?
- How many professional and managerial staff work on each shift?
- What is the client profile?
- What is the volume of activities, and are there any seasonal or other variations (evening, night and weekend)?
- What are the main paths – patients, visitors, staff, clean and dirty supplies?
- What are the main types of workplace accidents that occur?
- What technologies are used? Why were these particular technologies selected?
- What are their advantages and disadvantages, costs and benefits? How reliable are they? How much maintenance do they need?
- What is the logic of the layout design selected? Why was this particular architectural design selected?
- What works well? What are they proudest of?
- What would they improve if they could start over?

It is essential to document the visit, by taking notes, photos or a video, as needed. It is best to hold a group meeting immediately after the visit, so that members can identify the elements worth keeping and those that should be rejected, while their memories are fresh.

Producing a report

It is essential to produce a report of the visit. One person should be assigned specifically to this task, with the role of gathering the information collected by the group members. The report is then kept in the project file.

Here is *an example* of the visits organized for a group of medical imaging users involved in the construction of a new hospital on the outskirts of Montreal (Canada). In this case, it was particularly important to organize several site visits, because the service would be expanding considerably with the introduction of leading-edge technologies. In addition, several new functions were to be added, and the current players' knowledge of them was limited: nuclear medicine, magnetic resonance, mammography and angiography. There were also plans to introduce digital medical imaging (PAX), which would involve radical changes in the practices of all the players. Neither users nor professionals were familiar with the installation conditions required for hazardous materials such as radioactive products and electromagnetic waves [19].

Field observations and links analysis – Case Study No. 3: Observation and link analysis

Based on: "Evidence-Based Design of Hospital Bed Spaces: Nursing Activities in Intensive Care Unit", by Lu, Hignett, Healthcare Ergonomics and Patient Safety Research Unit, Department of Human Sciences, Loughborough University, UK.

One of the basic methods used by ergonomists is to observe how existing facilities actually work in the field. This is the best way to see what works well and what does not. This information is essential in creating the best possible design for future facilities. The following study compares systematic field observations and links analysis in the evaluation of working space for the standardization of the intensive care room.

Study

This study reviews the NHS Estates recommendations for hospital space dimensions by looking at the highest repeating unit in hospital design, i.e. single bed space in single rooms and multi-bed bays based on acute adult wards and intensive care units (ICUs), in terms of four key clinical activities: manual handling, resuscitation, disability access and infection control.

Observations

In order to investigate how the tasks of manual handling, resuscitation, infection control and disability access were carried out and what was really going on in the bed space during such tasks, one week of observations were undertaken. The observation areas were sited in the cardiac intensive care unit (CICU) at University Hospitals of Leicester NHS Trust, an acute hospital of around 2500 beds which provides a range of in-patient, day case and outpatient services. The subjects were nursing staff and post-operative patients on the CICU.

Simulations

The simulations were designed to test the different bed space layouts and dimensions measured at hospitals during site visits. They were undertaken in full-scale mock-ups built at Healthcare Ergonomics and Patient Safety Research Unit Laboratory of Loughborough University. The observational data were used to develop the task scenarios to be "enacted" in the mock-ups (Fig. 15.9). The participants were the nursing staff from CICU at UHL.

Three sessions were run over three days with 15 participating nurses. Each session had two groups of nurses, giving a total of six groups testing the layouts by repeatedly performing three tasks. Data collection was undertaken by multi-directional video recording for further analysis.

Link analysis

Link analysis was used to record the (1) movements of components, i.e. nursing staff, equipment/devices and furniture; and (2) participants' (nursing staff's) movements among equipment/devices, furniture and the simulation mannequin (patient) according to video footage. AutoCAD was used to draw the link diagrams as output to convey spatial information, and

FIGURE 9. Photo of the mock-ups.

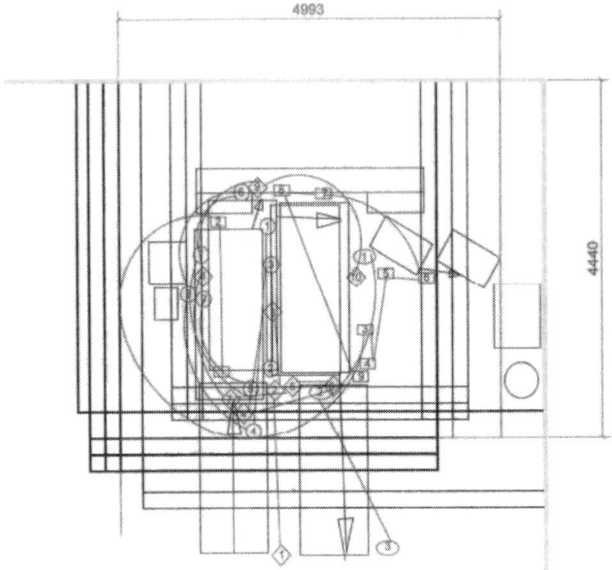

FIGURE 10. Link analysis of transferring a patient from bed to bed.

the result of each layout was tested. Figure 10 shows the link analysis result of a bed space for the task of transferring a patient from one bed to another.

Results

The average space used for washing and dressing a patient and then moving him from bed to wheelchair by a hoist is 4970 mm (width) × 4920 mm (length); for transferring a patient from one bed to another the average space is 5040 mm (width) × 5090 mm (length); for resuscitating a patient the space is 4810 mm (width) × 4847 mm (length). So the bed space needed to accommodate all three tasks is 5040 mm × 5090 mm, which is smaller than the room measured at the hospital, but larger than the NHS Estates recommendation (4500 mm × 4500 mm).

INFLUENCE OF ERGONOMICS

Partnership with public agencies

One of the most promising approaches involves the development of a partnership with government agencies to produce construction standards for hospital buildings. To forge a successful partnership, we must use concrete examples to show how ergonomics adds value to architectural projects.

Over the years, ASSTSAS, for example, has been able to develop good relationships with the government agencies that manage construction projects and develop standards. Our organization is now involved in construction standards review committees for healthcare facilities. Occupational health and safety and ergonomic concerns are now criteria used to

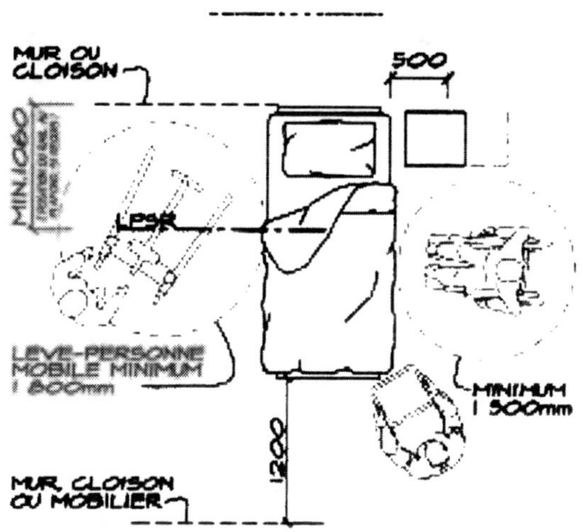

FIGURE 11. New construction standard for a typical bedroom.[2]

develop and update standards (see example in Fig. 11). Professional ergonomists are already involved in the pre-design phase of two university megahospital projects. The approach and methods described earlier are fully implemented during the design process with complete involvement of planners, architects and medical staff.

A different way to develop a partnership with a government agency is presented in case study 4. In this case, there are no state construction standards to refer to. Guidelines were published and distributed to every healthcare institution to help project managers implement an ergonomic review process.

Case Study No. 4: Ergonomic review process

Based on: "Hospital Design with Ergonomics in Mind", by Duffy, Ontario Safety Association for Community and Healthcare, Ontario, Canada.

Ergonomic review process

The ergonomic review must be integrated into the standard design process. The primary focus is to guide the users in identifying where musculoskeletal risk factors currently exist and how the future work environment can be enhanced to minimize or eliminate the risks. The individual conducting the review may be a health and safety professional, departmental manager and staff, or a facilities planner. The ergonomic review outlined is a seven-step process (see Fig. 12). As each step of the review is completed, feedback needs to be communicated to the project planner or coordinator. If a more in-depth assessment is necessary to resolve ergonomic concerns, an ergonomist should be called upon.

A Guidebook is provided to provide the ergonomic review process with relevant ergonomic information and a checklist for 13 hospital service areas. An anthropometric "Ergo wheel" is also available.

---

[2] CHQ, *Cadre de référence normatif*, Gouvernement du Québec 2006, www.chq.gouv.qc.ca.

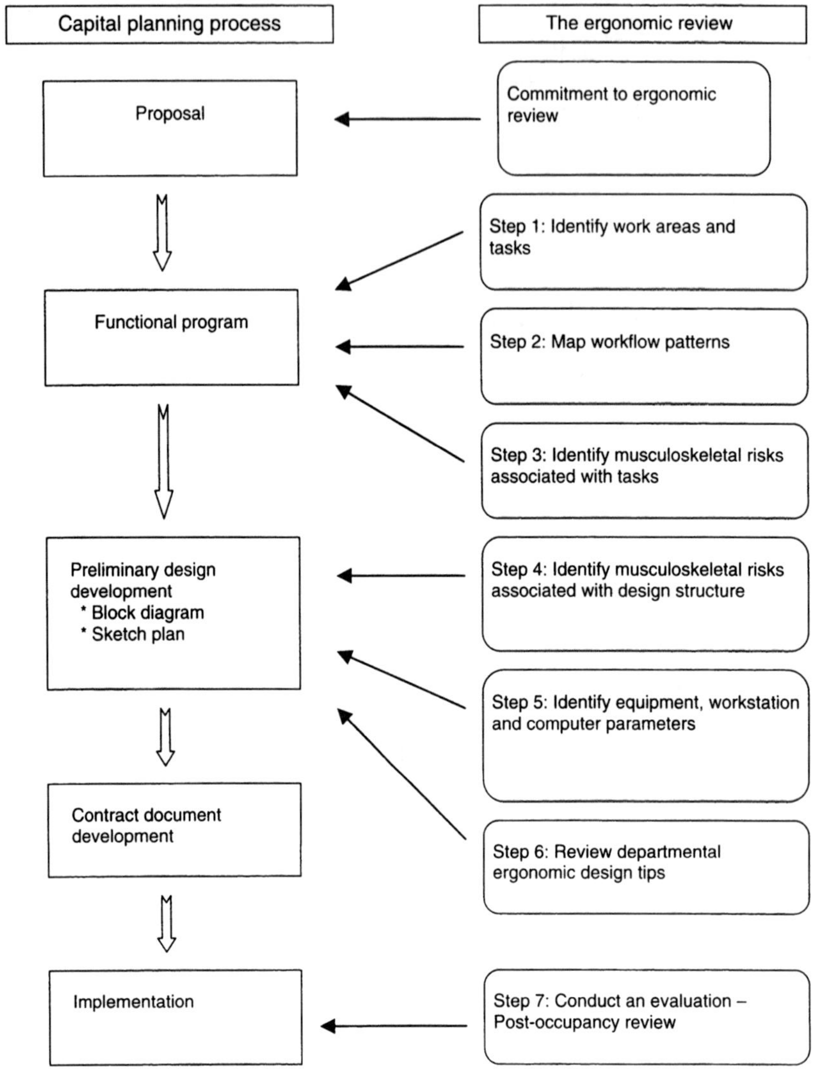

FIGURE 12. Integrating an ergonomic review into the capital planning process.

Marketing ergonomics

Efforts should be made to market our Ergonomic services to health facilities, particularly to hospital management and trade unions, explaining the importance of integrating ergonomics with health and safety concerns right from the initial design phase of architectural projects. This is an investment at a primary prevention level in line with the mission of health and safety organizations. Some material has been published to promote ergonomics expertise in this domain: a variety of subjects are discussed in a collection of practical guidelines for

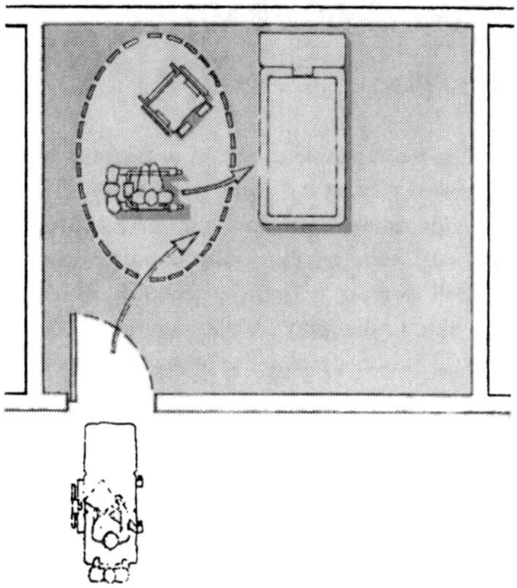

FIGURE 13. Bathroom layout.[3]

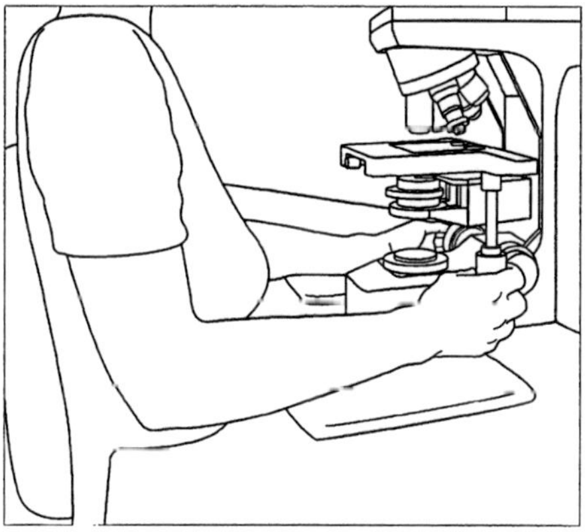

FIGURE 14. Working with a microscope.[4]

non-experts in the construction field, including project management and the design of typical work places such as patient rooms, bathrooms, dish washing rooms and office workstations. (Fig. 13).

A collection of data sheets on the ergonomic design of laboratory workstations was also published this year, in response to many reported complaints about musculoskeletal problems (Fig. 14). Articles are published regularly in the magazine, *Objectif Prévention*, which is published five times a year.

---

[3] Bertrand, G. (2004). *Aménagement des salles de bain*, brochure PARC, No. 7, ASSTSAS.

[4] Villeneuve, J. and Gambin, C. (2005). *Le travail au microscope*, fiche technique, ASSTAS.

LESSONS FOR ERGONOMIC PRACTICE

Limitations of participation

While the participation of the main people involved is essential to the design of the project, it does entail certain limitations that must be managed carefully.

For most of these individuals, this will be their first and only design experience. It is to be expected that they may have trouble understanding some of the information being conveyed and find it difficult to make helpful suggestions, at least early in the process. One common problem is being able to interpret architectural plans. It is, therefore, very important to introduce individuals to the process gradually, by simplifying technical language and using familiar spatial references and three-dimensional models to help them formulate opinions about the proposed facilities. Referring to the existing premises or organizing full-size simulations is extremely helpful, as earlier case studies have shown.

Another problem to avoid is people's tendency to give opinions in reaction to what is wrong in the existing facilities. The project leader should carefully explain the future vision of the project so that user representatives can be creative rather than simply reactive. Visits to reference sites are very useful for showing other ways of doing things, particularly when introducing new working methods.

The choice of user representatives and their regular attendance at meetings is also important to ensure the harmonious development of the project and to avoid any backtracking.

From micro to macro ergonomics

Micro ergonomics, the conventional ergonomics of workstation analysis, is the starting point for most practising ergonomists. But a medium- to large-scale project encompasses far more. The organizational questions asked at the design phase, for instance, concern the project as a whole: its mission, the characteristics of current and future patients, the implementation of new clinical approaches, new medical technologies, new information technologies, etc. These considerations have a direct impact on the architecture itself: overall spatial organization, building volumetric, the location of the various units, the surface area of each department, the optimal layout of the premises, and lastly, the position of each workstation. In other words, each workstation can be seen as a "micro system" which is an integral part of a larger system.

In this sense, the definition of ergonomics proposed by the IEA is very appropriate because all of the activities described below are applied in an architectural design process. "Ergonomists contribute to the planning, design and evaluation of tasks, jobs, products, organizations, environments and system in order to make them compatible with the needs, abilities, and limitations of people".

Multidisciplinarity

While the expertise of each professional discipline is specific, they also have skills that are complementary. Architects focus their attention primarily on buildings, engineers on technologies, and ergonomists on the well-being of the building's occupants, but a multidisciplinary approach is essential (Fig. 15). The ultimate aim is to design a hospital that dispenses quality services as efficiently as possible in a pleasant, comfortable, safe environment for patients, staff, visitors and suppliers.

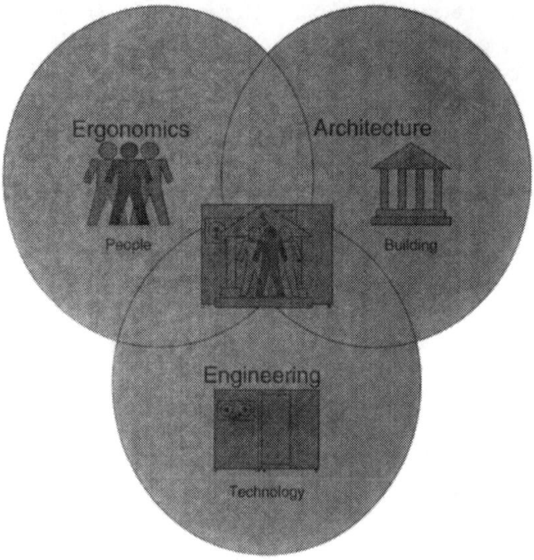

FIGURE 15. Multidisciplinary approach.

Role of the ergonomist

The role of the ergonomist in this field is a recent one, and is not always well understood. The ergonomist may be seen as an "outsider" by design professionals or as an "occasional expert" who can offer an opinion about furniture or specific issues such as the design of computer workstations. All too often, the ergonomist is called in at the end of the project, to develop the final plans. In this case, analysing activities and work processes sometimes leads to recommendations that challenge the design concepts when it is too late to make changes.

Ideally, the ergonomist should be fully involved in the planning team, from the initial design phase to the post-occupancy evaluation. One paper on the subject that attracted a great deal of attention described a situation in which a professional ergonomist was an integral part of the planning team in a regional health and safety department [20]. As an example of the outcomes, guidelines have been published to prevent violence to caregivers in psychiatric wards and emergency rooms [21].

CONCLUSION

Ergonomics as a discipline has a major role to play in the design process of hospital buildings; overlap or competition with traditional design professions is not an issue. Ergonomics provides useful, complementary expertise that improves end results. Participatory ergonomics seems to be a very effective method for improving the architectural design process. Before and after assessments in the field show that caregivers and top management are very pleased. In fact, there appear to be far fewer post-construction changes.

Ergonomic intervention in design projects is a relatively recent phenomenon. The public and building professionals, in general, are not very familiar with this discipline or its scope, and most have a very limited view of our profession. Advances are now being made and results are being achieved, but it is not easy to change building trade culture.

Intervention in architectural design projects challenges the scope of traditional ergonomic practice as limited to the workstation. The complexity of hospital architectural projects, the large number of parties involved and the financial and political stakes all require a much broader outlook and the development of new methods.

Relationships with hospital technical services and leading architectural firms must be productive; the added value of ergonomics is now starting to be recognized. Partnership with government agencies could lead to tangible results, but there is still a need for more marketing so that people can learn what ergonomics can do to improve the work environment.

REFERENCES

[1] Villeneuve, J. and Lebeau, G. (1994). *The PARC Program: Ergonomics and Architecture in Hospital.* IEA Proceedings, Toronto.

[2] Villeneuve, J. (1997). *Ergonomics in Hospital Design.* Occupational Health for Health Care Workers, 3rd International Congress, Edinburgh, ICOH/CIST, Ecomed.

[3] Morissette, L. (2006). *Better Living in a Child's World.* IEA Proceedings, Maastricht.

[4] Duffy, A.E. (2006). Hospital design with ergonomics in mind. In: Pikaar, R.N., Koningsveld, E.A.P., Settels, P.J.M. (eds), *Meeting Diversity in Ergonomics.* Proceedings IEA 2006 Congress. ISSN 0003-6870, Elsevier Ltd.

[5] Gamble, L. (2006). Essential ergonomics in the healthcare facility design process. In: Pikaar, R.N., Koningsveld, E.A.P., Settels, P.J.M. (eds), *Meeting Diversity in Ergonomics.* Proceedings IEA 2006 Congress. ISSN 0003-6870, Elsevier Ltd.

[6] Pikaar, R.N., et al. (1998). Ergonomics in process control rooms. *Part 1: Engineering Guideline; Part 2: Design Guideline.* The Hague: WIB International Instrument Users' Association.

[7] Remijn, S.L.M. (1993). *Mensgericht bouwen.* Enschede, The Netherlands: ErgoS Ergonomics and Engineering.

[8] Remijn, S.L.M. (2006). Integrating ergonomics into the architectural design processes: Tools for user participation in hospital design. In: Pikaar, R.N., Koningsveld, E.A.P., Settels, P.J.M. (eds), *Meeting Diversity in Ergonomics.* Proceedings IEA 2006 Congress. ISSN 0003-6870, Elsevier Ltd.

[9] Lu, J. and Hignett, S. (2006). Evidence-based design of hospital bed spaces: Nursing activities in intensive care unit. In: Pikaar, R.N., Koningsveld, E.A.P., Settels, P.J.M. (eds), *Meeting Diversity in Ergonomics.* Proceedings IEA 2006 Congress. ISSN 0003-6870, Elsevier Ltd.

[10] LESC. (1997). *L'ergonome, le Maître d'ouvrage et la Maîtrise d'œuvre.* Actes des Journées de Bordeaux sur la pratique de l'ergonomie, Université Victor Sagelan, Bordeaux 2.

[11] Querelle, L. and Escouteloup, J. (2006). From the path of surgical instruments to the design of operating theatres. In: Pikaar, R.N., Koningsveld, E.A.P., Settels, P.J.M. (eds), *Meeting Diversity in Ergonomics.* Proceedings IEA 2006 Congress. ISSN 0003-6870, Elsevier Ltd.

[12] Cardoso, V.M.B., et al. (2006). Ergodesign service: Ergonomic contribution to users of an university hospital. In: Pikaar, R.N., Koningsveld, E.A.P., Settels, P.J.M. (eds), *Meeting Diversity in Ergonomics.* Proceedings IEA 2006 Congress. ISSN 0003-6870, Elsevier Ltd.

[13] Villeneuve, J. (2006a). Ergonomic intervention in hospital architecture projects. In: Pikaar, R.N., Koningsveld, E.A.P., Settels, P.J.M. (eds), *Meeting Diversity in Ergonomics.* Proceedings IEA 2006 Congress. ISSN 0003-6870, Elsevier Ltd.

[14] Villeneuve, J. (2006b). The contribution of ergonomics to the design of hospital architecture. In: Pikaar, R.N., Koningsveld, E.A.P., Settels, P.J.M. (eds), *Meeting Diversity in Ergonomics.* Proceedings IEA 2006 Congress. ISSN 0003-6870, Elsevier Ltd.

[15] Huddy, J. (2002). *Emergency Department Design.* American College of Emergency Physicians, p. 23.

[16] Martin, C. (2000). *Dialogue: Maîtrise d'ouvrage et maîtrise d'œuvre ....* Octarès.

[17] Unger, B. (2002). In *Emergency Department Design* (J. Huddy, ed.) American College of Emergency Physicians, p. 221.

[18] Zimring, C. (1997). Site visits. In *Healthcare Design* (S.O. Marberry, ed.) John Wiley & sons.

[19] Villeneuve, J. and Thibault, B. (1999). *Les utilisateurs sur la planche à dessin,* 22 (4), Objectif Prévention.

[20] Gamble, L. (2006). Essential ergonomics in the healthcare facility design process. In: Pikaar, R.N., Koningsveld, E.A.P., Settels, P.J.M. (eds), *Meeting Diversity in Ergonomics.* Proceedings IEA 2006 Congress. ISSN 0003-6870, Elsevier Ltd.

[21] Gamble, L., et al. (2006). Facility design considerations to reduce violent encounters in emergency and psychiatric departments. In: Pikaar, R.N., Koningsveld, E.A.P., Settels, P.J.M. (eds), *Meeting Diversity in Ergonomics.* Proceedings IEA 2006 Congress. ISSN 0003-6870, Elsevier Ltd.

Meeting Diversity in Ergonomics (2007)
Pikaar, Koningsveld, and Settels (Eds)

16

CCTV Ergonomics: Case Studies and Practical Guidance

John Wood

CCD Design & Ergonomics Ltd, 95 Southwark Street, London SE1 OHX, UK

Abstract. Two case studies are used to illustrate the more important features to be considered in CCTV ergonomics. The examples are drawn from rail and road transportation, both of which involve safety-critical features. In each case the experiments conducted are described, along with how the information derived was then used in the CCTV systems design process. The article finishes with some practical guidelines covering such aspects as job design, picture and image specification, target sizes, monitor numbers and recruitment and selection.

Keywords: CCTV ergonomics, case studies, safety-critical contexts, practical guidelines

BACKGROUND

Whilst our understanding of the rules which govern how easily we can see characters on display screens are well established, those concerning CCTV images are far less well developed [1–6]. Well-documented failures include missing vans dangerously parked on level crossings or security images clearly showing violations which go unchallenged. In both cases replaying the recordings shows the failure is a human one – there is no question about the visibility of the targets – there just seems to have been no human reaction. In response, CCTV system specifiers have aimed to 'automate out' the operator – by movement detection for example – but usually *not* by paying more attention to underlying ergonomic factors.

The declining cost of CCTV systems, coupled with improvements in picture transmission capacity, has resulted in ever-larger CCTV schemes. Whereas 20 years ago the use of CCTV was relatively limited, it is now to be found offshore, in nuclear processing plants, controlling motorways, monitoring city centres and as an essential tool in security and surveillance systems. Other systems exploit the tiny size of the cameras to examine sewers or carry out internal medical examinations. In all cases the information is presented to the viewer on a screen.

The main ergonomic factors which can be manipulated to maximize performance are illustrated in Fig. 1. How the ergonomist can manipulate and control some of the performance-shaping factors is illustrated in the following case studies.

The first case study examines the use of a CCTV system for checking whether the safety lane on a motorway is clear so that it can be used to relieve traffic congestion. This study examined the effect of using different picture presentation display formats on accuracy and timings. The second case study examined the potential impact of changing display technology on the operation of railway level crossings. In this instance accuracy and speed were examined for different types of targets located on the railway line.

The chapter closes with some practical guidance which can be considered when taking account of the human factor issues arising from a CCTV programme.

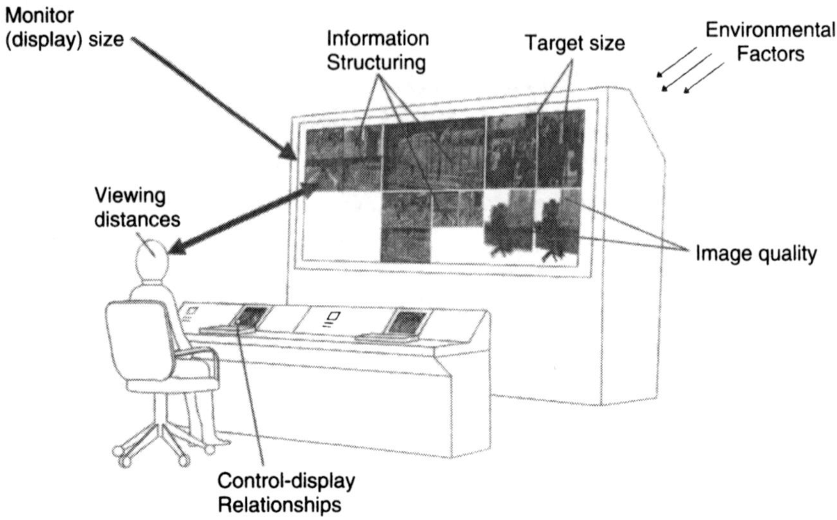

FIGURE 1. Human factors and CCTV system operation.

CASE STUDY 1: THE USE OF CCTV SYSTEMS FOR CHECKING MOTORWAY SAFETY LANES

Introduction and background

The UK Highways Agency's 'Active Traffic Management' (ATM) pilot project was designed to use the motorway hard shoulder (safety lane) as a temporary traffic lane to increase road capacity during periods of high demand. The hard shoulder was to be manually opened to traffic by control room operators using gantry-mounted traffic control signals. CCTV was proposed as the principal method of checking whether the hard shoulder was clear of obstructions and vehicles prior to opening.

Under the ATM operating concept the motorway was considered as a series of 'links', each of which corresponded to the length of carriageway between two junctions. The links were sub-divided into a number of 'sections', nominally 500 m in length, each of which had a signal gantry at its upstream end (see Fig. 2).

The hard shoulder was to be opened for traffic by setting an associated overhead speed sign on each of the section signal gantries in sequence, starting at the downstream end of the link (just before the downstream junction) and working back towards the upstream end (just after the upstream junction). Before each section of hard shoulder was signalled as open, the control room operators would scan it via CCTV to make sure that there were no stopped vehicles which might pose a hazard to traffic.

The CCTV system was being designed to provide 100% continuous coverage over the whole length of hard shoulder to be opened and was based upon an operational requirement that a detectable target (defined as an object with a height of at least 1.6 m) appearing in a CCTV camera image would have an apparent height of at least 10% of the total picture height [5]. To meet these requirements each section of hard shoulder would have to be covered by several cameras with overlapping fields of view (see Fig. 3).

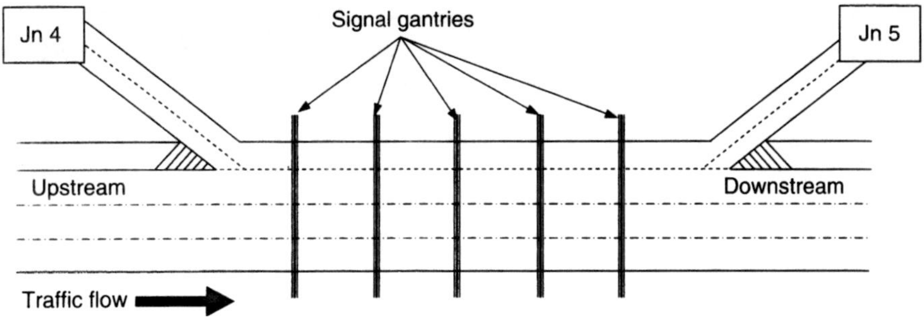

FIGURE 2. Schematic plan of an ATM pilot project Link sub-divided into four sections.

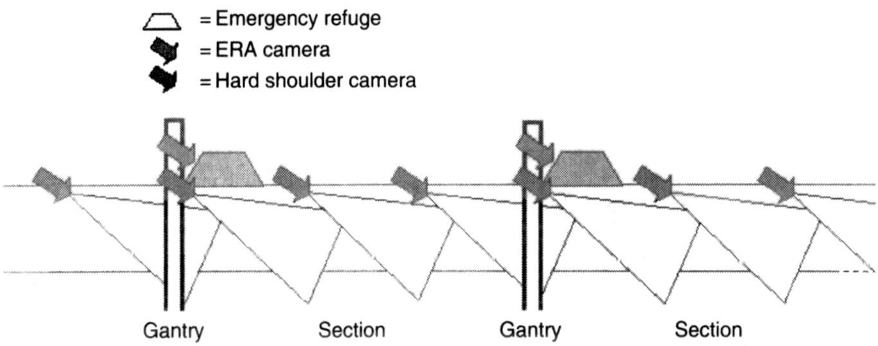

FIGURE 3. Schematic plan of an ATM hard shoulder section demonstrating hard shoulder and ERA camera positions and fields of view.

It was established that the maximum number of cameras required to cover a hard shoulder section would be 5, with a sixth camera covering Emergency Refuge Areas (ERA) associated with each section.

The project safety case determined that control room operators had to achieve a level of 98% reliability when scanning the hard shoulder for stopped or broken down vehicles prior to opening it.

The design of the scheme raised a number of important questions concerning the design of the CCTV system:

- How should the CCTV images be presented to operators in order to support optimal task performance?
- How many monitors can operators be expected to check reliably and efficiently?
- Should images be presented concurrently or sequentially?
- How big should the display monitors and images of the hard shoulder be?
- Can operators be expected to achieve the requisite level of reliability in checking CCTV images?
- How long might it take to reliably open sections and links?

A review of Human Factors standards and published literature was undertaken. This established the very limited extent of published guidelines available for the design and operation of CCTV systems and determined that these questions could not be satisfactorily answered with existing knowledge [2–4]. Therefore, a pragmatic user trial was undertaken to investigate CCTV presentation options in order to provide practical guidance for the design of the ATM control system.

Method

The following sizes of CCTV image were compared in the trial at a 700 mm viewing distance (all at 4:3 aspect ratio):

- smallest: 30 mm × 40 mm;
- standard: 41 mm × 54 mm;
- nontuple: 91 mm × 121 mm;
- native: 163 mm × 217 mm;
- full screen: 284 mm × 380 mm.

To compare how operator performance might be affected by presentation format (i.e. single image or multiples), two multiplexed layouts were developed (see Fig. 4). The same image size (91 mm × 121 mm) was used to compare single, quadruple (2 × 2) and nontuple (3 × 3)

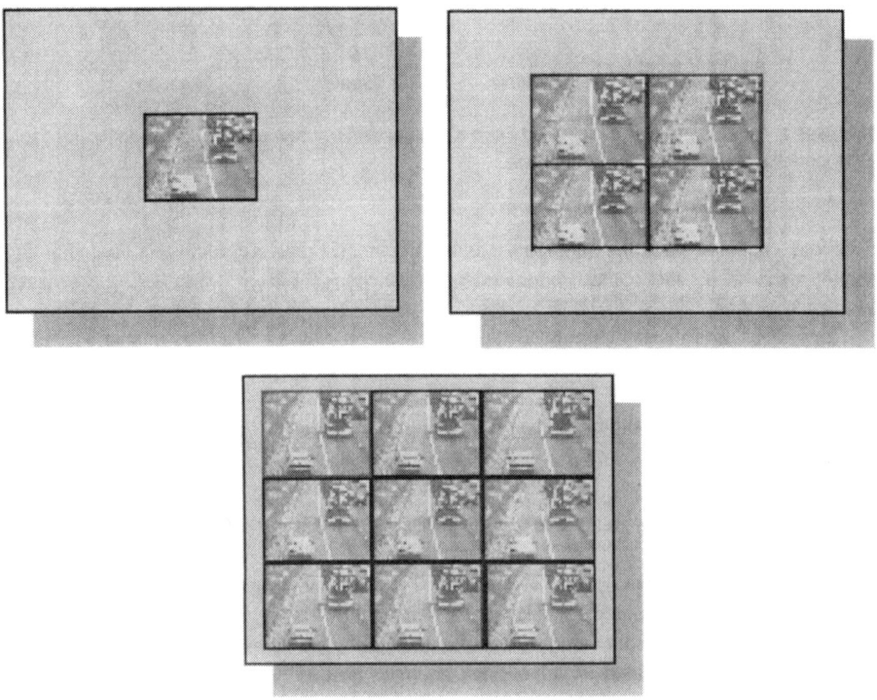

FIGURE 4. Image presentation formats showing single, quadruple and nontuple image formats.

image formats. To enable a comparison of performance between different display types, all conditions were repeated for all subjects on both a 19″ CRT monitor and a flat screen 19″ TFT monitor.

In order to compare performance with different input devices, the subjects repeated all conditions using both a mouse and a keyboard. A realistic target detection task was used to compare operator performance with different image sizes, presentation formats and display types. The subjects were asked to inspect several series of CCTV images and simply decide whether or not a target vehicle on the hard shoulder was present in each.

Live CCTV feeds from appropriate CCTV cameras, with fields of view specified for ATM, were not available and hence representative video footage was captured during day and night (under motorway lighting) conditions. The video was captured using a digital camcorder, transferred to PC and edited into 20 s clips. Two types of clips were captured:

- without a vehicle on the hard shoulder; and
- with a vehicle on the hard shoulder (stimulus clips).

The position of the target vehicle on the hard shoulder varied but was shown mainly at the far point of the field of view – the worst case scenario where the target is at its smallest. In practice, the position of vehicles within the camera field of view would be entirely random.

Operator performance was primarily measured by assessing error rate, i.e. failures to correctly identify whether an image included a vehicle on the hard shoulder or not.

Additionally, time taken to complete conditions was recorded in order to provide a basis upon which the time required to open sections and links was estimated.

A repeated measures study design was employed, in which each subject experienced each of 56 test conditions. The presentation sequence was randomized by means of a Latin Square in order to minimize potential order effects.

Each subject inspected 30 CCTV video clips for each of the 56 test conditions. As previously described there were two types of clip:

- non-stimulus (empty hard shoulder); and
- stimulus (clip includes a vehicle stopped on the hard shoulder).

A bespoke software program was developed to manage the video clip presentation and randomization. Stimulus clips were presented randomly in the sequence of non-stimulus clips at an approximate ratio of 1:10. This ratio for target images is probably higher than might be expected in reality. However, with a practical need to capture real data, it was essential to present a sufficient number of targets under each test condition.

Subjects were required to view each image presented and judge whether the hard shoulder was clear or whether a target (vehicle) was present. Subject responses and errors were recorded by the computer system controlling the presentation.

With a total of 56 different test conditions per subject, and with each condition requiring 30 video clips, each subject viewed a total of 1680 clips. The large number of clips presented to the subject in one sitting was designed to force errors; by presenting lots of stimuli, the chances for error were increased, thereby ensuring that some error scores were available for analysis.

Ten full-time operators (eight male and three female aged 40–64) from a District Council CCTV control centre took part in the study. Whilst unfamiliar with motorway monitoring and management, these subjects were familiar with searching CCTV images in a professional capacity and represented a relatively homogenous group in terms of training and expertise.

To provide a level of task realism, the video clips were presented in clusters of 10 as an approximate match to the number of images that might be required to open sections. Following a familiarization run, each subject ran through the full trial sequence. The trial was divided into two sessions, each consisting of 28 randomly presented conditions, with one session using the TFT monitor and the other the CRT monitor. The order in which monitors were used was randomized between subjects.

Subjects inspected each video clip in turn (highlighted by the software with a green 'bounding box') and decided whether or not a vehicle was present on the hard shoulder (Fig. 5), confirming their decision by keyboard button press or mouse click (depending on trial conditions).

The overall time for each full trial varied for each subject, but averaged approximately 2.5 hours.

Results

Overall

A mean of 3.7 errors (SD 0.74) per subject, per condition was recorded, which gives an overall error rate of approximately 1.2%, which lies within the 98% reliability requirement for the CCTV operation.

FIGURE 5. A 3×3 multiplex condition during a trial. The first two images on the top row have already been 'cleared' (bounded in red); the green bounding of the third image shows that this is the next one to be checked.

Individual scores, however, varied between 0% and 17% across operators and conditions. To determine the source of this variation the results were scrutinized in more detail, looking at the influences of image size and image format and examining 'dangerous errors' (i.e. failure to detect a target when present).

Image size

Errors are compared with image size in Fig. 6, which shows that the two smaller image sizes were the source of most errors. This was supported subjectively by the operators, who felt that they were too small to work with.

An acceptable rate of error is shown to be achieved with the three larger image sizes.

Image format

The effect of image format on errors is shown in Fig. 7.

There was little difference between formats. However, a more interesting result is demonstrated in Fig. 8, which compares image format with 'dangerous errors'.

Clearly, operators are more reliable in searching for vehicle targets when shown a single image than with multiple images.

Other factors affecting performance

Night Time images induced more errors than daytime and keyboard input resulted in more errors than mouse input. There were no significant differences between display types.

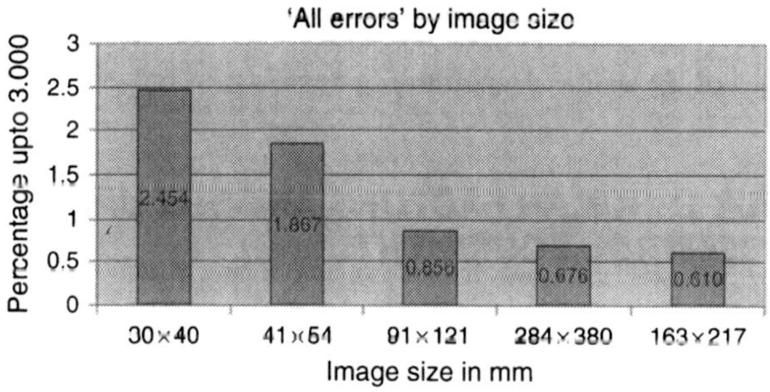

FIGURE 6. Errors vs. image size.

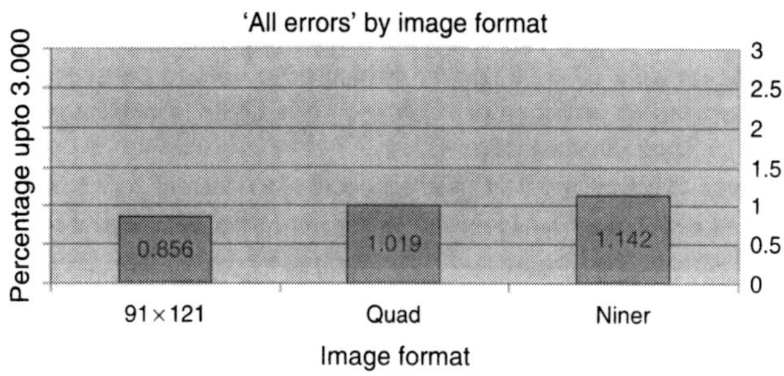

FIGURE 7. Errors vs. image format.

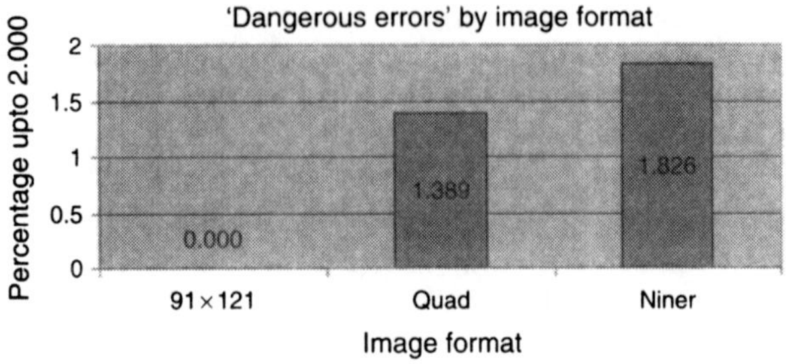

FIGURE 8. Dangerous errors vs. image format.

Tests were slower with single images than with multi-image formats, but not by much (e.g. average of 60 and 45 s, respectively). Times were marginally slower still with night images.

Conclusions

- Operators are able to achieve a suitable level of reliability with CCTV monitoring provided that the larger image formats are used.
- Operators are more reliable when they check a single image at a time.
- Serial presentation is preferred over multi-image formats. Serial inspection is a little slower.
- Either CRT or TFT (flat screen) displays are acceptable.
- Mouse input is preferred over keyboard input.
- Operators do need the ability to revisit their judgements.

CASE STUDY 2: HUMAN FACTORS AND THE UPGRADING OF RAILWAY CONTROL EQUIPMENT

Introduction and background

In the UK, selected level crossings are monitored remotely with CCTV cameras mounted adjacent to the track. This allows a number of sites to be controlled by a single operator at a suitably equipped workstation.

The original specification for these CCTV systems was based around black and white images. These pictures, transmitted to a remotely located level-crossing operator, are used to check whether the crossing is 'clear' before allowing the train to continue its approach. The existing displays used in signal boxes are black and white CRTs, which are getting increasingly less reliable and are costly to maintain.

The UK rail infrastructure operator, Network Rail, had decided to replace trackside black and white cameras with colour equivalents. As regards the remote operators' equipment, the unit costs of 'flat screens' had been steadily decreasing whilst the quality of images had been improving. The Human Factors team were asked to determine whether there would be any significant degradation in the performance of the operator if the CRT displays, as currently used in signal boxes, were replaced with their flat screen equivalents.

Preparatory work and experimental design

It was decided that actual level-crossing pictures would be used – preferably taken at a number of different types of crossings – and that judgements would be made by experienced level-crossing operators. It was also agreed that the object of the experiment was to compare the performance of the existing CRT displays with flat screens – it was not to establish what the smallest target might be that could be identified through the CCTV system. Effectively it was a comparison between current and future display technologies.

Getting hold of sample video from existing cameras proved to be more difficult than expected. This was solved using a specially adapted van, with a twin cameras mounted on an extendable pole, which allowed the collection of suitable images (Fig. 9).

Pilot trials were conducted at CCD offices using the edited video material. These trials allowed experimental protocols to be tested and refined and it was agreed that the following variables would be examined – with appropriate recordings being made at selected crossings:

FIGURE 9. Mobile camera van.

Factor	Variables to be examined
Crossing type	Single vs. multiple track
Lighting conditions	Daylight vs. dusk vs. nightime
Target	Wheelchair vs. child vs. 'sheep'[a]
Image type	Black and white vs. colour presentation

[a] Initial selection included a full-size, paper-mache replica of 'sheep'.

Both Network Rail and the ergonomics team reconfirmed that the primary objective was to identify whether there would be any significant degradation in operator performance when a black and white CRT display was replaced by a flat screen.

The existing requirement to 'check that the crossing is clear' sets neither an upper nor a lower limit on the nature of the object to be seen. In deciding targets to be used, the ergonomics team selected ones which were at the smaller end of the spectrum of possible targets and also of such a size that they *would* induce errors, thus allowing for comparisons between displays to be made. Two targets were finally selected: (a) a full-scale paper mâché model of a 3-year-old child and (b) an empty wheelchair (Fig. 10).

The specially adapted van included video recording equipment which allowed for records to be made independently from the Network Rail CCTV infrastructure. At each of the selected sites a series of video records was taken (1–6) with each type of target and under different lighting conditions. Figure 11 presents the locations used for target location.

The recording of the raw video was spread over 3 days including one overnight session. These video recordings were edited into 50 pairs of 10 s clips in both monochrome and colour. The 50 clips included a mix of all targets (including some without a target present), every level-crossing type, all lighting conditions and all test locations on the crossing.

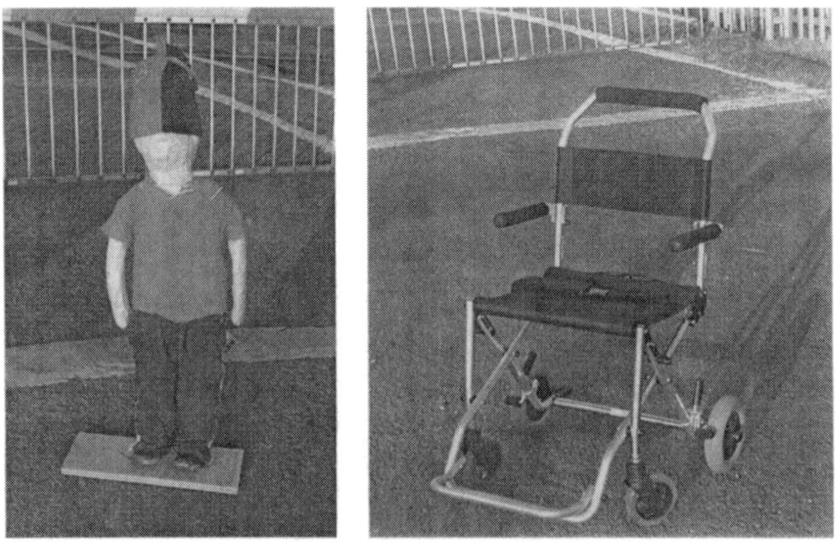

FIGURE 10. Targets used for trials.

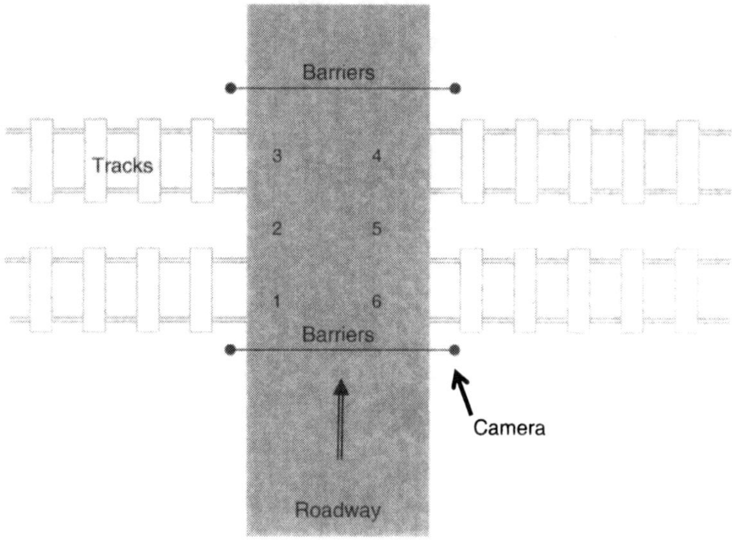

FIGURE 11. Locations used on crossings for test targets.

Experimental set-up and conduct of trials

The experiments were carried out in a signal box which maximized the number of available, experienced level-crossing operators with CCTV experience. These trials were run over two shifts involving a total of 21 signalling staff.

Following a standardized briefing session, subjects were given some sample tests for familiarization. Subjects were presented with two screens – a black and white CRT monitor and a flat screen which could present either colour or black and white images – and a small keypad for recording their decisions.

Once having familiarized themselves with the experimental procedure, subjects were asked to look at a sequence of images and determine whether they were clear or not of targets. Where a target was noted, subjects had to indicate what the target type was on a colour-coded keypad. Where no button was depressed, and thus no target judged to be present, the video sequence automatically moved onto the next scene after a set time. The experimental protocol followed the process presented in Fig. 12. Errors and response times were automatically recorded.

Subjects continued until they had completed the entire sequence of 50 tests, following which two more repeats were undertaken so that presentations on CRT (black and white), black and white flat screen and colour flat screen had been completed. At the end of the trials they were asked for their subjective views about the alternative screen technologies and were posed a more general, open-ended question asking for any other comments.

Results

The data collected was subjected to a battery of statistical tests including repeated-measures ANOVA, Friedman's test, Bonferroni Post-hoc test, Wilcoxon Signed Ranks test and '*T*'

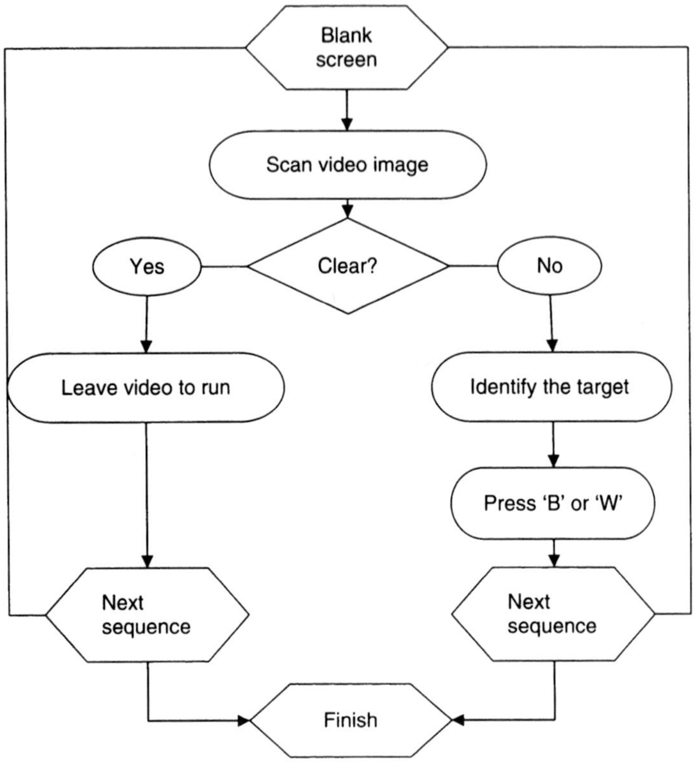

FIGURE 12. Experimental protocol used during trials.

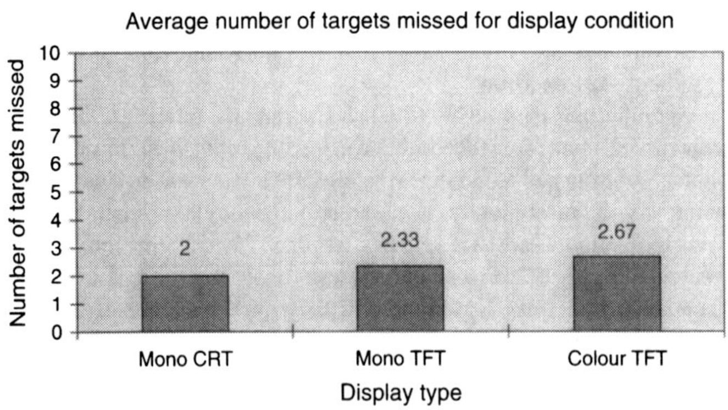

FIGURE 13. Targets missed vs. display type and image.

test. The main conclusion was that there was no significant difference between the different displays and thus the replacement of the CRT with flat screens would not result in any significant degradation in signaller performance. The key results are presented in the bar charts (Figs. 13 and 14).

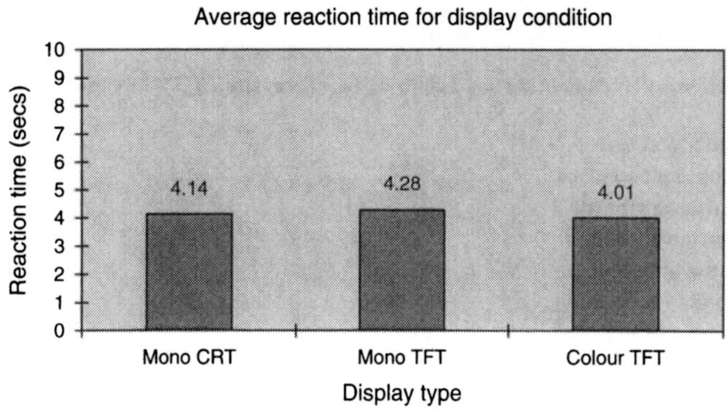

FIGURE 14. Reaction time vs. display type and image.

The targets missed results were tested statistically. Although the bar chart shows a difference between the error scores, this was not statistically significant. The number of targets missed, in general, was not affected by display type.

The performance with the mono TFT is marginally worse than the existing condition, mono CRT, and the full colour condition is marginally better. Statistical tests showed the differences to be significant. However, in real terms a reaction time difference of a few tens of milliseconds would not be a significant differentiator: the nature of the operational task means that operators are required to take the time necessary to satisfy themselves that the crossing is clear – thus the operator is not under time pressure.

In general, participants felt that the colour clips presented on the TFT monitor provided better definition for the images.

Conclusions and discussion of results

The overall conclusion was that there would be no reduction in safety by replacing the existing CRT monitors with flat screens showing either black and white images or colour ones.

When performance against type of targets were assessed it was noted that there was a difference between the 'boy' and the 'wheelchair'. No boy targets were missed. The wheelchair was difficult to see in some positions where its visibility was marginal. Using targets on the threshold of visibility raises wider issues about the minimum size of target that a signaller might be expected to see.

PRACTICAL GUIDELINES

The following section presents some practical guidelines which may be applied by Human Factors engineers and CCTV systems designers. These recommendations combine lessons learnt in the security industry with those gleaned from road and rail [6–11].

Recruitment and selection

Some of the positive characteristics exhibited by successful CCTV operators relate to:

(1) good attention span;
(2) minimum reaction time;
(3) multi-tasking ability;
(4) good memory skills;
(5) object tracking ability;
(6) hand/eye co-ordination.

For security-related tasks it has been found that successful CCTV operators typically exhibit the ability to:

(1) remain calm under pressure;
(2) work with minimal supervision;
(3) work as part of a team;
(4) make sense of information quickly;
(5) ignore distractions;
(6) make precise adjustments when moving camera controls quickly and accurately.

Training and management

When developing training programmes in relation to CCTV operation the following are some of the considerations which should be applied:

(1) Operators should be provided with a clear and detailed understanding of their tasks, role and responsibilities within the control room, including what responses and actions are required from them.
(2) Training should cover recognition of when fatigue occurs and when to take suitable rest breaks.
(3) Management should undertake performance reviews for operators and offer feedback; operators typically feel more motivated when their work is acknowledged.
(4) A structured shift handover should be built into the working process (i.e. during shift time) for briefings and transfer of information between operators.

Job design

Whilst the monitoring of a CCTV system may sometimes appear superficially to be a low-demand task, this is by no means universally true. Security monitoring, for example, can switch from 'general observations' to intense scrutiny when specific threats or targets are being sought. The earlier case studies also provide examples of where close, intense attention is required over short periods. This section presents some of the ergonomic considerations which should be applied when designing the CCTV task:

(1) Severe task underload can be a serious potential problem with CCTV monitoring and should be minimized to avoid monotony and boredom. Extreme underload is typically indicated by increased errors, loss of attention, increased boredom and easier distraction.
(2) Where the CCTV task involves periods of low workload, non-conflicting secondary tasks can usefully be introduced.

(3) Automatic incident detection systems may help to free operators from the more boring and tedious aspects of CCTV system operation.
(4) Typically operators remain more attentive when working in 20-min blocks with 5 min rest breaks in between. However, certain types of task may warrant even shorter periods, e.g. in level-crossing operation high levels of detection performance are more likely to be maintained if the total duration for image inspections is less than 5 min.
(5) Some sources suggest 3 hours to be the maximum continuous period over which most people can reliably carry out general surveillance tasks. Furthermore, they recommend a cumulative total of 6 hours per day dedicated to CCTV tasks as a maximum figure.
(6) Methods for directing operator attention to particular screens should be considered in situations where several monitors are present or when operators may be involved in secondary non-CCTV tasks. 'Attention grabbing' is maximized when screens that are normally blank are automatically tripped to show a picture.

Pictures and images

As a general rule auto-cycling of images should be avoided. Peripheral vision is useful for detecting movement, or status changes, on banks of monitors; the use of auto-cycling undermines this ability.

The importance of 'image complexity' is illustrated in Fig. 15. The requirement to identify an individual is obviously much more difficult where the scene is busy than when the same scene is quiet.

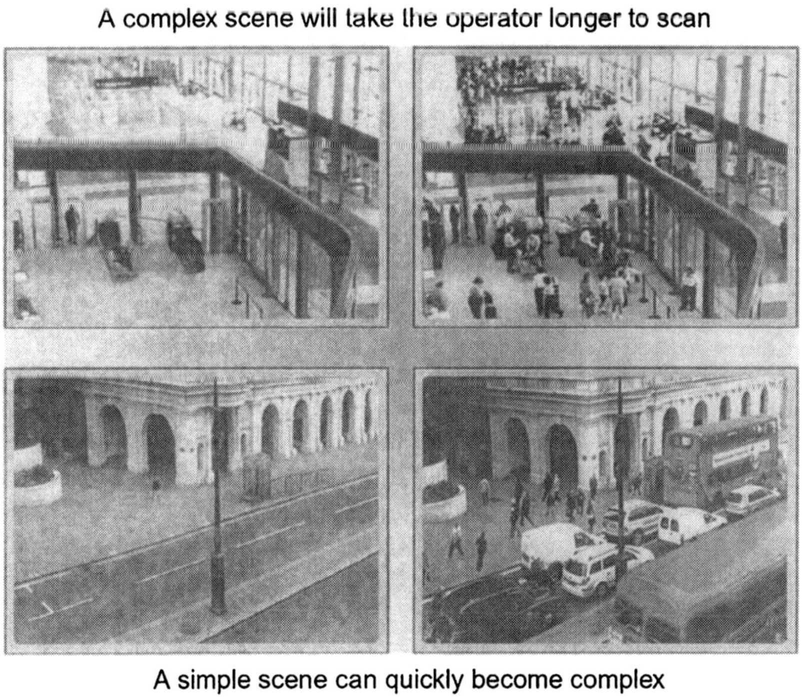

FIGURE 15. Examples of varying image complexity.

Although it is easy to make a subjective judgement about the level of complexity of an image, the development of a suitable metric is not trivial and is reported elsewhere [9].

Target size

The minimum acceptable size of targets largely depends on the nature of the task and the requirement for reliability and accuracy of target detection and recognition. The following guidance (Fig. 16) is offered when target sizes are being considered and has been derived from the security industry [5]:

(1) For monitoring a known presence, the target size should not be less than 5% of screen height for operators to have a reasonable chance of detecting it.
(2) For reliable detection tasks (and when target presence is not assured), the target size should not be less than 10% of screen height.
(3) For recognition tasks (e.g. where target type or number of targets is important), the target size should not be less than 30% of screen height, or not less than 50% for reliable recognition.
(4) For identification tasks (e.g. determining the identity of an individual), the image size should not be less than 120% of screen height.

Number of monitors and operator performance

The number of monitors provided should be appropriate for the task and related to performance requirements – ideally kept to a minimum. Research shows a significant reduction in detection rates when increased numbers of CCTV monitors are viewed (Table 1).

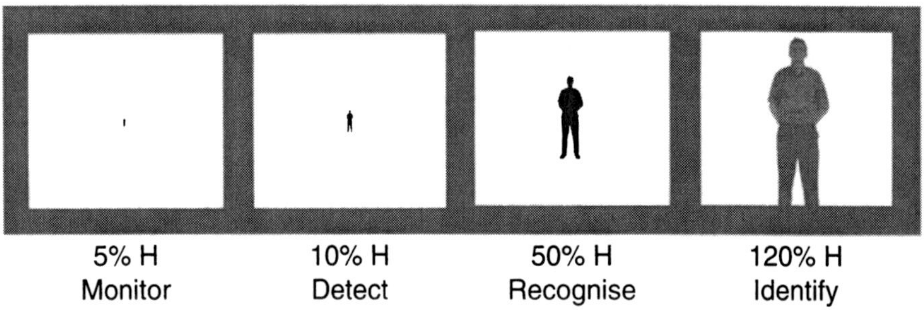

FIGURE 16. Target height as a percentage of screen height 'H' for different task requirements.

TABLE 1. Monitor numbers and target detection accuracy

Monitor numbers	Accuracy scores (%)
1	85
4	74
6	58
9	53

General recommendations on the number of monitors operators can reliably handle are as follows:

(1) An operator required to scan images to detect targets reliably (e.g. level-crossing operation) should concentrate attention on one active monitor, leaving the rest blank until activated by an alarm or event.
(2) Where pictures display considerable movement, and the task primarily involves general surveillance, no more than nine simultaneously displayed pictures should be observed by an individual.
(3) Where pictures display little movement, and the task consists essentially of general surveillance and observation, no more than 16 simultaneously displayed pictures should be observed by an individual.

ACKNOWLEDGEMENTS

For the case study in the railways, CCD would like to thank Network Rail for their support and assistance in the conduct of these trials.

REFERENCES

[1] Wood, J. (1997). *The Ergonomics of Closed Circuit Television Systems*. Proceedings of the 13th Triennial Congress of the International Ergonomics Association. Finland: Tampere.
[2] Tickner, A.H. and Poulton, E.C. Monitoring up to 16 synthetic television pictures showing a great deal of movement. *Ergonomics*, 16 (4), 381–401.
[3] Tickner, A.H., Poulton, E.C., Copeman, A.K., and Simmonds, D.C.V. Monitoring 16 television screens showing little movement. *Ergonomics*, 15 (3), 279–291.
[4] Tickner, A.H. and Poulton, E.C. Remote monitoring of motorways using closed-circuit television. *Ergonomics*, 11 (5), 455–466.
[5] Aldridge, J. *Operational Requirement Analysis – A New Approach to Effective Security*. Home Office, Police Scientific Development Branch, UK, PSDB Publication 17/94, ISBN 85893, 1995.
[6] Wallace, E., Diffley, D., Baines, E., and Aldridge, J. (1997). *Ergonomic Design Considerations for Public Area CCTV Safety and Security Applications*. Proceedings of the 13th Triennial Congress of the International Ergonomics Association. Finland: Tampere.
[7] Diffley, D. and Wallace, E. (1998). CCTV: Making it work. In *Training Practices for CCTV Operators*. Home Office, PSDB No. 9/98.
[8] *CCTV Ergonomics Good Practice: Guidance Notes*. (2004). Network Rail.
[9] Cole, H. (2006). *The Relationship Between Image Complexity and Task Difficulty as Judged by Novice and Expert Viewers*. Proceedings of the International Ergonomics Association, Maastricht.
[10] Keval, H.U. and Sasse, M.A. (2006). *Man or a Gorilla? Performance Issues with CCTV Technology in Security Control Rooms*. Proceedings of the International Ergonomics Association, Maastricht.
[11] Wood, J. (2008). *Planning and Design of Control Centres*. CRC Press (submitted for publication).

Meeting Diversity in Ergonomics (2007)
Pikaar, Koningsveld, and Settels (Eds)

17

Facing Diversity When Designing and Evaluating Driver Support Systems

Laura K. Thompson

Mechanical and Industrial Engineering, University of Toronto, 5 King's College Road, Toronto, Ontario, Canada M5S 3G8

Marcus Tönnis

Fachgebiet Augmented Reality, Technische Universität München, Boltzmannstraße 3, D-85748 Garching bei München, Germany

Abstract. An increasing amount of computing technology is being integrated into vehicles in the form of ADAS and in-vehicle information systems. In approaching the dream of fully automated cars, these driver support systems must be able to handle the diversity of individual drivers and driving situations. Consequently, the focus of research should be on design of adaptive and cooperative systems. In particular, the design process should focus on developing an integrated driver–vehicle interface instead of individual functions. Most importantly, the safety and comfort benefits of each driver support system need to be thoroughly evaluated as to minimise unsafe behaviour in the form of behavioural adaptation, driver distraction or high workload.

As an example of such an evaluation, this chapter describes a study concerning the design of minimally distractive control devices for an adaptive cruise control (ACC) system. ACC is designed to support the primary task of driving, in that it controls both vehicle speed and headway. However, adjusting the settings should not lead to driver distraction. Therefore, two driver–vehicle interface concepts for an ACC system were designed and tested in a fixed-base driving simulator. Adjusting both settings at once had the most detrimental effect on driving performance and visual distraction. Dividing the controls between the steering wheel and dashboard also increased visual distraction. The results are used to derive recommendations for the design of minimally distractive driver–vehicle interfaces.

Keywords: Vehicle ergonomics, driver support systems, advanced driver assistance systems, in-vehicle information systems, driver distraction, ACC, HUD

INTRODUCTION

Technological development and market competition compel the steady increase of in-vehicle information systems (IVIS) and advanced driver assistance systems (ADAS) available on the market. Car drivers must cope with increasingly more sophisticated technology in the vehicle [1]. With the aim of increasing the safety and comfort in vehicles, ADAS are designed to avoid collisions and reduce the severity of injuries. These systems assist the driver in the lateral and longitudinal control of the vehicle. Anti-lock braking (ABS), electronic stability (ESP) and adaptive cruise control (ACC) are examples of such systems. Furthermore, the implementation and use of market and after-market IVIS are also increasing. These infotainment systems provide the driver and occupants with information (e.g., satellite navigation, traffic and news reports), entertainment (e.g., music and video) and communication (e.g., mobile telephony and email) services. In fact, some systems are a combination of both ADAS and IVIS systems, such as satellite navigation, vehicle diagnostics, pedestrian detection and driver impairment monitoring [2].

However, the more functions that are integrated into the car, the more distractive the in-car environment can become. Completing secondary tasks with in-vehicle systems can have a detrimental effect on vehicle performance [3]. In the 100-car study, those engaging in a complex secondary task tripled their near-crash/crash risk [4]. For moderate secondary tasks the risk doubled. In this naturalistic study, secondary task distraction also contributed to 22% of crashes and near-crashes. Although many driver support systems are designed to support the primary task of driving, drivers do engage in secondary tasks to adjust the settings. Therefore, these systems must be implemented in such a way to support but not distract the driver in the primary task of driving. In addition, the driver's workload should not become too high or low while driving and interacting with driver support systems [5].

As a consequence, there has been much research interest in vehicle ergonomics, as reflected by an increased number of contributions to the IEA 2006 congress in this area. This chapter will focus on the topic of driver support systems (both ADAS and IVIS systems). In particular, we will discuss the relevant research in vehicle ergonomics presented at the 2006 IEA congress. We will start with the overall idea of fully automated cars and then focus on cooperative systems. These are systems that can bridge the gap between fully automated driving and current state of the art in vehicle technology. Using such systems we explain to what extent driver support systems can assist drivers in the near future. To explore the design challenges for such driver support systems, we furthermore examine design and evaluation processes for in-car driver support systems. These systems have diverse requirements that must be addressed during the design process and targeted in follow-up evaluations.

Using our work as an example of designing and testing control devices for ACC,[1] we describe our experiences with such design and evaluation processes and give guidelines for the design of minimally distractive ACC control devices.

CAN AUTOMATION HANDLE THE DIVERSITY?

Driving situations on roads are diverse. Street and weather conditions change from hour to hour and day to day. Furthermore, car drivers are also diverse. Depending on factors of experience, mood and comfort, their reactions to the traffic situation vary. The magnitude of accidents involving vehicles strengthens the need for automation in vehicles. The idea is to replace human control by computer systems that rely on sensor data. However, for practical, personal and safety reasons, complete replacement of human operators is only possible for specific environments, such as trained drivers in dedicated lanes. There is no practical way, in the near future, for any kind of automation to be designed to handle such diversity. The driver must remain the adaptive element in the driver-vehicle system in order to handle the abnormal situations unforeseen by designers.

The transition from the driver performing the task of driving to having a system perform the task implies a shift from manual control to supervisory control. This has implications for, among others, the cognitive requirements of the human operator. Those systems still need the driver to monitor and control the actions of the automation. In general, normal conditions can be dealt with automatically, but abnormal conditions still will require the driver to resume control. Similar to issues in the aviation domain, the driver could suffer from *out of the loop* performance problems when performing supervisory tasks with vehicle automation [6]. There

[1] The ACC system is also known as active, automatic or autonomous cruise control.

are also issues concerning how and when the driver should resume control, and how the driver can communicate with the automated system.

The design of driver support systems faces various challenges. One of these is to what extent the system is automated. A common theme of the research of Hoc and Young [7] suggests that it is not the level of automated control that is important, but rather the level of authority and communication between human and technology. They suggest that the best form of cooperation treats automation as a team member. That puts a different perspective on design, but is perfect for staying within the systems perspective of ergonomics – both human and machine are working in harmony towards a common goal. However, this has yet to be achieved.

With respect to the level of automation in driving, Brookhuis and Waard [8] conducted a simulator experiment in which bus drivers could switch between three different assistance modes. In their experiment, concerning automated public transport on the road, bus drivers could switch between fully automatic control, semi-automatic control and fully manual control. The semi-automatic control was analogous to driving a tram (i.e., the driver had only longitudinal control). Drivers were told to drive in the automatic mode but were allowed to switch to the other modes if they preferred. In addition, one portion of the route did not have designated lanes and so drivers were required to take over manual control. In order to test if drivers reclaim control in critical situations, four critical scenarios were added to the course, such as cyclists crossing the road or cars driven illegally in the designated lane. While operating in (semi-)automatic mode, seven drivers (28%) collided with the bicyclist on the first encounter. On subsequent encounters, drivers were able to regain manual control in time, with only one collision occurring. Generally, drivers trusted the automation, but training on unexpected events is necessary to facilitate fast takeover and avoid negative consequences of a first time encounter on the road.

This experiment was designed for public transport with a majority of designated lanes that allow for automated driving. What would happen if automated driving were introduced into personal vehicles? Here, a much wider variety of situations should be considered, such as the interactions between manually controlled and automated cars in traffic. Thus it will certainly take longer until such systems can be introduced into private cars. Only semi-automated driving can be realised in near future.

With respect to truck driving, Skottke et al. [9] also investigated the effects of automation on driving. They examined the potential of semi-automated truck convoys, in which the lead truck is driven manually. The remainder of the convoy follows automatically, except in certain situations, such as construction zones, bridges, tunnels and when exiting the highway. In these situations, the convoy drivers must resume manual control. A particular concern in the research of Skottke et al. was driver adaptation to the short headways (0.3 s) used by the automated driving system. Using a driving simulator study, they were able to demonstrate that there is a carry-over effect. Most drivers maintained an unsafe headway (<0.8 s) for more than 20 km after resuming manual control. This carry-over effect was seen not only after long periods of automated driving but also after short periods. These results show the possible negative consequences of automated driving that must be addressed before such systems are implemented.

Another aspect of diversity relates to the accuracy of driver support systems. Such systems rely on noisy sensor data that may trigger false alarms, as well as algorithms that may not match a driver's expectations and thus produce non-useful notifications. Lees et al. [10] investigated the influence of false alarms and non-useful warnings for a rear-end collision warning system. They evaluated the effect of such alarms and warnings on trust and reliance. The results suggest that non-useful alarms encourage drivers to exhibit greater caution, whereas false alarms diminish trust in the system such that drivers become less likely to respond to warnings. One explanation for these differences is that drivers understand non-useful alarms by linking contextual cues to alarm onset. These findings suggest that compared to false alarms, non-useful alarms do not diminish trust and leave drivers more prepared to respond to uncertain situations. Drivers receiving false alarms tended to disregard alarms and delay responding until they could verify threats more carefully.

Up to now, we have seen that automated driving is a goal of the future. In the meantime, we have to deal with issues concerning driver support systems that cooperate with the driver and adapt to the driving situation.

DESIGN PROCESSES

Design issues for driver support systems have also reached a level of maturity, so that developers have to deal with a wide variety of design aspects. There are many design philosophies and methods that can be used for the design of driver support systems.

The work of Lee et al. [11], for instance, considered the applicability of ecological interface design (EID) to driver support systems. The focus of many traditional systems is to provide discrete warnings that alert drivers to threatening situations. These warnings are often binary rather than graded and provide little information regarding the type of threat or how it is evolving over time. EID-inspired displays however, show information at multiple levels of abstraction and leave the determination of what constitutes a threat up to the driver. Lee et al. developed three EID-inspired displays to support the judgement of safe gaps in a lane change situation. In two experiments, they compared these three novel displays to a traditional display. Participants adapted with a similar speed to both the traditional and EID-inspired displays. With short viewing times, one EID-inspired display outperformed the other displays. With long viewing times, one EID-inspired display and the traditional display similarly supported the judgement of safe lane change situations. However, this EID-inspired display promoted a more precise calibration between judgement accuracy and confidence in the judgement. In terms of the applicability of EID, it is also important to note that not all of the EID-inspired displays outperformed the traditional display. Not every design philosophy is applicable to every situation and so it is important to consider the context in which a design will be used and also to evaluate it in this context.

A successful design process is one that incorporates information from many sources and uses an iterative cycle of analysis and design. An excellent example is the BIONIC[2] project [12], which developed 'eyes-free' driver controls for the secondary and auxillary systems in the vehicle (navigation, climate and entertainment systems), with the aim of minimising visual demands within the car. The design process was quite comprehensive, incorporating 13 phases and 5 studies. A novel preliminary phase in this project consisted of conducting interviews with visually impaired people to determine tactile cues (i.e., what specific characteristics facilitate non-visual use). Other phases included literature reviews, focus groups, usability trials in market vehicles, tests with cardboard cut-outs, CAD evaluations (SAMMIE), driving simulator trials with non-functional prototypes, and road trials with a functional prototype. The final design incorporated three separate pods, one each for satellite navigation, climate controls (HVAC) and in-car entertainment. When evaluated against the conventional controls in a Honda Civic, the eyes-off-road time was 17% less with the BIONIC interface (which

[2] Blind operatION of In-car Controls.

exceeded the target value of a 10% reduction). The research project also generated many design guidelines for 'eyes-free' interfaces, such as providing hand control reference points, avoiding tactile noise and providing a clear indication of modality.

There are other design issues, apart from 'eyes-free', that developers of driver support systems could take into account. This short overview gives an idea about the diversity of the design space for driver support systems.

EVALUATION PROCESSES

To what extent can, and should, driver support systems be evaluated? Which tools and metrics should be used to assess the driver's behaviour and the performance of the driver support system?

First, there are many human factors methods that can be used to evaluate driver support systems [13]. However, not all methods have been used specifically in the driving domain. Some methods can be applied early in the design phase, such as task analyses, checklists, heuristics and focus groups. For example, Porter et al. [12] used cardboard cut-outs and non-operational rapid prototypes. Ma et al. [14] investigated driver trust in automated driving aids by using a driving-like task (PC with gaming controls) and the 'Wizard-of-Oz' method, in which the experimenter acted as the navigation system and either spoke the directions on a cell phone or displayed them on a laptop screen. Other methods require more functional prototypes, such as usability and experimental studies with mock-ups and simulations. For example, Lee et al. [11] used a task simulation, in which displays were shown superimposed on a picture of a left side mirror on a PC monitor. By viewing the display, the participants had to judge whether a lane change could be safely made. Finally, there are the methods that use medium to high fidelity systems and are conducted in advanced driving simulators [8], closed test tracks [15] and on-road studies [12, 16]. The method selected depends on many factors, including project costs, safety risks and the ability to generalise results.

Driver support systems are frequently evaluated at a late stage in the project when fully functional prototypes are available. However, designers must be capable of conducting tests in the early stages of development. Therefore, Chen et al. [17] developed a test method to be used by in-car device designers to conduct usability tests during concept design. In this method, the usability tests are conducted using a rough mock-up of a vehicle and a touchscreen simulation of the proposed in-car device design. The mock-up consists of a video game steering wheel and pedals, an adjustable car seat and a computer monitor to display a simulated driving task. The method uses a toolkit of objective measures (task time, errors and glances) and subjective measures (ergonomics audit form and usability questionnaire) to uncover usability problems with the design. Industry feedback to the method was positive.

Another issue is how to 'measure' the comfort of a driver support system. Didier and Landau suggest a method that compares the comfort of two different systems to be evaluated [16], as to facilitate both relative and absolute measurements and judgements. They describe how the method was used to compare two different ACC systems. The results demonstrate that detailed questionnaires coupled with driving data give the most accurate representation of comfort. These questionnaires should not simply ask for overall comfort but should ask about detailed aspects of the system.

The driving simulation itself is also an issue when evaluating driver support systems [18]. There are many different types of driving simulators, ranging from simple lab settings using a PC and video game controls, to fully immersive driving simulators. There are so many possible variations that it becomes difficult to compare results between studies. In addition, the reporting of driving simulator studies needs to be improved or even standardised [19]. In order to replicate and validate results, Jonsson and Chen's review found five categories of information that could be useful if described as part of a driving simulator study: design and participants, simulator fidelity and vehicle, roadway scenario and driving environment, driving performance methods and driving behaviour measures. They also re-emphasise the need to determine the transfer of results from simulator studies to actual on-road performance.

A second crucial question is what measures should be used in the evaluation of driver support systems. Although these systems are designed to support the driver, they may lead to an increase in unsafe behaviour due to driver distraction or behavioural adaptation [2]. The aim of recent research has been to determine which measures are sensitive to levels of driver distraction or inattention. For example, the HASTE[3] Project [20] demonstrated that the steering wheel reversal rate is particularly sensitive to the amount of visual distraction due to an IVIS. Further research is required to validate which metrics should be included in the vehicle ergonomics 'tool box'. This, in turn, also requires the further development of tools to measure these metrics.

Measuring glance behaviour has been a popular method for assessing driver distraction. However, many systems are obtrusive and uncomfortable to the participant. In addition, analysis of the vast amounts of generated data is very time-consuming. Glance-tracking systems must be unobtrusive so that they do not themselves distract the driver. Two categories of eye-tracking systems can be classified. First, *outside-in* systems capture an exocentric view of the participant. Cameras are mounted in the vehicle to capture videos of the driver's head and eyes. *Inside-out* systems, on the other hand, capture the egocentric view of the participant. These head-mounted systems directly film the user's eye and the forward field of view into the environment.

In general, outside-in systems allow head movement that is less obtrusive than inside-out systems, which must be worn on the head. Outside-in systems that do not allow head movements, such as those used in laboratory studies with a PC and chin rest, are the most accurate but obtrusive and uncomfortable over time. The remaining *outside-in* systems are more comfortable because they allow freedom of movement, but this comes at the cost of a lower tracking area (the driver must stay in sight of the camera) and lower gaze tracking quality (when the eyes are obstructed). *Inside-out* systems enable a much greater tracking range, but rely heavily on image processing software and/or manual offline (post-hoc) analyses.

A popular *outside-in* system is FaceLAB by Seeing Machines [21]. This system unobtrusively tracks both head and gaze position. When the eyes are obstructed, it continues to track head position. An example of an *inside-out* head-mounted system is Dikablis[4] [22], developed online by Lange et al., which supports wireless gaze tracking over a large field of view. It is capable of pupil detection, but the glance analysis is performed ofline with post-recording software.

As shown in the examples above, there are many methods and measures used to evaluate driver support systems, from simple mock-ups to high-fidelity driving simulators and naturalistic field studies. Driver support systems are evaluated in terms of performance and safety measures along with comfort and usability measures. Thus the diversity seen in drivers and driving situations is also present in the methods and measures used to evaluate driver support systems.

---

[3] Human–machine-interface and the Safety of Traffic in Europe.

[4] DIgitale Kabelose BLIckerfassungs System (digital wireless gaze tracking system).

DESIGNING AND EVALUATING ACC CONTROLS

To illustrate the processes used in our project for the design and evaluation of a driver support system, we report on one of our experiments. In this experiment, we compared two different ACC driver control layouts [23]. We describe the concept and the methodology and compare it with the related work illustrated in the previous sections.

ACC, which is available in many luxury cars, is an ADAS that was designed to increase safety and driver comfort. Within operational boundaries, ACC not only controls the vehicle speed (as is the case with conventional cruise control), but also controls the headway to the vehicle ahead. Previous ACC studies have concentrated on behavioural adaptation while driving with ACC [e.g., 24–26], but did not specifically address the task of adjusting the ACC settings. Therefore, the aim of this study was to investigate the extent of driver distraction while interacting with an ACC system. Two different driver–vehicle control concepts were designed to explore the effects on task performance, driving performance and visual distraction. These concepts differed in the location of the controls (steering wheel vs dashboard) and the degree of haptic feedback (with or without click stops). The aim was not to develop a fully cooperative system [7], but rather to determine what factors aid or hinder the communication between the driver and the ACC system.

The following sections describe the two ACC driver–vehicle interface concepts and the experimental method used to evaluate the two designs. The results are then used to derive recommendations for the design of minimally distractive driver controls.

Concept

Driver controls

Two different ACC control concepts were developed and implemented in a static driving simulator. These were named the *divided* and *integrated* concepts. Both of these concepts were used to set the desired speed and the desired headway of the ACC system. For the *divided concept*, a speed selector knob was located on the dashboard to the left of the steering wheel and a headway slider was located on the left side of the steering wheel (Fig. 1). For the *integrated concept* both controls were combined on the left side of the steering wheel, with a barrel key for the headway surrounded by a selector ring for the speed (Fig. 2). All controls (except for the headway slider from the divided concept) were implemented with click stops in increments of 5 km/h for the desired speed and 0.1 s for the desired headway. The desired

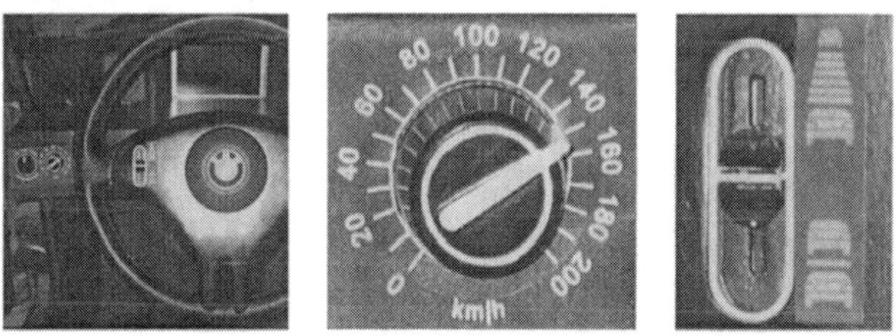

(a) Overall setup (b) Speed selector knob (c) Headway selector slider

FIGURE 1. Driver controls for the divided concept.

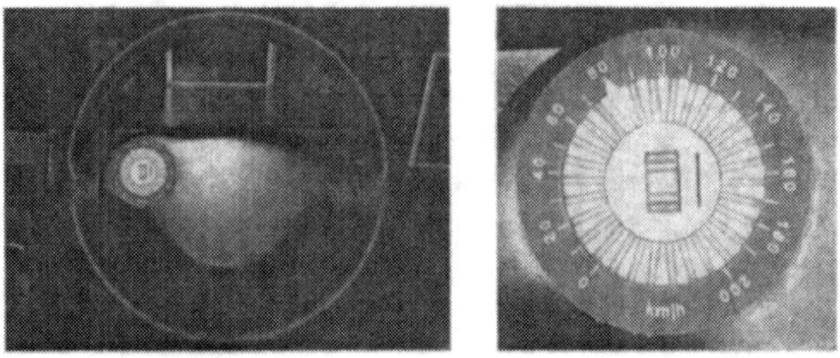

(a) Overall setup (b) Control devices

FIGURE 2. Driver controls for the integrated concept.

speed could be set between 30 and 200 km/h, and the headway between 0.9 and 2.0 s. These click stops follow the recommendations of Porter et al. [12] for multimodal feedback. Along with a visual indication of the current setting in the head-up display (HUD) and on the control itself, the driver can feel the adjustment.

Visual feedback in the HUD

Visual feedback about the ACC settings was shown in a virtual head-up display (Fig. 3) as analogue symbols with a digital value. Each complete symbol subtended an angle of 1.5° and was located 6.5° below the horizon. In the HUD, the desired speed was shown to the left of the current speed and the desired headway to the right. Placing the visual feedback in the HUD may reduce eyes-off-road time, but may also increase cognitive capture.

FIGURE 3. The driver's view of the head-up display. The desired speed is on the left, the desired headway is on the right and the current speed is in the centre.

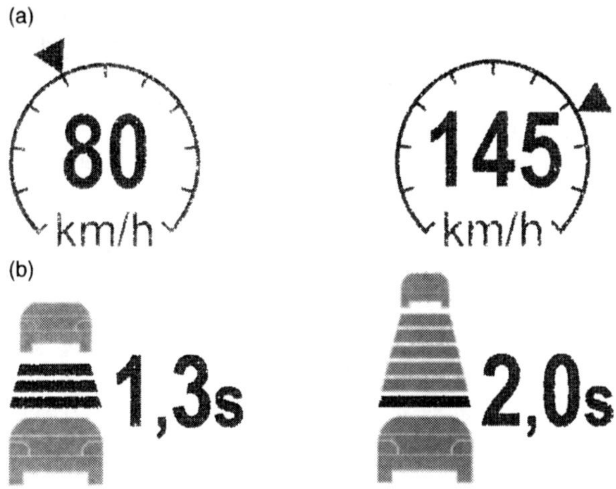

FIGURE 4. Visual feedback in the head-up display. (a) HUD symbols for the desired speed. (b) HUD symbols for the desired headway.

The symbols were designed in conjunction with the driver controls in order to maintain primary control–display compatibility [27]. In order to increase the desired speed, the driver rotated the control to the right. Correspondingly, the orange triangle also moved to the right around the speedometer icon (Fig. 4(a)). In order to increase the headway, the driver pushed upwards on the control. Correspondingly, the number of bars in the symbol increased and the upper car moved further away (Fig. 4(b)).

Method

Participants

Twelve drivers (six males, six females) participated in the study. They were between 19 and 53 years old ($M = 37$, $SD = 14$). All held a valid driver's license and were primarily staff or students at the Technische Universität München, Germany. Two drivers had participated in experiments in the driving simulator and in road trials and one other driver had participated in road trials. The volunteers were paid 30 Euros for the two hour experiment.

Apparatus

The experiment was run in a fixed-base driving simulator with a 40° field of view. Subjects drove a modified BMW convertible with automatic transmission and simulated motor sounds. The course was a two-lane rural road with long curves interspersed by some villages. The driving scene was projected at a resolution of 1024 × 768 pixels and 24-bit colour onto one screen located 3.5 m from the driver.

A driving simulator seemed to be suitable for this kind of experiment [19], because the roadway scenario and driving environment were appropriate to the situation where ACC is commonly used. The road course enabled various opportunities to encounter different traffic scenarios, as suggested by Didier and Landau [16]. The varying speed limits in the road course would require the driver to change the ACC speed settings.

Two interchangeable, wireless steering wheel inlays were developed to house the steering wheel controls. In addition, the speed control for the divided concept was mounted permanently in the dashboard. For the head-up display, DWARF[5] [28] was used as a component-oriented toolkit to easily implement and test prototypes of the driver-vehicle interfaces. Finally, the Dikablis [22, 29] helmet-mounted eye-tracking system was used to generate eye glance videos of the driver's right eye superimposed on the scene ahead.

Since this experiment was part of an exploratory study of future driver support systems, we chose to use mid-fidelity prototypes and simulations. This allowed us the flexibility to investigate driver-vehicle interactions, while maintaining full functionality of the ACC system but not endangering the drivers. The driver controls were physical prototypes built into the cockpit of an actual vehicle. However, the head-up display and driving scene were simulated.

Procedure

Upon arriving at the driving simulator, participants completed a demographic questionnaire and read a description of the driver controls for the ACC system. Once finished, they entered the vehicle, adjusted the seat and the eye-tracking system was calibrated. The drivers subsequently drove one practice round with eye-tracking (12 min). The ACC system was turned off and only the current speed was shown in the HUD.

For each concept, drivers were first trained and then completed the experimental tasks. The first part of the training focused solely on task completion while parked. Drivers were given verbal instructions from the experimenter to change the ACC settings. Practice tasks were repeated until the tasks could be completed without errors. In the second part of the training, drivers repeated practice tasks while driving. Training continued until they could adjust both ACC controls while driving at least 80 km/h without any lane departures and they felt confident to continue. Therefore, in some cases, the training lasted up to 15 min.

For the experimental tasks, the participants drove the same rural road course as in the practice trial, but were given 18 verbal instructions from the experimenter to change the ACC settings. The location and order of the tasks were identical for both concepts. After completing the experimental tasks with both concepts, the drivers were interviewed about their subjective opinions of the ACC controls.

Experimental design

A within-subject design was used, with all drivers using both ACC concepts. The order was counter-balanced based on age and gender. In order to examine only the task of changing the ACC settings and not the task of driving with ACC, the ACC functionality was disabled. Therefore, the drivers had to maintain proper speed (based on the traffic signs), headway and lane position.

The independent factors were *concept* (divided or integrated), *task type* (adjust desired speed, headway or both) and *task length* (small or large adjustments). See Table 1 for descriptions.

The dependent variables included task performance, driving performance, glance behaviour and subjective measures [23, 30]. The *task performance* measure was the elapsed time to complete the various tasks. The measurement of the task time started directly after the verbal instructions were given and ended when the correct value was shown in the HUD and the driver's left hand returned to the starting position.

The *driving performance* measures were an indication of how well the driver could maintain the correct speed (speed deviation and average speed compared to the posted speed) and lane position (lane deviation, lane departure time, steering angle variation and large steering

[5] Distributed Wearable Augmented Reality Framework.

TABLE 1. Description of the independent factors.

Factor	Level	Description
ACC concept	Divided	The desired speed is set on the dashboard and the headway is set on the steering wheel
	Integrated	Both the desired speed and headway are set on the steering wheel
Task type	Speed	Adjust the desired speed
	Headway	Adjust the desired headway
	Both	Adjust both the desired speed and headway
Task length	Small	A small adjustment to the control (e.g., 5 km/h faster or 0.1 s closer)
	Large	A large adjustment to the control (e.g., 40 km/h faster or 0.5 s closer)

corrections). Other than the measure of large steering corrections [31], these measures were defined in the HASTE Project [32]. For this experiment, large steering corrections were defined as sudden steering wheel movements greater than 3.7°.

The measures of *glance behaviour* were an indication of driver distraction and workload. For the analysis, the driver controls and displays were classified as 'off-road'. The total glance time 'off-road', the maximum glance duration 'off-road' and number of glances 'off-road' were calculated based on a manual offline analysis of the eye glance videos according to ISO 15007-1 [33]. Therefore, the glance time included the dwell time (fixations and saccades on an object) and the transition time to that object.

The *subjective measures* served to complement the objective data. The driver's opinion of their task performance (i.e., the usability of the controls) and their preference of ACC control and control location were asked in a semi-structured interview. The questions were of various formats, including ranking, scoring, short answer and long answer.

An initial analysis of the data indicated that all dependent measures were not normally distributed ($p < 0.05$ according to the Kolmogorov–Smirnov Z-test). In addition, some of the measures were categorical (e.g., usability scores). Therefore, non-parametric tests were used to find significance, in particular the Mann–Whitney U-test (the U values were converted to Z values) for testing the difference between two groups and the Kruskal–Wallis H-test (χ^2 values) for three or more groups.

Results and discussion

Task time

There was no statistical difference between the task times for the divided and the integrated concepts, $Z = -0.60$, $p = 0.55$. On the contrary, the task times differed significantly between the three task types, $\chi^2 = 114.65$, $p < 0.001$. The adjustment of both controls together took significantly longer than the individual adjustment of the speed, $Z = -8.94$, $p < 0.001$, or headway, $Z = -9.53$, $p < 0.001$, as can be seen in Fig. 5. With both concepts, the median time was 4.5 s to adjust the speed and 4.3 s to adjust the headway, whereas the adjustment of both settings required 9.1 s with the divided concept and 7.5 s with the integrated concept. The number of adjustments also had a significant effect on the task time; the tasks with large adjustments took approximately twice as long as the tasks with small adjustments, $Z = -10.48$, $p < 0.001$. In particular, this effect was quite strong for the adjustment of the headway using the *divided* concept. When considering both the task type and number of adjustments, the task times while drivers had to make large adjustments to both controls were considerably

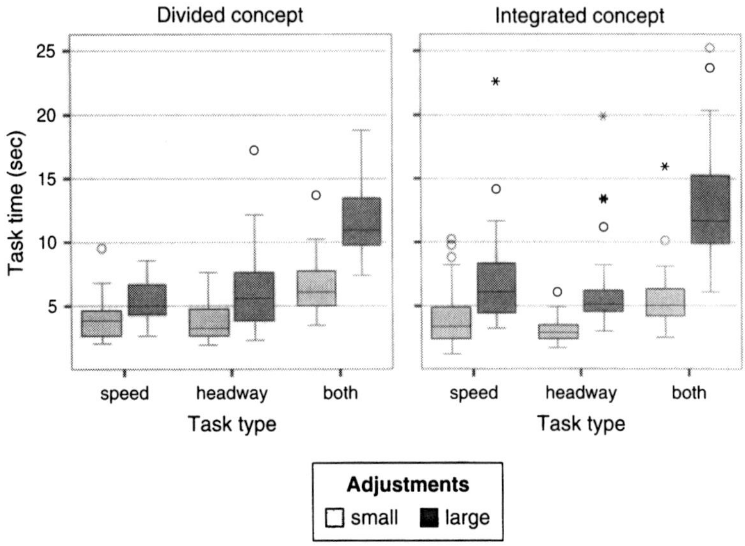

FIGURE 5. Task times per concept, task type and number of adjustments.

longer than all other tasks, with a median of 11.0 s for the divided concept and 11.7 s for the integrated concept.

Driving performance measures

The ACC concept did not influence the speed-keeping measures, but the type of task did have an effect on the speed deviations and average speed. Participants drove significantly slower when adjusting both the speed and headway controls compared to adjusting only one setting, $\chi^2 = 12.71, p < 0.01$. While adjusting the desired headway the mean speed was 1.5 km/h below the speed limit and while adjusting the desired speed it was 2.1 km/h below. However, while adjusting both controls the average speed fell 6.4 km/h below the speed limit. Contrary to other dependent measures, the number of adjustments (task length) did not affect the average speed.

Furthermore, the speed deviations were significantly larger when both controls were adjusted as compared to only one control, $\chi^2 = 24.63$, $p < 0.001$, and also significantly longer for tasks with large adjustments compared to the small adjustments, $Z = -5.67, p < 0.001$. This effect was particularly strong for the integrated concept. These findings are portrayed in Fig. 6.

The two concepts did not influence the driver's ability to maintain lane position; there was no significant difference between concepts for lane deviations and lane departure time. On the other hand, the type of task did have a significant effect. The lane deviation was significantly higher ($Mdn = 33$ cm) when both the speed and headway were set ($\chi^2 = 35.34$, $p < 0.001$) as compared to changing only one setting ($Mdn = 21$ cm for *speed* and $Mdn = 19$ cm for *headway* tasks). This coincides with a significantly higher lane departure time for this task compared to the *speed* and *headway* tasks, $\chi^2 = 12.64$, $p < 0.01$ (Fig. 7).

In addition, the lane deviations and lane departure times were also significantly greater for the tasks with large adjustments compared to the tasks with small adjustments, $Z = -5.12$, $p < 0.001$ and $Z = -4.76, p < 0.001$, respectively. However, this main effect is mostly caused by one of the tasks. As shown in Fig. 7, the lane departure times were much larger while drivers

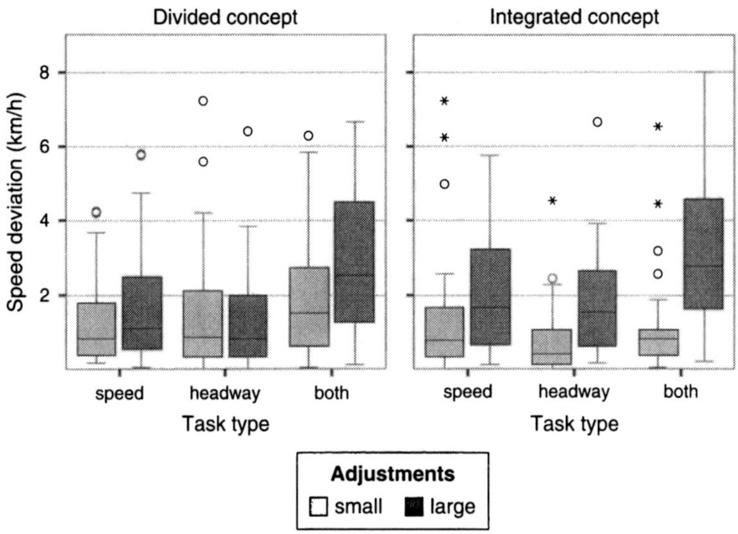

FIGURE 6. Speed deviation per concept, task type and number of adjustments.

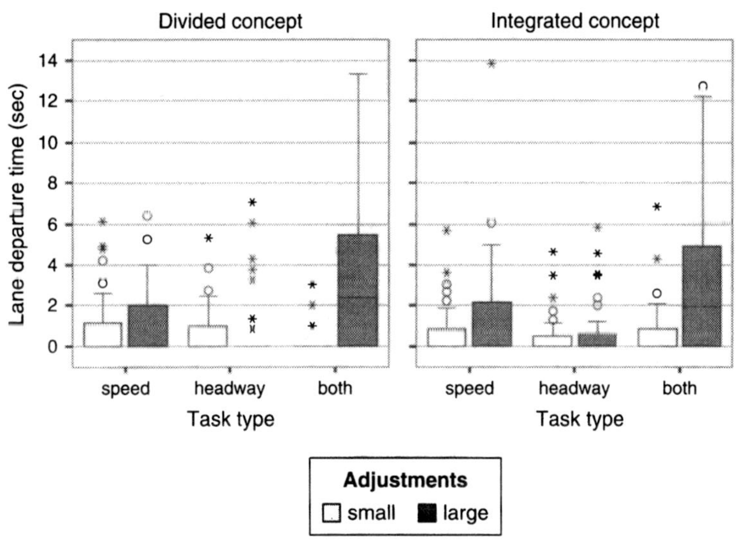

FIGURE 7. Lane departure time by concept, task type and number of adjustments.

made large adjustments to both controls ($Mdn = 2.1$). The task of making large adjustments to the desired speed was the only other task with a median above zero ($Mdn = 0.5$ s).

Similar to the lane-keeping measures, the steering measures also did not differ between the two ACC concepts, but differences were observed between tasks. Steering angle variation differed significantly between all three tasks, $\chi^2 = 22.95$, $p < 0.001$. This steering activity was the smallest when drivers adjusted the headway ($Mdn = 1.26°$), followed by the speed

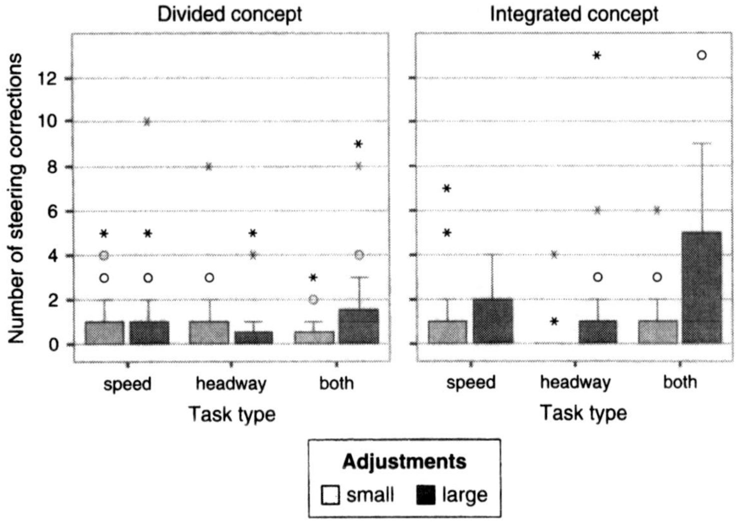

FIGURE 8. Number of large steering corrections by concept, task type and number of adjustments.

task ($Mdn = 1.64°$) and the largest when drivers adjusted both controls ($Mdn = 1.80°$). Drivers also had larger variations in steering and larger steering corrections for the tasks with large adjustments compared to the tasks with small adjustments, $Z = -3.50$, $p < 0.001$ and $Z = -3.01$, $p < 0.01$, respectively.

In addition, drivers had significantly fewer large steering corrections while adjusting the headway control compared to adjusting only the speed control or both controls, $\chi^2 = 13.97$, $p < 0.001$. This main effect was primarily caused by the integrated concept, as can be seen in Fig. 8. With the integrated concept, for example, 86% of the tasks with small changes to the headway were completed without any large steering corrections compared to only 39% of the tasks with large adjustments to both controls. Therefore, the steering control of the integrated concept appears to be the main cause of the large steering adjustments.

Glance behaviour measures

The number of glances 'off-road' depended significantly on the type of task, $\chi^2 = 90.75$, $p < 0.001$, as shown in Fig. 9. There were significantly more glances 'off-road' while adjusting both controls as opposed to adjusting only one of the controls. There was no significant difference between the ACC concepts, though there appears to be an interaction between the design of the driver controls and the type of task. For the speed task, the *divided* concept had the fewest glances, but for the headway task the *integrated* concept had the fewest glances. Both concepts required a similar number of glances to adjust both controls.

In addition, drivers required significantly more glances (approximately twice as many) to make large adjustments to the controls compared to small adjustments, $Z = -8.967$, $p < 0.001$. The number of glances for the tasks with small adjustments generally falls within the four glances recommended by Zwahlen et al. [34], but the number of glances required for the tasks with large adjustments do not. The most glances were observed while drivers used the integrated concept to make large adjustments to both controls; the number of glances ranged from 4 to 19 with a median of 8. With the divided concept, drivers required 3 to 12 glances with a median of 7.5 for the same task.

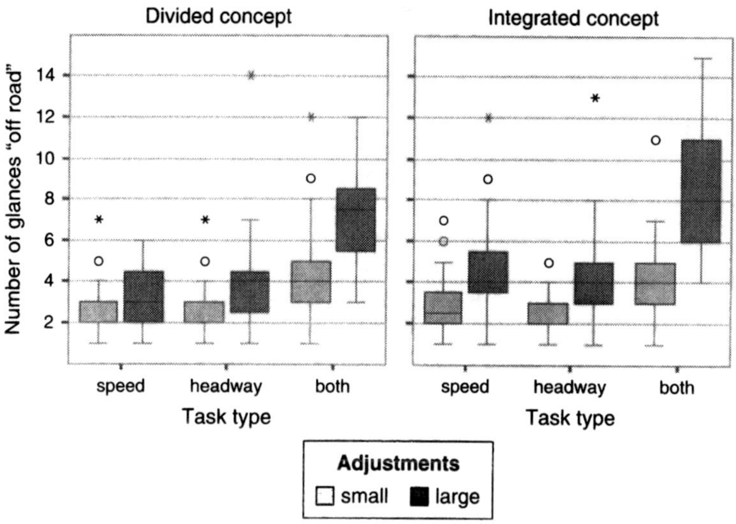

FIGURE 9. Number of glances 'off-road' by concept, task type and number of adjustments.

The total glance time 'off-road' was significantly smaller when drivers used the integrated concept, $Z = -2.21$, $p = 0.027$. It also depended on the type of task, $\chi^2 = 94.511$, $p < 0.001$, and the number of adjustments, $Z = -9.656$, $p < 0.001$, as is portrayed in Fig. 10. The total glance time was longer while adjusting both controls, particularly when large adjustments had to be made. Some drivers in particular had long glance times while using the *divided* concept to make large adjustments to the headway. For example, making large adjustments to both controls caused the longest total glance times with means of 7.2 s for the divided

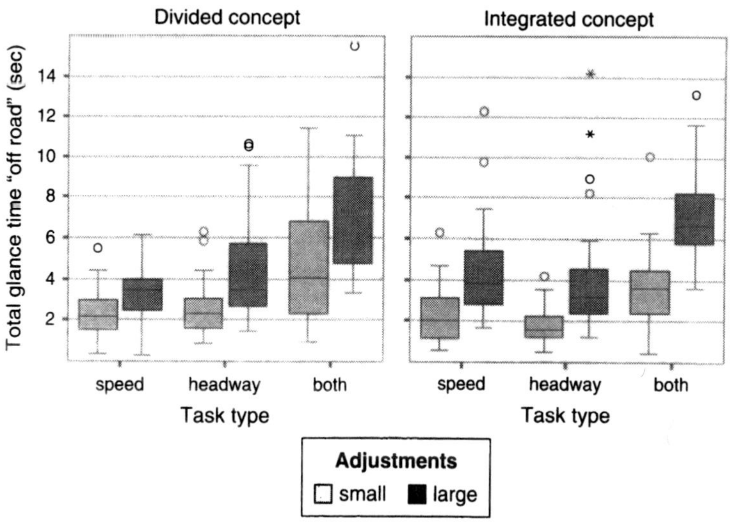

FIGURE 10. Total glance time 'off-road' by concept, task type and number of adjustments.

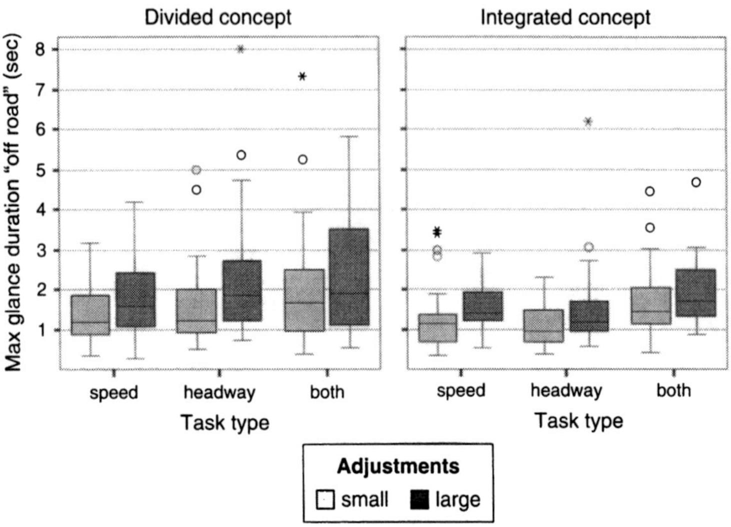

FIGURE 11. Maximum glance duration 'off-road' by concept, task type and number of adjustments.

concept and 7.8 s for the integrated concept. On the other hand, making small adju stments to the headway caused the shortest total glance times with means of 2.4 and 1.7 s for the two concepts, respectively.

The maximum glance durations followed a similar trend to the total glance time measure. The glance durations were longer with the divided concept, $Z = -2.78$, $p < 0.01$, longer while adjusting both controls compared to only one control, $\chi^2 = 19.697$, $p < 0.001$, and longer while making large adjustments to the controls, $Z = -4.54$, $p < 0.001$. These results are portrayed in Fig. 11. According to the AAM recommendations, the maximum glance durations should not exceed 2.0 s [35]. As can be seen in Fig. 11, both concepts fail to meet this criterion, in particular the divided concept. This has implications for traffic safety, since Klauer et al. [4] demonstrated that glance durations of more than 2 s at least double the crash risk compared with baseline driving.

Subjective measures

Subjectively, most drivers preferred the integrated concept; they rated the usability and appearance higher, and considered it to be less distracting and faster to use. Ten of 12 drivers wanted the ACC settings to be displayed in the HUD and 11 drivers wanted the controls to be located together on the steering wheel.

Most drivers preferred the *integrated* concept. On a scale from 1 to 6, where 1 is the best rating, drivers rated the usability higher for the *integrated* concept ($Mdn = 2$) compared to the *divided* concept ($Mdn = 4$), $Z = -1.76$, *ns*. These usability ratings are shown in Fig. 12(a). Most drivers mentioned that the click stops were helpful but should be improved in both designs since the control sometimes got stuck between two values or jumped past the desired value. Having to reach and locate the control on the dashboard was an additional criticism for the integrated concept. Drivers also generally preferred the appearance of the *integrated* concept ($Mdn = 2$) compared to the *divided* concept ($Mdn = 3$), $Z = -1.41$, *ns*. These appearance

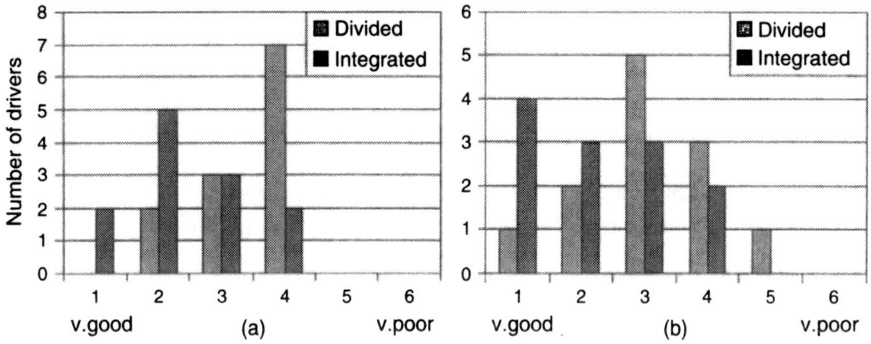

FIGURE 12. Subjective ratings for the ACC concepts based on usability (a) and appearance (b).

ratings are portrayed in Fig. 12(b). Some drivers mentioned that implementing a higher-fidelity prototype would increase this rating.

When asked about display location, 10 of 12 drivers wanted the ACC settings to be displayed in the HUD, 6 of 12 drivers wanted these to be displayed on the dashboard and 1 driver wanted a display on the steering wheel (multiple answers allowed). They all wanted the display symbols to be kept together. When asked about control location, all drivers preferred a speed control on the steering wheel and 11 drivers preferred a headway control on the steering wheel. Two drivers wanted a headway control on the dashboard and five drivers mentioned that both controls could also be located on a stalk control (e.g., the location of the current BMW 5 Series control). If the ACC system could automatically detect speed limits (and thus require less adjustments), only five drivers wanted the controls to remain on the steering wheel. Six drivers would then want the controls on the dashboard and one would want stalk controls.

Conclusions for the design of ACC driver controls

The ACC is designed to support the primary task of driving, however the adjustment of the ACC system's settings is in itself a secondary task that should not distract the driver. Task time, driving performance and glance behaviour were therefore used as measures of driver distraction to compare two ACC driver control concepts. These two concepts differed in control location (steering wheel vs dashboard) and haptic feedback. They both used a head-up display to provide feedback on the ACC settings and current speed. Based on the experimental results, several design recommendations can be supported for the design of minimally distracting ACC driver controls.

First, dividing the controls between the steering wheel and dashboard caused longer mean and maximum glance times and a lower glance frequency to the displays and controls. It appears that the separated control location also confused some drivers, since there were glances to the dashboard while adjusting the headway although the control was located on the steering wheel. Thus, the controls for an ACC system should be kept together, preferably on the steering wheel. This supports the UMTRI Guideline 3.2: 'controls used most frequently or for critical functions should be close to the predominate position of the hands' [36].

Second, most drivers looked longer at the speed control for the *integrated* concept compared to the *divided* concept (in particular while making large adjustments to the speed). The large steering corrections also increased while using the speed control of the *integrated* concept. During the post-trial interview, many participants mentioned that it was difficult to operate a rotating disk on the steering wheel. The frame of reference of the control rotates with the steering wheel and so they had to look at the control instead of being able to feel the location of the control. Therefore, caution is advised when using rotational controls on the steering wheel.

Third, the mean and total glance durations to the HUD were longer while using the headway slider (*divided* concept) compared to the scroll wheel with click stops (*integrated* concept). The task time was also longer. Since click stops facilitate 'eyes-free' operation, it is recommended that ACC controls provide this form of haptic feedback. These findings support the guidelines from the BIONIC project [12].

Finally, the type and length of task affect driving performance and glance behaviour. Adjusting the desired headway proved to be the least demanding task and adjusting both the desired speed and headway was the most demanding. Small adjustments to the driver controls were also less visually demanding than large adjustments. Both the driving performance measures and glance measures worsened as task time increased. Therefore, ACC controls should be designed so that drivers need only to make a few adjustments to one control at a time.

Future research should investigate how ACC driver controls can be incorporated into the complete cockpit and integrated with the other controls for the primary, secondary and auxiliary systems in the vehicle. A seamless integration with other ADAS, such as collision warning and headway control, is particularly important.

DRIVER SUPPORT SYSTEMS IN THE FUTURE

Various individual driver support systems are currently under development, with some already available in high-class vehicles, such as ACC and heading control. Drivers are also bringing after-market devices into their vehicles, such as MP3 players, satellite navigation systems and mobile phones. Up to now, these systems interfaced solely with the driver, and had their own control devices, displays and warnings. Future work in this field of research needs to investigate how these individual systems can be seamlessly integrated into one complete driver support system. Attention should be focused on the effective use of visual, auditory and haptic modalities so as not to overload or distract the driver. Only an integrated system can provide a consistent multi-modal driver–vehicle interface and act as an adaptive, cooperative partner, whose sole purpose is to support the driver.

ACKNOWLEDGEMENTS

This work was supported by the Natural Sciences and Engineering Research Council of Canada (NSERC), the German Academic Exchange Service (DAAD) and BMW Forschung und Technik, Munich. We would like to recognise the support of Prof. Heiner Bubb, Prof. Gudrun Klinker and Christian Lange. We are very grateful to Daria Popiv for spending many hours programming the driving simulator. We would also like to thank Alexander Peters and Johannes Güllich for their assistance with conducting the experiments and analysing the eye-tracking videos.

REFERENCES

[1] Freymann, R. (2006). *HMI: A Fascinating and Challenging Task*. Proceedings of the 16th World Congress on Ergonomics (IEA 2006). Maastricht, The Netherlands: Elsevier.

[2] Lenior, D., Janssen, W., Neerinex, M., and Schreibers, K. (2006). Human-factors engineering for smart transport: Decision support for car drivers and train traffic controllers. *Appl. Ergon.*, 37 (4), 479–90.

[3] Lansdown, T.C., Brook-Carter, N., and Kersloot, T. (2004). Distraction from multiple in-vehicle secondary tasks: Vehicle performance and mental workload implications. *Ergonomics*, 47 (1), 91.

[4] Klauer, S.G., Dingus, T.A., Neale, V.L., et al. (2006). *The Impact of Driver Inattention on Near-Crash/Crash Risk: An Analysis Using the 100-car Naturalistic Driving Study Data*, Washington, DC: National Highway Traffic Safety Administration, DOT HS 810 594.

[5] Neerincx, M.A., Hoedemaekera, M., and de Gier, E. (2006). *Adaptive In-Car User Interfaces Based on Personalized Work Load Estimation*. Proceedings of the 16th World Congress on Ergonomics (IEA 2006). Maastricht, The Netherlands: Elsevier.

[6] Endsley, M.R. and Kiris, E.O. (1995). The out-of-the-loop performance problem and level of control in automation. *Hum. Factors*, 37 (2), 381–94.

[7] Hoc, J.-M. and Young, M.S. (2006). *Driver Support Systems: How to Cooperate*? Proceedings of the 16th World Congress on Ergonomics (IEA 2006). Maastricht, The Netherlands: Elsevier.

[8] Brookhuis, K.A. and de Waard, D. (2006). *Consequences of Automation for Driver Behavior and Acceptance*. Proceedings of the 16th World Congress on Ergonomics (IEA 2006). Maastricht, The Netherlands: Elsevier.

[9] Skottke, E.-M., Wille, M., and Debus, G. (2006). *Automated Truck Driving on German Roadways: Simulator Data and Ergonomic Challenges*. Proceedings of the 16th World Congress on Ergonomics (IEA 2006). Maastricht, The Netherlands: Elsevier.

[10] Lees, M.N., Lee, J.D., and Brown, M.D. (2006). *Collision Warnings and Driver Brake Response to Critical and Noncritical Events*. Proceedings of the 16th World Congress on Ergonomics (IEA 2006). Maastricht, The Netherlands: Elsevier.

[11] Lee, J.D., Hoffman, J.D., Stoner, H.A., et al. (2006). *Application of Ecological Interface Design to Driver Support Systems*. Proceedings of the 16th World Congress on Ergonomics (IEA 2006). Maastricht, The Netherlands: Elsevier.

[12] Porter, J.M., Summerskill, S.J., Burnett, G.E., and Prynne, K. (2006). *Design of 'Eyes Free' Secondary Controls for Drivers*. Proceedings of the 16th World Congress on Ergonomics (IEA 2006). Maastricht, The Netherlands: Elsevier.

[13] Stanton, N. (ed.). (2005). *The Handbook of Human Factors and Ergonomics Methods*. Boca Raton: CRC Press.

[14] Ma, R., Kaber, D.B., Kim, S.-H., and Wang, X. (2006). *Effects of In-Vehicle Navigation Automation and Reliability on Driver Trust*. Proceedings of the 16th World Congress on Ergonomics (IEA 2006). Maastricht, The Netherlands: Elsevier.

[15] Bender, E., Landau, K., and Bruder, R. (2006). *Driver Reactions in Response to Automatic Obstacle Avoiding Manoeuvres*. Proceedings of the 16th World Congress on Ergonomics (IEA 2006). Maastricht, The Netherlands: Elsevier.

[16] Didier, M. and Landau, K. (2006). *Comfort Evaluation of Adaptive Cruise Control (ACC): A Comparison Between Two Different Systems*. Proceedings of the 16th World Congress on Ergonomics (IEA 2006). Maastricht, The Netherlands: Elsevier.

[17] Chen, C.-C., Woodcock, A., and Scrivener, S.A.R. (2006). *A Simulation Based Evaluation Method for In-Car Interface Designers*. Proceedings of the 16th World Congress on Ergonomics (IEA 2006). Maastricht, The Netherlands: Elsevier.

[18] Andersen, J., Ball, K., Caird, J., et al. (2004). *Critical Issues in Driving Simulation Methods and Measures*. Proceedings of the 48th Annual Meeting of the Human Factors and Ergonomics Society, pp. 2252–5.

[19] Jonsson, I.-M. and Chen, F. (2006). *How Big is the Step from Driving Simulators to Driving a Real Car?* Proceedings of the 16th World Congress on Ergonomics (IEA 2006). Maastricht, The Netherlands: Elsevier.

[20] Carsten, O.M.J., Merat, N., Janssen, W.H., et al. (2005). *HASTE final report*. Leeds: Institute for Transportation Studies, University of Leeds.

[21] Seeing Machines. (2006). FaceLAB version 4. Available from: http://www.seeingmachines.com/facelab.htm.

[22] Lange, C., Wohlfarter, M., and Bubb, H. (2006). *Dikablis (Digital Wireless Gaze Tracking System) – Engineering and Application Area*. Proceedings of the 16th World Congress on Ergonomics (IEA 2006). Maastricht, The Netherlands: Elsevier.

[23] Thompson, L.K., Tönnis, M., Lange, C., et al. (2006). *Effect of Active Cruise Control Design on Glance Behaviour and Driving Performance*. Proceedings of the 16th World Congress on Ergonomics (IEA 2006). Maastricht, The Netherlands: Elsevier.

[24] Rudin-Brown, C.M. and Parker, H.A. (2004). Behavioural adaptation to adaptive cruise control (ACC): Implications for preventive strategies. *Transportation Reseasrch Part F: Traffic Psychology and Behaviour*, 7 (2), 59–76.

[25] Sato, T., Akamatsu, M., Takahashi, A., et al. (2006). *Influence of ACC on Frequency of Situations Experienced While Driving on Highway*. Proceedings of the 16th World Congress on Ergonomics (IEA 2006). Maastricht, The Netherlands: Elsevier.

[26] Stanton, N.A. and Young, M.S. (2005). Driver behaviour with adaptive cruise control. *Ergonomics*, 48 (10), 1294–313.

[27] Bubb, H. (1993). Systemergonomische gestaltung. In *Ergonomie* (H. Schmidtke, ed.), 3rd ed. München: Hanser Verlag, pp. 390–420.

[28] Bauer, M., Bruegge, B., Klinker, G., et al. (2001). *Design of a Component-Based Augmented Reality Framework*. ACM Proceedings of ISAR, pp. 45–54.

[29] Lange, C. (2005). *The Development and Usage of Dikablis (Digital Wireless Gaze Tracking System)*. European Conference on Eye Movements (ECEM 2005), Bern.

[30] Thompson, L.K. (2005). *Development and Evaluation of a Driver Assistance Concept for Speed and Lane Assistance*. M.Sc. Thesis. Germany: Technische Universität München.

[31] Wierwille, W.W., Antin, J.F., Dingus, T.A., and Hulse, M.C. (1988). Visual attentional demand of an in-car navigation display system. In *Vision in Vehicles II* (A.G. Gale, M.H. Freeman, C.M. Haslegrave, P. Smith, S.P. Taylor, eds) Amsterdam: Elsevier, pp. 307–16.

[32] Roskam, A.J., Brookhuis, K.A., de Waard, D., et al. (2002). *HASTE Deliverable 1 – Development of Experimental Protocol*. Leeds: Institute for Transport Studies.

[33] ISO. (2002). *Road vehicles – Measurement of Driver Visual Behaviour with Respect to Transport Information and Control Systems – Part 1: Definitions and Parameters*. ISO 15007-1:2002, International Organization for Standardization.

[34] Zwahlen, H.T., Adams Jr, C.C., and DeBald, D.P. (1988). Safety aspects of CRT touch panel controls in automobiles. In *Vision in Vehicles II* (A.G. Gale, M.H. Freeman, C.M. Haslegrave, P. Smith, and S.P. Taylor, eds) Amsterdam: Elsevier Science B.V. (North-Holland), pp. 335–44.

[35] AAM. (2003). *Statement of Principles, Criteria and Verification Procedures on Driver Interactions with Advanced In-Vehicle Information and Communication Systems*. Draft version 3.0. Washington, DC: Alliance of Automobile Manufacturers.

[36] Green, P., Levison, W., Paelke, G., and Serafin, C. (1994). *Suggested Human Factors Design Guidelines for Driver Information Systems*. Technical Report, FHWA-RD-94-087. Ann Arbor, MI: The University of Michigan, Transportation Research Institute.

Index

CPSIA information can be obtained at www.ICGtesting.com
Printed in the USA
LVOW13*1031060814

397727LV00013B/235/P

9 780080 453736